D0064933

Bioremediation Engineering

Other McGraw-Hill Environmental Engineering Books of Interest

AMERICAN WATER WORKS ASSOCIATION • *Water Quality and Treatment*

BAKER • *Bioremediation*

CHOPEY • *Environmental Engineering for the Chemical Process Industries*

CORBITT • *Standard Handbook of Environmental Engineering*

FREEMAN • *Hazardous Waste Minimization*

FREEMAN • *Industrial Pollution Prevention Handbook*

FREEMAN • *Standard Handbook of Hazardous Waste Treatment and Disposal*

JAIN • *Environmental Impact Assessment*

LEVIN, GEALT • *Biotreatment of Industrial and Hazardous Waste*

KOLLURU • *Environmental Strategies Handbook*

MCKENNA, CUNNEO • *Pesticide Regulation Handbook*

MAJUMDAR • *Regulatory Requirements of Hazardous Materials*

NANNEY • *Environmental Risks in Real Estate Transactions*

WALDO, HINES • *Chemical Hazard Communication Guidebook*

Bioremediation Engineering

Design and Application

John T. Cookson, Jr.

McGraw-Hill, Inc.

New York   San Francisco   Washington, D.C.   Auckland   Bogotá
Caracas   Lisbon   London   Madrid   Mexico City   Milan
Montreal   New Delhi   San Juan   Singapore
Sydney   Tokyo   Toronto

Library of Congress Cataloging-in-Publication Data

Cookson, John T.
   Bioremediation engineering : design and application / John T.
Cookson, Jr.
      p.   cm.
   Includes bibliographical references and index.
   ISBN 0-07-012614-3
   1. Bioremediation. I. Title.
  TD192.5.C66   1995
  628.5′2—dc20
                                      94-26856
                                         CIP

  2 3 4 5 6 7 8 9 0 DOC/DOC 9 0 9 8 7 6 5

ISBN 0-07-012614-3

*The sponsoring editor for this book was Gail F. Nalven, the editing
supervisor was Peggy Lamb, and the production supervisor was Pamela
A. Pelton. It was set in Century Schoolbook by McGraw-Hill's
Professional Book Group composition unit.*

Printed and bound by R. R. Donnelley & Sons Company.

This book is printed on recycled, acid-free paper containing a minimum
of 50% recycled de-inked fiber.

Cover: *An aerial view of the French Limited Superfund Site in
Crosby, Texas. The 28 million gallons of contaminated sludge and soil
contained in and under this 7.3 acre lagoon (lower right corner) has
been successfully bioremediated. This is the largest and most complex
bioremediation project completed to date. About 70 million gallons of
hazardous industrial and commercial waste had been placed in this
sand mine by 1971. Bioremediation saved over $60,000,000 compared
to incineration.*

Courtesy: *The French Limited Task Group and ENSR Consulting
and Engineering, Houston, Texas.*

Dedication

This book is dedicated to Dr. Nathan C. Burbank, who possessed that rare quality of developing the best from those he taught. Except for parents, few of us have the fortune to have someone positively influence our life, particularly, in a way that significantly alters future years professionally and personally. I am one of those few, having been guided and motivated by Dr. Burbank (known to many as Burbie).

Dr. Burbank, as head of the Department of Civil and Sanitary Engineering at Washington University, St. Louis, Mo., guided my educational and professional career during my undergraduate years. He not only introduced me to environmental engineering, but he was solely responsible for directing my career toward graduate studies at the California Institute of Technology, Pasadena, Calif.

Although his guidance at the time appeared subtle, looking back one recognizes the profound scope of Burbie's influence. His ever-present optimism, enthusiasm, and joy of helping others provided the catalyst for their achievements. He nurtured and motivated many students during his teaching career that included:

Oklahoma State University, Okla. (1945–1956)
Washington University, St. Louis, Mo. (1958–1965)
University of Hawaii, Honolulu, Hawaii (1965–1977)
Pima Community College, Tucson, Ariz.

His professional achievements were numerous, including degrees from Harvard, Oklahoma State University, and Massachusetts Institute of Technology. He received the Arthur Bedell award of the Water Pollution Control Federation and the Resources Division Award of the American Water Works Association.

His support of and influence on students and many others went well beyond the requirements of academic excellence. His drive to aid and better the well being of others never ceased, as is illustrated by classes he conducted at the Arizona State Prison complex during his so-called "retirement" years. I am pleased to dedicate this book in honor of a great humanitarian.

Contents

Preface

The purpose of this book is to provide the fundamentals that support the degradation of hazardous compounds coupled with design and operational techniques for bioremediation systems. The book is intended for design engineers, regulatory officials, project managers, and students interested in remediation. It is assumed that they have some knowledge of conventional methods of wastewater treatment, design of biological processes, water chemistry, and microbiology. The title, *Bioremediation Engineering: Design and Application*, was chosen to emphasize the engineering aspects of using microorganisms for the remediation of contaminated soil, sludge, and groundwater.

This book has its origins in my application of biological processes to industrial wastewater treatment for the past 30 years, and bioremediation during the last 8 years. The contents reflect my teaching approach to technology transfer as a professor of environmental engineering for 10 years, and a practical approach to implementation from 18 years as head of an environmental engineering firm. The text integrates the scientific fundamentals with experience gained by many professionals in bioremediation. I have attempted to provide the latest design approaches and their applications.

It is difficult to write a book on a developing technology. Bioremediation has grown from an insignificant level to over 15 percent of all site remediations within 5 years. However, this growth highlights the need for a coherent and comprehensive presentation of the subject. A text of this nature is not only necessary for the application of bioremediation but is important for the acceptance of this technology. Biological processes have been perceived by some as unpredictable or unreliable. Apprehension can result from inappropriate applications or a lack of knowledge of the fundamentals of biological reactions and process design. I thank the many professionals who have attended my bioremediation training sessions for their motivation and requests that this text be written.

John T. Cookson, Jr.

Acknowledgments

This book could not have been accomplished without the support of my wife, Toni, who must have wondered where I disappeared to during the last 12 months. I thank her for understanding my one-track mind and inability to converse on normal everyday subjects. I also appreciate the support of my daughters, Valerie, Angela, and Christina. They made the duties of a father easy, allowing me the luxury to concentrate on this project. Finally, only one person could have interpreted my manuscript notes and maze of arrows and "insert here" diagrams. I am grateful to Joanne Meyer for her interpretation of my editing and assistance with a number of tasks in the preparation of this manuscript.

1

Application of Biological Processes

Introduction

We are blessed with what seemed, in earlier times, an unlimited abundance of land and resources. Today, however, every resource in the world shows in greater or lesser degree the carelessness and negligence to which each has been subjected. Thus we find ourselves confronted with the perplexing problem of reclaiming our national resources from hazardous chemicals. In the United States alone, the job is extraordinary, not to mention the imminent environmental problems of the previous Communist countries. European expenditures over the next 10 years are estimated at $1.33 trillion (ENTEC, 1993).

The Congressional Office of Technology Assessment (OTA) has projected cleanup of the nation's hazardous waste sites. Those under the jurisdiction of the Environmental Protection Agency (EPA) will require up to $500 billion and several decades to complete. OTA estimates that the EPA current list of waste sites will mushroom to as many as 9000 over the next 10 years (Table 1.1). Since 1980, EPA has identified 31,000 abandoned waste sites. Furthermore, these numbers do not include the thousands of industrial sites that have not made the EPA priority list. These additional sites that require cleanup are estimated at 19,000, plus 295,000 leaking underground storage tanks. The Department of Energy (DOE) is responsible for 3700 sites that are no longer active. Costs of remediation for DOE alone are estimated at $130 billion. To this must be added the 7300 sites under the Department of Defense (DOD) at an estimated cleanup cost of $25 billion.

Coupled with existing contamination is the continued release of hazardous chemicals. The U.S. EPA reported on the amount of toxic

TABLE 1.1   Sites Awaiting Cleanup as of 1993

Agency responsible	Number
U.S. Environmental Protection Agency*	
Superfund	1500–2000 sites
RCRA corrective action	1500–3000 sites
Underground storage tanks	295,000 units
U.S. Department of Defense*	7300 sites
U.S. Department of Energy*	4000 sites
States	19,000 sites
Potential New Sites	
U.S. Bureau of Land Management†	
Potential landfills†	3000 sites
Abandoned mine sites†	400,000 acres
U.S. National Park Service†	2400 sites
U.S. Bureau of Reclamation†	20–400 sites
U.S. Bureau of Indian Affairs†	
Abandoned mines	Not surveyed
Underground storage tanks	Not surveyed
U.S. Forest Service†	
Abandoned mine sites	2500 sites

*U.S. EPA, Cleaning Up the Nation's Waste Sites: Markets and Technology Trends, EPA 542-R-92-012.
†Majority Staff Report, U.S. House of Representatives, July 1993.

chemicals released to the environment during the year 1987 (U.S. EPA, June 1989). This information is a result of over 74,000 reports submitted to EPA by 19,000 industrial facilities employing at least 10 persons. The results are as follows:

- 9.6 billion lb of toxic chemicals were released to streams and U.S. waters.

- 1.9 billion lb were sent to municipal wastewater treatment plants

- 2.7 billion lb were put into landfills.

- 3.2 billion lb were injected into underground wells.

- 2.6 billion lb were sent to off-site treatment and disposal facilities.

Fifty percent of the reported releases were from manufacturing facilities that produce the chemical and the other 50 percent from businesses that use the chemicals.

During the next decade, bioremediation will play an expanding role in the remediation of contaminated soil, sludge, and groundwater. Within the last decade, bioremediation has grown from an unknown technology to one of the major treatment technologies considered for source control at Superfund sites. The basis for this growth is bioremediation's lower cost than technologies such as incineration and containment. Bioremediation also attracts interest because it destroys most organic wastes, eliminating health and ecological effects as well as future environmental liabilities.

Bioremediation is the application of biological treatment to the cleanup of hazardous chemicals. It requires the control and manipulation of microbial processes in surface reactors or in the subsurface, in situ treatment. Although bioremediation is less costly than other technologies, it is scientifically intense. The optimization and control of microbial transformations of organic contaminants requires the integration of many scientific and engineering disciplines. A greater degree of data as well as professional disciplines are necessary for assessing site applicability to bioremediation.

It was not long ago that many hazardous compounds were considered recalcitrant (inherent resistance to any biodegradation). In the last decade, significant knowledge has been gained on microbial metabolism of xenobiotic compounds (unnatural or synthetic compounds). If organisms can degrade these compounds, why does long-term contamination of sites exist? The answer to this question is the key to successful applications of bioremediation, and to answer this question requires adequate site assessments.

Hazardous compounds persist because environmental conditions are not appropriate for the microbial activity that results in biochemical degradation. The optimization of environmental conditions is achieved by understanding the biological principles under which these compounds are degraded, and the effect of environmental conditions on both the responsible microorganisms and their catalyzed reactions. One of the first engineering evaluations is determining the limiting environmental conditions at a specific site. The final design must provide the controls to manipulate this environment for enhancing biodegradation of the target compounds. Target compounds are those hazardous chemicals that are to be remediated by bioremediation.

During the initial efforts of developing innovative technologies for hazardous waste treatment, the application of biological systems received little attention. Bioremediation was given limited attention for several reasons. First, many hazardous compounds were believed to be resistant to biodegradation. Second, major deficiencies existed in the research on the biochemistry of microbial interactions. Third, the engineering profession contributed to this problem because of a lack of knowledge on biological processes. Unfortunately, this lack of knowledge supported the misconception that biological systems are uncontrollable and unpredictable. This attitude is not new but has always hindered the environmental profession, leading to the purely physical-chemical advanced wastewater treatment plants of the 1970s. These plants not only were costly but did not achieve destruction of the contaminants. Pollutants were only transferred from one medium to another.

Historical Beginnings

People have used biological systems since the recording of history. Wine production is illustrated in ancient myths, attributing the origin of wine to divine intervention rather than microorganisms. Producing raised breads has a more recent history. The use of yeast originated in Egypt about 6000 years ago.

People originally used indigenous yeast flora on grapes as the inoculum for fermentation. Indigenous microorganisms are the inoculum for most wastewater treatment and sludge composting processes. The first bioengineering process that used specific flora was yeast for raising bread and producing beer. In both processes, the inoculation is a specialized strain of yeast. Baker's yeast is used in the baking industry and brewer's yeast in the beer industry.

The basic practice of controlling water dates back to the Roman Empire, and the application of biological processes to benefit humans has existed since the recording of history. However, the design of biological processes for waste treatment is recent. The impetus for the application of biological processes to environmental improvement was and still is the concern for preserving human health.

The subject of pollution and disease received only occasional attention until 1832 and again in 1848 when the Asiatic cholera epidemic erupted in London. In 1849, John Snow reported on the relationship between fecal pollution of waters and disease. These findings were before the acceptance of Louis Pasteur's germ disease theory that provided the basis for public health and sanitation professions.

By 1860 scientists realized that a relationship between microorganisms and chemical change existed. The pioneer on these studies was Louis Pasteur. Pasteur studied the spoilage of beer and wine that he showed to be caused by growth of undesirable organisms. Pasteur referred to this spoilage as diseases of beer and wine.

The relationship between pollution and health was not recognized until the latter half of the nineteenth century. This era of sanitation awareness was ushered in by epidemic destruction across the world and the age of medical bacteriology. This era swept over Britain and hurdled the Atlantic in the middle of the nineteenth century. In America, Boston became the focal point for the sanitary awakening of America.

The development of the first engineered biological process in America had its foundations in the 1886 activities of the Massachusetts State Board of Health. This agency was organized under Dr. Henry Pickering Walcott, a physician aware of the public health and water pollution relationships. He established the Lawrence Experiment Station to determine the best practical methods for the purification of sewage or disposal of refuse. From 1890 to 1910, four principal investigators were responsible for developing a biological process for waste treatment, known as the activated sludge process. These investigators were:

Allen Hazen, an engineer-chemist in charge of the Lawrence Experimental Station

George C. Whipple, an MIT graduate in Civil Engineering

Harrison P. Eddy, a Boston consultant

William R. Nichols, a chemistry professor at MIT

Bioremediation is the extension of the principles of biological process design to treating hazardous chemicals. The treatment of hazardous chemicals is a greater challenge than treating domestic sewage because the chemical transformations must be precisely controlled. Sophisticated bioengineering, however, has a long history and success record. The biological process for ripening of curds is a complex process with many process variables. Process control of specific variables determines the nature of cheese produced. For cheddar cheese the process is controlled to convert only 25 to 35 percent of the protein to soluble products. Almost all the protein is converted for soft cheeses such as Camembert and Limburger.

Unlike early wine production, cheese requires a specific metabolic pathway with specific microorganisms. A lactic acid bacterium is used for hard cheeses that releases hydrolytic enzymes, known as exoenzymes. Exoenzymes function outside of the microbial cells. For soft cheeses, enzymes from yeast and other fungi that grow on the surface are desired. Frequently a very specific organism is used to impart a unique flavor to the cheese. Roquefort cheese gains its unique taste from the reactions catalyzed by a blue-colored mold, *Penicillium roquefort*. In the manufacturing of Swiss cheese, a species of the *Propioni bacterium* is used which produces carbon dioxide, yielding the characteristic holes of Swiss cheese.

Bioengineering

Industry has successfully applied biological processes for the production of many chemicals and food products. Engineers design and operate microbial processes to produce numerous high-quality products. Typical food items commercially produced by biological processes are cheese, soy sauces, yogurt, wine, beer, and alcohol. Microbial processes produce many industrial chemicals, including organic acids, solvents, polymers, specialized enzymes, and insecticides. A listing of some key industrial chemicals is given in Table 1.2. Many medical products are produced by biological processes, including antibiotics, steroids, and genetically engineered cancer-fighting compounds (Table 1.3).

Economics motivates the application of biological processes for the production of commercial products. The use of microorganisms for synthesis of organic compounds is frequently cheaper than nonbiological synthesis. This is even true for very simple chemical structures such as gluconic acid, citric acid, acetic acid, and ethanol.

TABLE 1.2   Industrial Chemicals Produced by Biological Processes

Acids	Alcohols and solvents
Citric	Acetone
Fumaric	Butanol
Gluconic	Ethanol
Glutamic	Fusel oil
Isoascorbic	Glycerol
Itaconic	
Lactic	
Lysine	
5'-Nucleotides	
Oxalic acid	
Tartaric acid	

Miscellaneous	Enzymes
Brewer's yeast	Amino butyric transaminase
Diacetyl	Amylase
Dihydroxyacetone	Catalase
Ephedrine	Cellulase
Ergot	Collagenase
Erythorbic acid	Diastase
	Glucose oxidase
	Glutamic acid decarboxylase
	Hemicellulase
	Invertase
	Lipase
	Pectinase
	Penicillinase
	Proteases
	Streptokinase-streptodornase

SOURCES: Stanier, 1986; Aiba, 1965.

TABLE 1.3   Medical Products Produced by Biological Processes

Antibiotics		
Amphotericin	Hygromycin	Ristocetin
Bacitracin	Kanamycin	Streptomycin
Blasticidin-S	Leucomycin	Tetracyclines
Cephalosporins	Neomycin	Thiostrepton
Colistin	Novobiocin	Trichomycin
Chloramphenicol	Nystatin	Tylosin
Cycloheximide	Oleandomycin	Tyrocidine
Cycloserine	Paromomycin	Tyrothricin
Erythromycin	Penicillins	Vancomycin
Fumagillin	Polymyxin	Viomycin
Gramicidin		

Steroids	Vitamins
Cortisone	Riboflavin (B_2)
Hydrocortisone	Vitamin B_{12}
Prednisolone	
9α-Fluoro-16β-methylprednisolone	
9α-Fluoro-16β-methylprednisolone	
6α-Methylprednisolone	

SOURCES: Stanier, 1986; Aiba, 1965.

A second reason is that biological processes can produce very complex organic molecules. Many of these, such as vitamin B_{12} and riboflavin, virtually preclude a commercially feasible chemical synthesis process (Fig. 1.1).

Economics is also the motivation for the use of biological processes to treat hazardous waste, as it was in the 1970s for wastewater treatment. In the late 1960s and early 1970s, the EPA initiated a major research effort to develop advanced wastewater treatment technology. One approach was to replace the biological processes in wastewater treatment with physical-chemical processes.

The motivation for reducing the dependency on biological processes from wastewater treatment was the misconception that biological processes are unreliable and unpredictable. Several full-sized physical-chemical plants were built. Soluble metals were precipitated, suspended solids coagulated and settled, dissolved gases stripped, and dissolved organic compounds adsorbed. Operation of these physical-chemical plants resulted in a very short design history. They were very costly to operate. They generated a large volume of sludge which had significant disposal costs. Finally, these treatment facilities only changed the phase of the contamination from water to sludge and/or air. On the other hand, biological processes convert a significant percentage of suspended and dissolved organic compounds to carbon dioxide, methane,

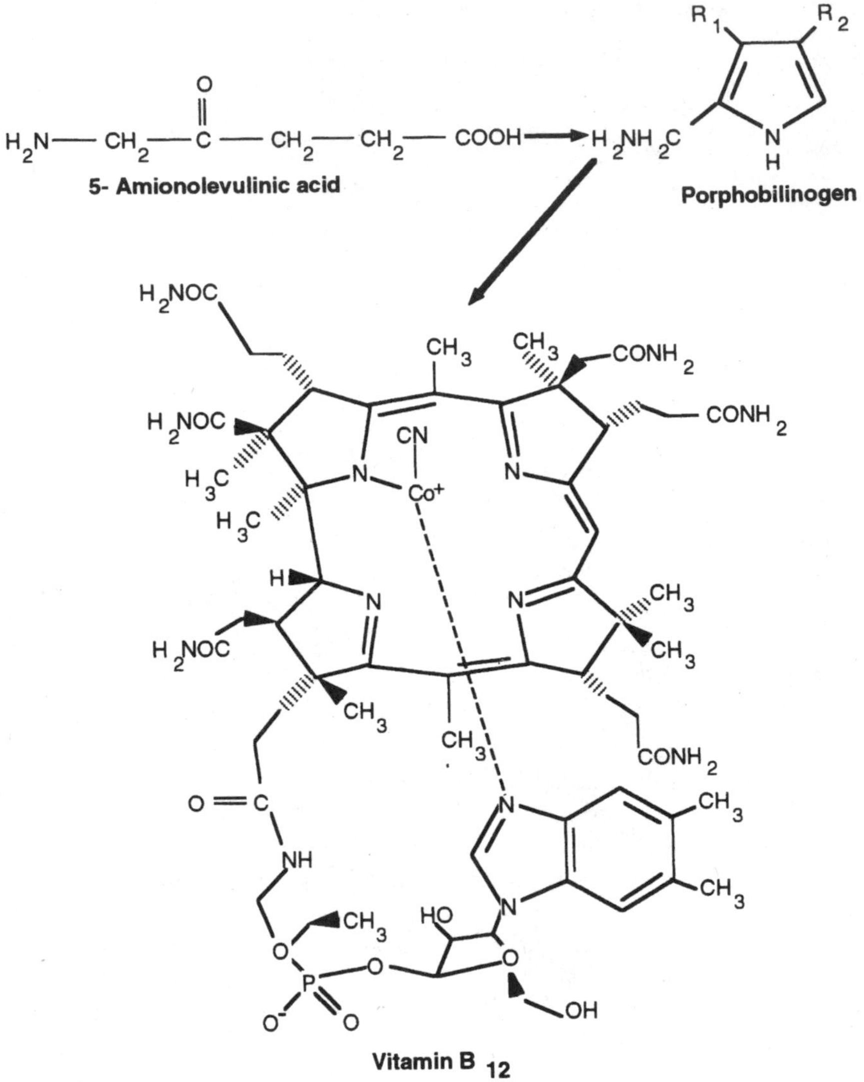

Figure 1.1 Biosynthesis of vitamin B$_{12}$ by *Pseudomonas denitrificans*. (*O'Sullivan*, 1991.)

nitrogen, hydrogen, and water. Biological wastewater processes have been developed for specific treatment goals (see Table 1.4).

To some degree, the cleanup of hazardous waste sites parallels that of wastewater treatment. Initial cleanup efforts were directed at physical and chemical processes. It is now clear that restricting cleanup to these processes is too great a cost burden for our economy. Bioremediation is often less costly (Table 1.5). Second, it is the only

TABLE 1.4   Biological Processes Designed for Specific Treatment Functions

Activated sludge and fixed film processes	General organic compound removal by conversion to carbon dioxide and water
Nitrification process	Conversion of nitrogen organic compounds and ammonia to nitrate
Denitrification process	Conversion of nitrate to nitrogen gas
Anaerobic digestion process	Conversion of more stable organic sludges to organic acids, alcohols, methane, hydrogen, and carbon dioxide
Alternate anoxic oxic process	Promotion of phosphate removal

TABLE 1.5   Cost-Effectiveness of Bioremediation*

Method	First year	Second year	Third year
Incineration	$530*	None	None
Solidification	$115	None	None
Landfill	$670	None	None
Thermal desorption	$200	None	None
Bioremediation	$175	$27	$20

SOURCE: Bioremediation Report, King Publishing Group, Washington, D.C., 1993.
*Annual cost for treatment per cubic yard.

practical solution for complete destruction or mineralization of an organic compound outside of incineration. Mineralization is the transformation of an organic compound to an inorganic form.

Biological processes are not unreliable when the process is controlled and the biochemistry and necessary environmental conditions are understood. The key to process performance in the commercial manufacturing of products is control of specific microorganisms under specific environmental conditions. This is also the key to successful bioremediation, process control through environmental control to bring about specific microbial transformations. Transformation is used in the same context as degradation. It simply means that the organic compound was changed. Neither transformation nor degradation implies mineralization. However, mineralization implies that transformation or degradation of the organic compound occurred.

Bioremediation Engineering

Bioremediation is the application of biological process principles to the treatment of groundwater, soil, and sludges contaminated with hazardous chemicals. There is little difference between the design principles of wastewater biological processes and those for bioremedi-

MICROBIAL CATALYZED REACTIONS

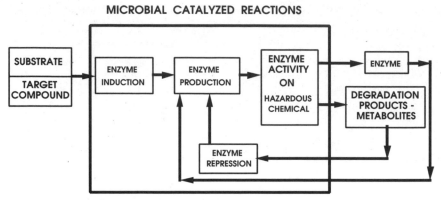

Figure 1.2 Generalized principles of microbial metabolism.

ation of hazardous chemicals. These processes are typical of those that environmental and chemical engineers have employed. Biological processes are catalyzed chemical processes. Bioremediation is more intricate because it uses a catalyst (enzyme) that is supplied by microorganisms to catalyze the destruction of a specific hazardous compound (Fig. 1.2). The hazardous compound (target compound) may or may not be the substrate. These catalyzed chemical reactions are conducted within a modular unit (the "cell") or outside of the cell. The principal reactions are oxidation-reduction reactions that are necessary for the generation of energy for the microorganisms.

The application of catalyzed oxidation-reduction reactions would be reactively simple if it weren't for the very specialized nature of microorganisms and their biochemistry. Meticulous environmental control is necessary for the catalyst production and the desired reaction. Thus, for successful bioremediation, the process must be controlled with the presence of a suitable energy source, an electron donor-acceptor system, and nutrients. Because of these complexities, process design and operation is all too frequently undertaken without complete knowledge of the chemical reactions and biological principles involved. This leads to uncertainty in the design and response of the system. Theoretical knowledge on all chemical aspects is never complete. Thus successful bioremediation frequently requires pilot testing. As knowledge about the mechanisms, kinetics, and microbial interactions for the transformation of hazardous chemicals expands, continued improvement will be made in the design and operation of biological remediation.

Many hazardous compounds that were considered recalcitrant are degradable. These compounds are better defined by the term persistent. Compounds resistant to biodegradation under a specific set of conditions are called persistent. In 1952, E. F. Gale stated, "It is probably not unscientific to suggest that somewhere or other some organism exists which can, under suitable conditions, oxidize any substance which is theoretically capable of being oxidized." If this principle is true, then why do contaminated sites exist with no or little self-purification? Understanding this is the foundation to designing a successful bioremediation project.

The extensive time that hazardous chemicals can exist in the ground is similar to the existence of humus and peat bogs. Humus, which makes up the dark brown or black color of fertile soils, is a very complex mixture of persistent organic compounds. It consists principally of degradation products of stable components of woody plants. Radiocarbon dating has established humus from certain soils to be thousands of years old. Peat is another natural organic material more persistent than many organic compounds. Peat bogs also will preserve human flesh and hair. On display in the British Museum is a well-preserved body that predates the Roman occupation of Britain.

The persistence of organic compounds relates directly to the environmental conditions that support microbial activity. For example, the high moisture and organic content that exists in peat bogs causes oxygen depletion. Oxygen is a necessary electron acceptor for the effective degradation of some compounds. In the absence of oxygen, other compounds serve as the electron acceptor, which leads to the accumulation of the breakdown products of fermentation metabolism. These compounds are mainly alcohols and organic acids. The accumulation of these compounds inhibits microbial activity, significantly reducing the diversity of microorganisms. This introduces two important concepts for bioremediation. First, the desired reactions may require a specific electron acceptor. Second, any single microbial species is extremely limited in what it can transform.

Organisms are highly specialized and play very specific roles in the mineralization of organic compounds. The ability of the microbial world to break down complex organic compounds is primarily a reflection of the metabolic versatility of its individual members. Recognition of this concept leads to the conclusion that a proper consortium of mixed microbial species is necessary. The principles that result in the preservation of peat and the more easily degraded organic matter are the same for hazardous chemicals. Environmental conditions are not suitable for microorganisms and their catalyzed reaction. Correct this condition, and you have biodegradation. Bioremediation requires containing, optimizing, and controlling nature's biological and chemical systems.

Some of the persistent chemicals are the nonoxygenated aromatic compounds. One of the first documented field investigations that illustrated natural bioremediation of a nonoxygenated aromatic compound

TABLE 1.6   Chemical Contamination at Pensacola, Fla., Site

Creosote wastes
  Nonoxygenated nitrogen heterocycles of:
    2,4-Dimethylpyridine
    Quinoline
    2-Methylquinoline
    Acridine
  Sulfur heterocycles of:
    Benzothiophene
    Dibenzothiophene
  Pentachlorophenol
  Diesel fuel

was for a contaminated aquifer in Pensacola, Fla. (Pereira, 1987). The site, located directly north of Pensacola Bay, was an 18-acre wood treatment facility. Waste materials were discharged into two unlined lagoons. Chemicals consisted of creosote with nonoxygenated nitrogen heterocycles, sulfur heterocycles, and pentachlorophenol solubilized in diesel fuel (Table 1.6). Although this contamination existed for 80 years, little mineralization of the organic compounds occurred. However, monitoring data illustrated that an anaerobic microbial system was transforming these compounds.

Field monitoring confirmed that indigenous microorganisms brought about this in situ bioremediation response. The geological cross section and monitoring well locations are illustrated in Fig. 1.3. The well's labels have increasing numbers with increasing distance from the lagoons. Microbial activity was demonstrated by total cell counts of 10^6 cells per gram of soil and by the production of methane (Table 1.7). The methanogenic microbial system yields methane as a degradation product.

Biochemical transformation of the contaminating chemicals is illustrated by the disappearance of isoquinoline with distance from the lagoons and the presence of 1[2H]-isoquinolinone (Fig. 1.4). Each monitoring location had multiple sampling wells with depths labeled in meters. The compound 1[2H]-isoquinolinone is a transformation product or metabolite of isoquinoline. The degradation pathway for this compound is discussed in Chap. 4 (Fig. 4.30). The same response is illustrated for quinoline, the original pollutant, and its metabolite quinolinone (Fig. 1.5).

Treatability studies documented that the indigenous microbial community could completely mineralize the nitrogen heterocycles under methanogenic conditions. Enhanced biodegradation occurred with the addition of nutrients and a carbon source such as acetate or propionate. The fact that 80 years passed with little recovery of this aquifer from contamination is explained when one addresses the

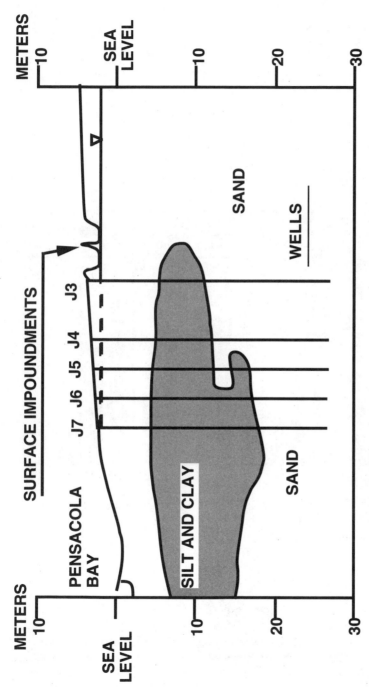

Figure 1.3 Geologic cross section of aquifer and well sites J3 through J7, Pensacola, Fla., site. (*Pereira, 1987.*)

13

TABLE 1.7   Microbial Activity at the Pensacola, Fla.,
Site

Microbial population
   10^6 total cells per gram of soil
Microbial type
   Facultative anaerobes
   Anaerobes
   Methanogens
Microbial metabolism
   Fermentation
   Methane fermentation
Microbial location
   From the surface of the lagoons to 150 m downgradient

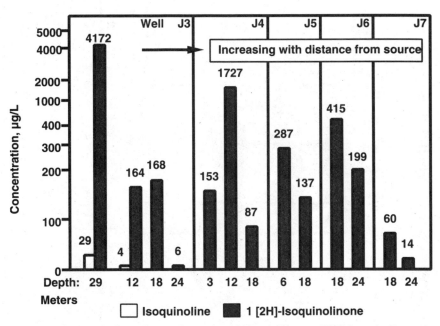

Figure 1.4   In situ biological transformation of isoquinoline to 1[2H]-isoquinolinone in a Pensacola, Fla., aquifer. (*Pereira, 1987.*)

requirements for bioremediation. The environmental conditions were not optimized for successful bioremediation.

The requirements for biodegradation are presented in Fig. 1.6 in descending order of importance. Of prime importance are microorganisms capable of producing enzymes that will degrade the hazardous chemical (target compound). Enzymes degrade compounds through exploitation of the organism's energy needs. Of prime importance is an energy source and electron acceptor, since microorganisms gain their energy through oxidation-reduction (redox) reactions. Also

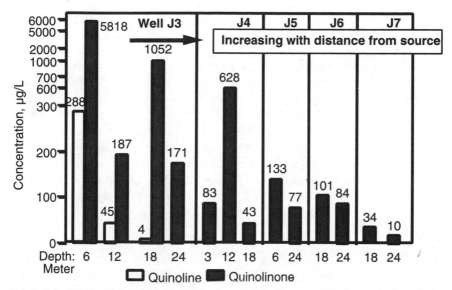

Figure 1.5 In situ biological transformation of quinoline to quinolinone in Pensacola, Fla., aquifer. (*Pereira, 1987.*)

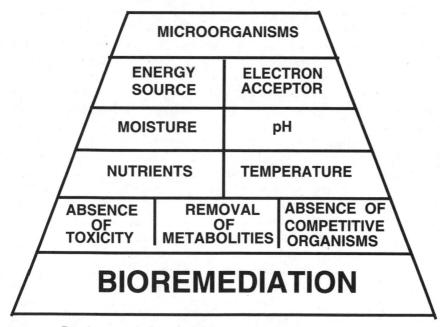

Figure 1.6 Requirements for bioremediation.

important are adequate moisture, pH, temperature, and nutrients for cellular growth. These requirements are discussed in Chap. 3.

To achieve bioremediation, the engineer must establish the limiting environmental conditions and then control these conditions for optimized bioremediation. Unfortunately, knowledge about the biotransformation of hazardous chemicals and the appropriate process controls is not widespread. One must start with the basic chemical reactions (biochemistry) and then include the interrelationships between the microbial consortiums, their energy and environmental needs, and process control. Successful process control requires knowledge on mass transfer, hydrogeology, and materials handling. Attention to these fundamentals results in successful and predictable bioremediation.

Chemical Nature of Hazardous Spills

A significant percentage of hazardous spills are petroleum-based, volatile solvent–based, or consist of polynuclear aromatic hydrocarbons (PAHs). Sites contaminated with pesticides, polychlorinated biphenyls (PCBs), and other specialty compounds represent a much smaller percentage. At one time industrial waste was classified according to the nature of the industry. However, industries use chemicals and generate by-products that include most of the chemical categories. Even the service industries use many chemicals that as a waste are considered hazardous. Waste-handling facilities and landfills have received waste from multiple sources. Society's broad use of chemicals and chemical toxicity has resulted in the need to identify specific chemicals.

Legislative action has provided the emphasis for identifying specific chemicals. U.S. EPA's formulation of the Hazardous Substance List (Appendix A) has propelled pollution control technology and waste management to recognize the chemical species over the industrial category. For these reasons, characterization of a site's contamination is based on historical chemical use records coupled with sophisticated instrumentation, including gas chromatography and mass spectrometry.

The widespread usage and storage of petroleum fuels has made petroleum hydrocarbons the most prevalent soil and groundwater contaminant. The petroleum industry has generated hydrocarbon mixtures with varying chemical characteristics. Sources of contamination include chemical storage tanks, oil-water separators, refining facilities, crude oil and fuel storage, drilling muds, oil field brine, and service stations. The composition of petroleum mixtures and sludges varies according to their origin, storage, treatment, and weathering conditions. Petroleum contamination, regardless of its source, frequently contains a large mixture of hydrocarbons. Jet fuels can contain over 300 different hydrocarbons (Swindell, 1988).

Many petroleum products contain aliphatic hydrocarbons, alicyclic hydrocarbons, aromatic hydrocarbons, and heterocyclic compounds. For example, fuel oil (No. 2) is a mixture containing approximately 30 percent alkanes, 45 percent cycloalkanes, and 25 percent aromatic compounds (Arvin, 1988). Fuel oil normally contains small amounts of heterocyclic compounds that contain the atoms of nitrogen, sulfur, and oxygen. Crude oils contain as much as 22 percent of these heterocyclic compounds.

Gasoline makes up most of the stored fuels. Gasoline is not a simple mixture but is a complex combination of hydrocarbons and other additives. One typical hydrocarbon mixture for gasoline is illustrated in Table 1.8. The seven predominant chemicals are isopentane, p-xylene, n-propylbenzene, 2,3-dimethylbutane, n-butane, n-pentane, and toluene (Table 1.9). These compounds make up more than 53 percent of the mixture.

A hydrocarbon spill will change in composition with time. Such factors as volatility, solubility, and biotic and abiotic reactions bring about weathering. After weathering, the composition of the gasoline spill changes significantly (Table 1.10). The major portion of chemicals change from that found in fresh gasoline (Table 1.11). This weathering action is a result of three mechanisms. Most of the short-chain aliphatic hydrocarbons volatilize. The more soluble compounds are transported from the site as a result of groundwater movement, and other compounds undergo biological transformations. Biodegradation of chemical components of hydrocarbon mixtures occurs simultaneously, but at different rates. These rates vary significantly, leading to sequential disappearance of individual compounds. These principles apply to the weathering of all chemical spills.

Creosote contamination is the second most prevalent hazardous waste problem (U.S. EPA, 1991). Creosote is a complex mixture of over 200 major chemicals. Creosote consists of approximately 85 percent polynuclear aromatic compounds (PAH), 12 percent phenolic compounds, and 3 percent heterocyclic compounds. Since many of the larger PAH compounds are not soluble in water, the chemical distribution changes in contaminated groundwater. The water phase becomes enriched with 45 percent phenolic compounds, 38 percent heterocyclic compounds, and only 17 percent PAH compounds (Arvin, 1988). The major mass of PAH compounds stay in the soil or at the site of the original spill or lagoon.

Volatile organic compounds represent the third most prevalent hazardous waste contamination. One portion of these volatile chemicals

TABLE 1.8   Composition of a Regular Gasoline

Component number	Chemical formula	Mole weight, g	Mass fraction	Mole fraction
1. Propane	C_3H_8	44.1	0.0001	0.0002
2. Isobutane	C_4H_{10}	58.1	0.0122	0.1999
3. n-Butane	C_4H_{10}	58.1	0.0629	0.1031
4. trans-2-Butene	C_4H_8	56.1	0.0007	0.0012
5. cis-2-Butene	C_4H_8	56.1	0.0000	0.0000
6. 3-Methyl-1-butene	C_5H_{10}	70.1	0.0006	0.0008
7. Isopentane	C_5H_{12}	72.2	0.1049	0.1384
8. 1-Pentene	C_5H_{10}	70.1	0.0000	0.0000
9. 2-Methyl-1-butene	C_5H_{10}	70.1	0.0000	0.0000
10. 2-Methyl-1,3-butadiene	C_5H_8	68.1	0.0000	0.0000
11. n-Pentane	C_5H_{10}	72.2	0.0586	0.0773
12. trans-2-Pentene	C_5H_{10}	70.1	0.0000	0.0000
13. 2-Methyl-2-butene	C_5H_{10}	70.1	0.0044	0.0060
14. 3-Methyl-1,2-butadiene	C_5H_8	68.1	0.0000	0.0000
15. 3,3-Dimethyl-1-butene	C_6H_{12}	84.2	0.0049	0.0055
16. Cyclopentane	C_5H_{10}	70.1	0.0000	0.0000
17. 3-Methyl-1-pentene	C_6H_{12}	84.2	0.0000	0.0000
18. 2,3-Dimethylbutane	C_6H_{14}	86.2	0.0730	0.0807
19. 2-Methylpentane	C_6H_{14}	86.2	0.0273	0.0302
20. 3-Methylpentane	C_6H_{14}	86.2	0.0000	0.0000
21. n-Hexane	C_6H_{14}	86.2	0.0283	0.0313
22. Methylcyclopentane	C_6H_{12}	84.2	0.0000	0.0000
23. 2,2-Dimethylpentane	C_7H_{16}	100.0	0.0076	0.0093
24. Benzene	C_6H_6	78.1	0.0076	0.0093
25. Cyclohexane	C_5H_{12}	84.2	0.0000	0.0000
26. 2,3-Dimethylpentane	C_7H_{16}	100.2	0.0390	0.0371
27. 3-Methylhexane	C_7H_{16}	100.2	0.0000	0.0000
28. 3-Ethylpentane	C_7H_{16}	100.2	0.0000	0.0000
29. 2,2,4-Trimethylpentane	C_8H_{18}	114.2	0.0121	0.0101
30. n-Heptane	C_7H_{16}	100.2	0.0063	0.0060
31. Methylcyclohexane	C_7H_{14}	98.2	0.0000	0.0000
32. 2,2-Dimethylhexane	C_8H_{18}	114.2	0.0055	0.0046
33. Toluene	C_7H_8	92.1	0.0550	0.0568
34. 2,3,4-Trimethylpentane	C_8H_{18}	114.2	0.0121	0.0101
35. 2-Methylheptane	C_8H_{18}	114.2	0.0155	0.0129
36. 3-Methylheptane	C_8H_{18}	114.2	0.0000	0.0000
37. n-Octane	C_8H_{18}	114.2	0.0013	0.0011
38. 2,4,4-Trimethylhexane	C_9H_{20}	128.3	0.0087	0.0065
39. 2,2-Dimethylheptane	C_9H_{20}	128.3	0.0000	0.0000
40. p-Xylene	C_8H_{10}	106.2	0.0957	0.0858
41. m-Xylene	C_8H_{10}	106.2	0.0000	0.0000
42. 3,3,4-Trimethylhexane	C_9H_{20}	128.3	0.0281	0.0209
43. o-Xylene	C_8H_{10}	106.2	0.0000	0.0000
44. 2,2,4-Trimethylheptane	$C_{10}H_{22}$	142.3	0.0105	0.0070
45. 3,3,5-Trimethylheptane	$C_{10}H_{22}$	142.3	0.0000	0.0000
46. N-Propylbenzene	C_9H_{12}	120.2	0.0841	0.0666
47. 2,3,4-Trimethylheptane	$C_{10}H_{22}$	142.3	0.0000	0.0000
48. 1,3,5-Trimethylbenzene	C_9H_{12}	120.2	0.0411	0.0325
49. 1,2,4-Trimethylbenzene	C_9H_{12}	120.2	0.0213	0.0169

TABLE 1.8   Composition of a Regular Gasoline (Continued)

Component number	Chemical formula	Mole weight, g	Mass fraction	Mole fraction
50. Methylpropylbenzene	$C_{10}H_{14}$	134.2	0.0351	0.0249
51. Dimethylethylbenzene	$C_{10}H_{14}$	134.2	0.0307	0.0218
52. 1,2,4,5-Tetramethylbenzene	$C_{10}H_{14}$	134.2	0.0133	0.0094
53. 1,2,3,4-Tetramethylbenzene	$C_{10}H_{14}$	134.2	0.0129	0.0091
54. 1,2,4-Trimethyl-5-ethylbenzene	$C_{11}H_{16}$	148.2	0.0405	0.0260
55. n-Dodecane	$C_{12}H_{26}$	170.3	0.0230	0.0129
56. Naphthalene	$C_{10}H_8$	128.2	0.0045	0.0033
57. n-Hexylbenzene	$C_{12}H_{20}$	162.3	0.0000	0.0000
58. Methylnaphthalene	$C_{11}H_{10}$	142.2	0.0023	0.0015
Total			0.9969	1.0000

SOURCE: Johnson, P. C., Kemblowski, M. W., and Colhart, J. D., "Quantitative Analysis for the Clean-up of Hydrocarbon-Contaminated Soils by In-Situ Soil Venting," *Groundwater*, Vol. 28, No. 3, pp. 413–429, 1990.

TABLE 1.9   Major Constituents of Fresh Gasoline

Constituent	Percent
Isopentane	10.5
p-Xylene	9.6
n-Propylbenzene	8.4
2,3-Dimethylbutane	7.3
n-Butane	6.3
n-Pentane	5.9
Toluene	5.5
Total	53.5

find their source from petroleum hydrocarbon spills. These are the aromatic volatiles of benzene, toluene, ethylbenzene, and xylene (BTEX compounds). Other major volatile organic compounds are the halogenated solvents that are used as solvents, as cleaning solutions, and for chemical synthesis. In a survey of contaminated water supplies, the most prevalent halogenated contaminants were (Semprini, 1988):

- Trichloroethylene
- 1,1,1-Trichloroethane
- Tetrachloroethylene
- *cis*- and *trans*-1,2-Dichloroethylene
- 1,1-Dichloroethane

TABLE 1.10   Composition of a Weathered Gasoline

Component number	Chemical formula	Mole weight, g	Mass fraction	Mole fraction
1. Propane	C_3H_8	44.1	0.0000	0.0000
2. Isobutane	C_4H_{10}	58.1	0.0000	0.0000
3. n-Butane	C_4H_{10}	58.1	0.0000	0.0000
4. trans-2-Butene	C_4H_8	56.1	0.0000	0.0000
5. cis-2-Butene	C_4H_8	56.1	0.0000	0.0000
6. 3-Methyl-1-butene	C_5H_{10}	70.1	0.0000	0.0000
7. Isopentane	C_5H_{12}	72.2	0.0200	0.0296
8. 1-Pentene	C_5H_{10}	70.1	0.0000	0.0000
9. 2-Methyl-1-butene	C_5H_{10}	70.1	0.0000	0.0000
10. 2-Methyl-1,3-butadiene	C_5H_8	68.1	0.0000	0.0000
11. n-Pentane	C_5H_{12}	72.2	0.0114	0.0169
12. trans-2-Pentene	C_5H_{10}	70.1	0.0000	0.0000
13. 2-Methyl-2-butene	C_5H_{10}	70.1	0.0000	0.0000
14. 3-Methyl-1,2-butadiene	C_5H_8	68.1	0.0000	0.0000
15. 3,3-Dimethyl-1-butene	C_6H_{12}	84.2	0.0000	0.0000
16. Cyclopentane	C_5H_{10}	70.1	0.0000	0.0000
17. 3-Methyl-1-pentene	C_6H_{12}	84.2	0.0000	0.0000
18. 2,3-Dimethylbutane	C_6H_{14}	86.2	0.0600	0.0744
19. 2-Methylpentane	C_6H_{14}	86.2	0.0000	0.0000
20. 3-Methylpentane	C_6H_{14}	86.2	0.0000	0.0000
21. n-Hexane	C_6H_{14}	86.2	0.0370	0.0459
22. Methylcyclopentane	C_6H_{12}	84.2	0.0000	0.0000
23. 2,2-Dimethylpentane	C_7H_{16}	100.0	0.0000	0.0000
24. Benzene	C_6H_6	78.1	0.0100	0.0137
25. Cyclohexane	C_5H_{12}	84.2	0.0000	0.0000
26. 2,3-Dimethylpentane	C_7H_{16}	100.2	0.1020	0.1088
27. 3-Methylhexane	C_7H_{16}	100.2	0.0000	0.0000
28. 3-Ethylpentane	C_7H_{16}	100.2	0.0000	0.0000
29. 2,2,4-Trimethylpentane	C_8H_{18}	114.2	0.0000	0.0000
30. n-Heptane	C_7H_{16}	100.2	0.0800	0.0853
31. Methylcyclohexane	C_7H_{14}	98.2	0.0000	0.0000
32. 2,2-Dimethylhexane	C_8H_{18}	114.2	0.0000	0.0000
33. Toluene	C_7H_8	92.1	0.1048	0.1216
34. 2,3,4-Trimethylpentane	C_8H_{18}	114.2	0.0000	0.0000
35. 2-Methylheptane	C_8H_{18}	114.2	0.0500	0.0468
36. 3-Methylheptane	C_8H_{18}	114.2	0.0000	0.0000
37. n-Octane	C_8H_{18}	114.2	0.0500	0.0468
38. 2,4,4-Trimethylhexane	C_9H_{20}	128.3	0.0000	0.0000
39. 2,2-Dimethylheptane	C_9H_{20}	128.3	0.0000	0.0000
40. p-Xylene	C_8H_{10}	106.2	0.1239	0.1247
41. m-Xylene	C_8H_{10}	106.2	0.0000	0.0000
42. 3,3,4-Trimethylhexane	C_9H_{20}	128.3	0.0250	0.0208
43. o-Xylene	C_8H_{10}	106.2	0.0000	0.0000
44. 2,2,4-Trimethylheptane	$C_{10}H_{22}$	142.3	0.0000	0.0000
45. 3,3,5-Trimethylheptane	$C_{10}H_{22}$	142.3	0.0250	0.0188
46. n-Propylbenzene	C_9H_{12}	120.2	0.0829	0.0737
47. 2,3,4-Trimethylheptane	$C_{10}H_{22}$	142.3	0.0000	0.0000
48. 1,3,5-Trimethylbenzene	C_9H_{12}	120.2	0.0250	0.0222
49. 1,2,4-Trimethylbenzene	C_9H_{12}	120.2	0.0250	0.0222
50. Methylpropylbenzene	$C_{10}H_{14}$	134.2	0.0373	0.0297

TABLE 1.10   Composition of a Weathered Gasoline (Continued)

Component number	Chemical formula	Mole weight, g	Mass fraction	Mole fraction
51. Dimethylethylbenzene	$C_{10}H_{14}$	134.2	0.0400	0.0319
52. 1,2,4,5-Tetramethylbenzene	$C_{10}H_{14}$	134.2	0.0400	0.0319
53. 1,2,3,4-Tetramethylbenzene	$C_{10}H_{14}$	134.2	0.0000	0.0000
54. 1,2,4-Trimethyl-5-ethylbenzene	$C_{11}H_{16}$	148.2	0.0000	0.0260
55. n-Dodecane	$C_{12}H_{26}$	170.3	0.0288	0.0181
56. Naphthalene	$C_{10}H_8$	128.2	0.0100	0.0083
57. n-Hexylbenzene	$C_{12}H_{20}$	162.3	0.0119	0.0078
58. Methylnaphthalene	$C_{11}H_{10}$	142.2	0.0000	0.0000
Total			1.0000	1.0000

SOURCE: Johnson, P. C., Kemblowski, M. W., and Colhart, J. D., "Quantitative Analysis for the Clean-up of Hydrocarbon-Contaminated Soils by In-Situ Soil Venting," *Groundwater*, Vol. 28, No. 3, pp. 413–429, 1990.

TABLE 1.11   Major Constituents of Weathered Gasoline

Constituent	Percent
2,2-Dimethylheptane	12.4
Toluene	10.5
2,3-Dimethylpentane	10.2
n-Propylbenzene	8.3
n-Heptane	8.0
Total	49.4

Applications of Bioremediation

A number of chemical contaminants and waste categories have been remediated by bioremediation (Table 1.12). Applications include the three most common hazardous waste occurrences (Table 1.13). The majority of bioremediation applications have been to fuel spills and fuel storage facilities. The U.S. EPA has recently reported the distribution of bioremediation activities on Superfund sites. Bioremediation activities by chemical category are 33 percent petroleum, 28 percent creosote, 22 percent solvents, 9 percent pesticides, and 8 percent other (Fig. 1.7). The other waste group includes industrial facilities with mixed chemical contributions. These bioremediation activities include above-surface and subsurface technologies. These are summarized in Table 1.14 and discussed in Chaps. 7 to 9.

Bioremediation has definite advantages and disadvantages that must be considered when selecting a technology (Tables 1.15 and

TABLE 1.12   Chemical Compounds and Wastes That Have Been Bioremediated

Acetone	Industrial wastes
Acrylonitrile	Isopropyl acetate
Animal fats and grease	Methanol
Anthracene	Methylene chloride
Benzene	Methylethyl ketone
Benzopyrene	Methylmethacrylate
t-Butanol	2-Methylnaphthalene
Butylcellosolve	Monochlorobenzene
Chrysene	Naphthalene
Coal tar	Pentadecane
Crude oil	Petroleum hydrocarbons (miscellaneous)
2,4-D	Phenanthrene
1,2-Dichloroethane	Phenols
Diesel fuel	Polynuclear aromatic hydrocarbons (PAHs)
Dodecane	Pyrenes
Ethylacrylate	Stoddard solvent
Ethylbenzene	Styrene
Ethylene glycol	Tetrahydrofuran
Fatty amines	Toluene
Fluoranthene	Trichloroethylene (TCE)
Gasoline	1-Tridecene
Hexadecane	Xylene
Hexane	

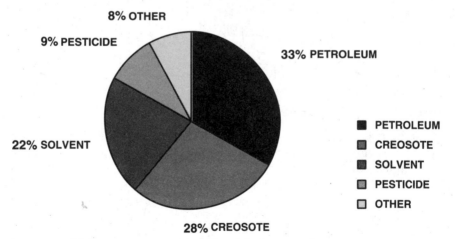

Figure 1.7   Major waste types remediated by bioremediation. (*U.S. EPA/540/ N-93/001.*)

1.16). Bioremediation can treat chemicals to acceptable levels as frequently as alternate technologies. The two major advantages are lower capital cost and the ability to perform the task on site.

Disadvantages include the scientific intensity of bioremediation design and operation (Table 1.16). With bioremediation, one must

TABLE 1.13   The Most Common Hazardous Waste
Occurrences

No. 1   Petroleum hydrocarbons
   Major contaminants:
      BTEX compounds
         Benzene
         Toluene
         Ethylbenzene
         Xylene
      Aromatics
      Heterocyclic aromatics

No. 2   Halogenated aliphatic compounds
   Major contaminants:
      Trichloroethylene
      1,1,1-Trichloroethane
      Tetrachloroethylene
      cis-trans-1,2-Dichloroethylene
      1,1-Dichloroethane

No. 3   Industrial waste lagoons and wood preservation
   Major contaminants:
      Creosote
      Polynuclear aromatics
      Pentachlorophenol
      Nitrogen heterocyclic aromatics
      Aromatics

TABLE 1.14   Bioremediation
Systems

Solid-sludge wastes:
   Designed lagoon
   In situ lagoon
   Land farming
   Composting
   Slurry reactors
Liquid wastes:
   Designed lagoon
   In situ lagoon
   Surface reactors
   Surface fixed film reactors
   In situ
      Vadose zone
      Saturated zone

manage the system extremely well. This, however, is offset by the
lower capital intensity of the process. Bioremediation activities may
prolong the presence of hazardous materials on site versus removal.
This provides for potential worker liability and accidental release.
However, for other technologies liabilities are associated with trans-

TABLE 1.15   Advantages of Bioremediation

- Can be done on site
- Permanent elimination of waste
- Biological systems are cheaper
- Positive public acceptance
- Long-term liability risk eliminated
- Minimum site disruption
- Eliminates transportation cost and liability
- Can be coupled with other treatment techniques

TABLE 1.16   Disadvantages of Bioremediation

- Some chemicals cannot be bioremediated
- Extensive monitoring needs
- Site-specific requirements
- Toxicity of contaminates
- Scientific intensive
- Potential production of unknown by-products
- Perception of unproved technology

portation and the ultimate disposal system. Thus many of the disadvantages that can be listed for bioremediation can be equally said of other technologies. A frequent question is can one guarantee the results. There is hardly any cleanup activity dealing with hazardous waste for which the treatment program can be guaranteed.

A significant concern of some engineers, scientists, and regulators is that bioremediation is an unproved technology. This concern is based on the lack of extensive field experience. However, the number of sites at which bioremediation has been completed is rapidly expanding. The U.S. EPA Superfund Records of Decision (ROD) is including bioremediation treatment for source control. The first ROD at a Superfund remedial site was signed in 1987 at Seymour Recycling in Indiana (Dean, 1992). EPA has listed 159 sites that as of 1993 plan to use or have used bioremediation (EPA/540/N-93/001). Through fiscal year 1991, bioremediation represented 9 percent of 498 site remedial actions (Table 1.17) (Dean, 1992).

Another concern is the possibility that unknown by-products can go undetected during the cleanup program. Adequate field monitoring, knowledge of the degradation products, and treatability studies can provide the information to prevent movement of undesired by-products. Laboratory testing of biological systems is less transferable to field activities than that of physical-chemical processes. However, for contaminated sites, laboratory testing doesn't accurately predict field

TABLE 1.17   Alternative Treatment
Technologies Applied To Superfund
Remedial Actions*

Technology	Percent of actions
Solidification and stabilization	26
Off-site incineration	17
On-site incineration	13
Soil vapor extraction	17
Bioremediation	9
Surface	5
In situ	4
Thermal desorption	6
Soil washing	3
In situ flushing	3
Dechlorination	2
Solvent extraction	1
Other	<3

SOURCE: Dean, 1992.
*Data derived from 1982–1992 Records of
Decision and anticipated activities as of
February 1992.

results for many processes. This is due to the matrix in which contamination exists as well as the variance in concentrations and chemical species. Differences within a site's geology and soil chemistry
make laboratory studies, and even field pilot studies, less successful
in predicting full-scale results. This problem exists regardless of the
treatment alternate being evaluated. A final disadvantage is that not
all chemicals can be bioremediated.

2

Managing a Bioremediation Project

Introduction

Groundwater and soil remediation are both a technical and management challenge. Although this can be said for any engineering project, bioremediation of hazardous substances demands the highest of skills. The number of scientific and technical disciplines that must be coordinated with several regulatory agencies and citizens requires the best that management can deliver. For a successful project, management is as important as technology, and they are so tightly integrated that it is sometimes difficult to separate their components.

A successful project is the net result of four major components: good negotiation, good project management, good engineering, and good implementation of field operations. In this section, the project management of bioremediation is discussed. The balance of this text addresses the engineering and field operations.

The project manager orchestrates the necessary inputs from many disciplines. This requires adequate technical knowledge on bioremediation so appropriate limits are placed on the scope of each effort, while ensuring adequate data for successful design and operation. The capabilities of the technology and its limitations must be understood, and the potential regulatory areas for negotiation recognized. Not all negotiated items may be favorable for the client, or even the cleanup effort. You must recognize those that are vital for the success versus those that can be offered as your part of the mediation. It may be unwise to be unbending on a request costing several thousand dollars when you need to negotiate decisions that influence the project's success or the cost by several hundred thousand dollars. The final approach to environmental projects is a result of negotiation.

Most projects are performed in phases, but these phases often overlap and one impacts on the other. From a discussion standpoint, a bioremediation project can be grouped in the following phases:

Defining the project and goals

Assembling the project team

Initial review of remediation options

Performing site characterization

Screening and selecting remediation alternatives for detail evaluation

Developing a design concept

Developing the supporting elements of the project

Designing and conducting laboratory treatability testing

Designing and conducting field pilot testing

Revising process design

Obtaining final approvals

Performing field remediation activities

Obtaining site closure

Defining the Project and Goals

One of the first aspects of any project is to recognize the players and establish the client requirements and needs. Frequently the client requirements may not be those that you believe they need or those the regulatory agency believes. Because of this multiplayer interaction, the management of remediation projects requires as much skill as any engineering program that one will ever administrate. To succeed in management, first recognize who are the players that make up the remediation effort. Remediation projects are not a simple two-player (client and consultant/contractor) arrangement. Most remediation projects have three main players who interact with a fourth (Fig. 2.1). These include the client or the potential responsible parties, the regulatory agency (one or more may be involved at various stages), the consultant/contractor, and the public. This triparty triangle is linked through bonds varying from weak to strong, and they operate within the circle of public domain (Fig. 2.1).

The client-consultant/contractor bond is strong. The bonds between the client and regulatory agency may be strong or weak, depending on how the client has organized their interaction rela-

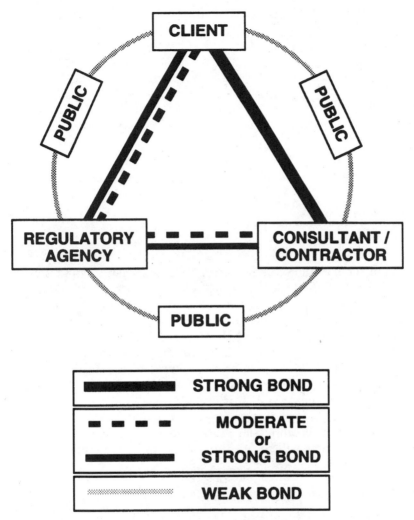

Figure 2.1   Interaction diagram of remediation project players.

tive to the consultant. As a result, the bond between the consultant and the regulatory agency is either weak or strong. The public's interaction is a weak interaction that influences all components of the triangle.

The project manager must establish the nature of these interactions and who will be responsible for communications to each party. One and only one person of each party should be the contact for communication flow. After communications and responsibilities have been established, determine the nature and extent of the environmen-

TABLE 2.1   Information Needs for
Problem Definition

Establish client requirements
Delineate regulatory requirements
Establish preliminary project goals
Establish contamination:
   Nature
   Extent
Delineate project sensitivity:
   Health and safety
   Ecological
   Political
Determine site constraints
Establish schedule requirements
Define potential solutions
Determine information needs

tal problem. This initial data gathering includes scientific and nonscientific information (Table 2.1).

The client's requirements are integrated with regulatory requirements, and each may be influenced by the public. From these interactions one establishes the goal of the remediation program. The program's goal defines the extent and hopefully the degree of cleanup. Cleanup levels may be delineated in general terms or they can be very specific. Most projects, however, are initiated with significant negotiations continuing toward this end.

Defining a cleanup level is a complex and potentially time-consuming process. It involves all of the players in the remediation program (Fig. 2.1). Some states will set cleanup levels for target compounds in terms of concentrations. Other states will not provide a specified concentration but strive to achieve the best attainable level by the selected technology. This level is not defined until a decision is made by the regulatory agency that cleanup has been accomplished. Thus the acceptable cleanup may not be known until the remediation effort has been completed to the best ability of the consultant/contractor. For these states, the screening, design, and project implementation proceeds with regulatory input and guidance, but no defined cleanup limits.

Agencies may require cleanup to background or drinking water levels. These requirements may result in excessive costs or impractical operations that provide little improvement in overall environmental conditions or risk. In such cases, consideration should be given to the use of risk-based cleanup levels. Performing an environmental risk assessment for establishing risk-based cleanup levels can achieve a balance between remediation cost and the potential risks present at a particular site. For large remediation projects, there are two up-front studies that initially appear costly: risk assessments and treatability studies. However, both risk assessments and treatability studies can provide a return of 2 to 10 times their actual cost. Project management must recognize when these studies are appropriate and adequately present the potential cost savings to the client.

The engineering goal is to achieve the program goal without risking health, ecological impacts, or litigation exposure. From a theoretical standpoint, this requires the collection of necessary data for design and operation within a given budget. From a practical standpoint, achieving this is quite complex. First is the unknown degree of site heterogeneity and even the extent of contamination. Data needs change and expand as an investigation proceeds, making a fixed budget difficult to work with. Second, the scope of each phase is negotiated with all involved parties. The project's scope may be established because of political reasons as much as health or environmental risk. These negotiations are not free of influence by the fourth player. The public's interests and potential for interaction should not be underestimated. Sensitive projects frequently result in the interaction with citizens' committees, review boards, community groups, or even congressional leaders.

An important part of good project management is maintaining the professional image of the team. This impression is frequently conveyed by the individual responsible for communication to the players. A consultant/contractor team is retained for a remediation job because they sold their capabilities to complete a successful project. This probably resulted from a proposal, oral presentation, or experience with the client. In a multiparty project, this effort does not end after job selection. The regulatory officials and citizens are of equal importance and must be sold on the consultant/contractor's ability.

Anticipating likely questions and being prepared to respond immediately rather than having to prepare the response after the challenge significantly aids the team's image. Be prepared to address known issues that will be raised before they are raised by the regulatory agency or citizens. Frequently asked questions of bioremediation are tabulated in Table 2.2. The remediation action plan should address each of these in a clear and detailed presentation. Those areas requiring further studies should be so indicated and a preliminary approach stated for obtaining answers.

The Project Team

Any engineering project is only as successful as its management and the capabilities of the project team. Good management requires one designated head, the project manager, who is responsible for the total

TABLE 2.2   Frequent Questions of Regulators and Citizens

How do you know these compounds can be biologically degraded?
Are secondary by-products produced?
What is the toxicity of the secondary by-products?
How will the secondary by-products be controlled?
Does the bioremediation process increase the migration of contaminants?
Does the process increase volatilization?
How can you confirm degradation?
How will you document cleanup?
Where has this technology been applied before?
Are the microorganisms natural?
What are the biological dangers of these microorganisms?
How long will the project take?

project. The project manager must have a clear definition of the project, its goal, and how the principal parties interact. This knowledge is vital to the proper configuration of the project team. The nature and extensiveness of the project team are a direct function of the project's scope. It can include many professional disciplines. The project team can be divided into categories related to job functions and project phases. Potential team job functions are provided in Table 2.3.

Initial Review of Remediation Options

Potential remediation techniques influence the type of data that must be obtained for screening and selection of remediation alternatives. An initial review of remediation options is important before site characterization. Potential remediation options should be listed and the data requirements of each delineated. For bioremediation, this step takes on a different scope versus other technologies.

It is at this stage that the preliminary feasibility of bioremediation must be determined. Evaluating bioremediation as an alternate remediation technology starts with an assessment of the degradation potential of the contaminants. This assessment includes the ease or difficulty of degradation, the ability to achieve total mineralization, as well as the environmental conditions necessary for mineralization. These principles are discussed in Chaps. 3 to 5.

Site Characterization

A well-managed bioremediation project is achieved only when site evaluations have been properly performed. Site characterization is the first and most important stage in the design of any bioremediation effort. Since bioremediation is a scientific intense technology, site

**TABLE 2.3   Job Functions and
Professions for a Bioremediation Team**

Contracting officer
Project manager:
  Regulatory coordinator and analysts
  Community relations coordinator
  Quality assurance officer
  Health and safety officer
  Permit coordinator
  Risk assessment coordinator
  Design engineer chief:
    Environmental engineer
    Civil and structural engineer
    Mechanical engineer
    Electrical engineer
  Scientific team leader:
    Geologist and hydrogeologist
    Microbiologist
    Industrial hygienists
    Chemists
  Field operations chief:
    Trainer
    Site engineer
    Electrician
    Carpenter
    Welder
    Plumber
    Equipment operator
    Equipment maintenance technician
    Sampling technician
    Security officer
  Analytical laboratory coordinator:
    Field analytical technicians
    Data validation analyst
    Chemist

characterization requires an intense investigation, particularly for in situ treatment. Each discipline must be integrated to its proper extent during site characterization and throughout the bioremediation program.

The site characterization program collects data for evaluating alternatives and supporting design. It must define the potential biological systems and the site characteristics that impact on biological reactions and must provide information for process control. Site data include physical, chemical, and biological parameters as well as hydrogeological data. Hydrological studies must provide a thorough understanding of the subsurface flow characteristics. Biological and chemical data must provide information on the site conditions important to chemical transformation of the hazardous compounds (target compounds). Chapter 6 discusses the nature of these data, how they are obtained, and their application.

Screening and Selecting Remediation Alternatives

After site characterization, the potential remediation alternatives are screened. The screening may be performed in several phases. The first phase eliminates the obvious. Based on the preliminary screening, one or more are selected for detailed evaluation. Key evaluation factors include technical feasibility, cleanup schedule, regulatory response, and site-specific constants. Although cost is important, it is difficult to project reasonable costs during the first screening. Several alternatives may be selected for more detailed evaluations, including treatability studies and cost projections. The review, screening, and selection of the preferred remediation generally follow the approach provided in Table 2.4. However, each project is unique, as are the players. Regulatory agencies are not similar nor do they follow the same definition of cleanup.

Developing the Design Concept

The most appropriate remediation alternative is formulated into a preliminary design. The first step is determining if the project will be in situ, surface, or a combination. This decision is usually governed by the location of the contaminants, the site's hydrogeological conditions, and site constraints. If surface treatment is selected, the system is either a pump and treat or excavate and treat depending on the media: liquid or soil. If in situ is selected, the location of the contamination and the required metabolism mode establish the feasible

TABLE 2.4   Components for Screening and Selecting a Proposed Remedial Action*

- Screen technologies for remedial action and select appropriate technologies for review
- Develop remedial action alternatives
- Preliminary screening of alternatives and selection of alternatives for a detailed analysis
- Detailed analysis of selected alternatives including an assessment of the technology feasibility, environmental and public health impact, ability to meet cleanup levels, institutional acceptability, and cost
- Selection of the preferred alternative and development of a remedial action plan

*Modified from California state guidance document, California Department of Health Services, 1985.

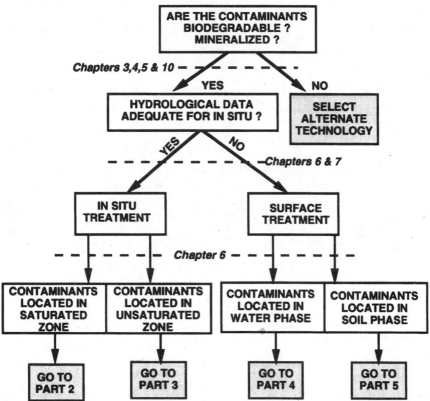

BIOREMEDIATION ?

Figure 2.2   The bioremediation decision tree, part 1.

approach. The conceptual design approaches for bioremediation can be guided by the decision trees of Figs. 2.2 through 2.8. Factors that determine which path to select on these decision trees are discussed in the remaining chapters of this text. Pertinent chapter numbers are located next to each major decision.

Supporting Elements

It is the project manager's responsibility to see that all supporting elements of a project are scheduled and properly developed. Many of these programs must be submitted to regulatory agencies, and in some cases permits must be obtained. Permitting requirements of local and state agencies must not be forgotten. Address all regulatory requirements promptly for maintaining the project's schedule and working relationships.

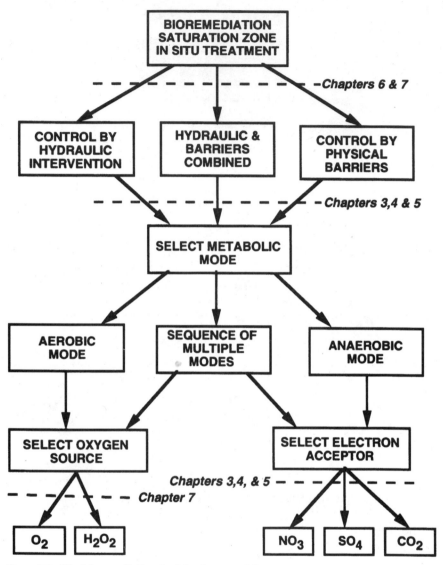

Figure 2.3 The bioremediation decision tree, part 2.

Supporting elements for remediation projects are listed in Table 2.5. Each is important for a successful and safe program. Most of these elements are distinct projects requiring a program leader and a well-defined plan. They are usually formulated by an interdisciplinary team of technical experts. Many of these elements must be submitted to regulatory agencies for approval before field remediation activities.

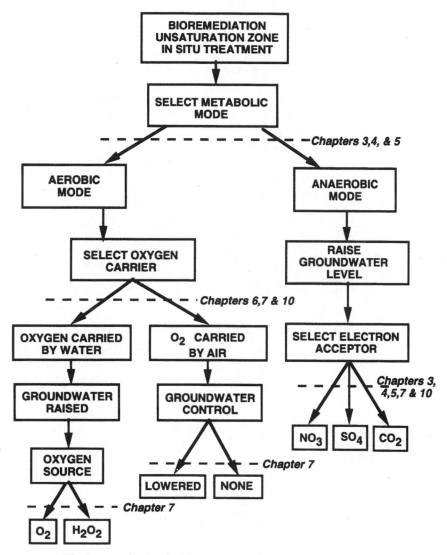

Figure 2.4 The bioremediation decision tree, part 3.

Health and safety plan

The health and safety plan must consider a broad range of potential health and safety problems. These include occupational exposures to hazards associated with the waste and use of equipment. A listing of considerations is provided in Table 2.6.

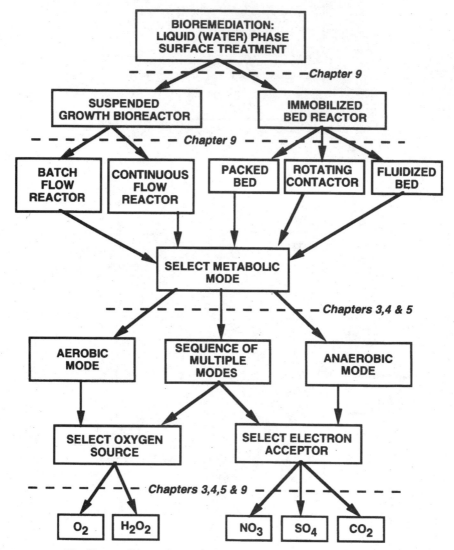

Figure 2.5 The bioremediation decision tree, part 4.

Spill and contingency plan

An important support element of any hazardous waste cleanup action
is a contingency plan. The contingency plan is additional insurance to
protect against the conditions that may threaten health and safety of
site workers, the surrounding environment, and people. The contingency plan sets forth responsibilities and the sequence of activities for
an emergency.

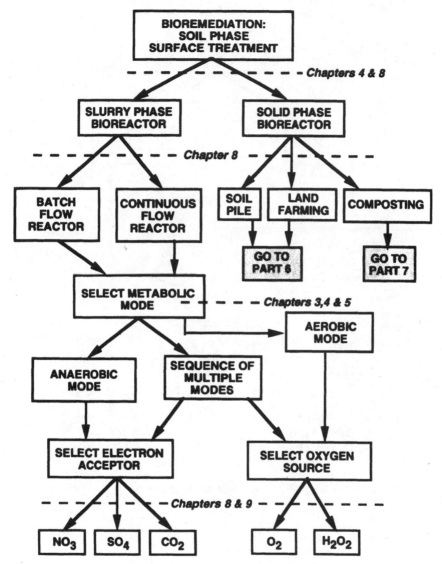

Figure 2.6 The bioremediation decision tree, part 5.

Monitoring plan

Monitoring programs must be designed for multiple purposes. These include monitoring for personnel safety and medical screening, monitoring for process operation, monitoring for site emissions, and monitoring for site closure. Each monitoring program has a specific goal and must be designed to properly characterize this goal. Monitoring is an impor-

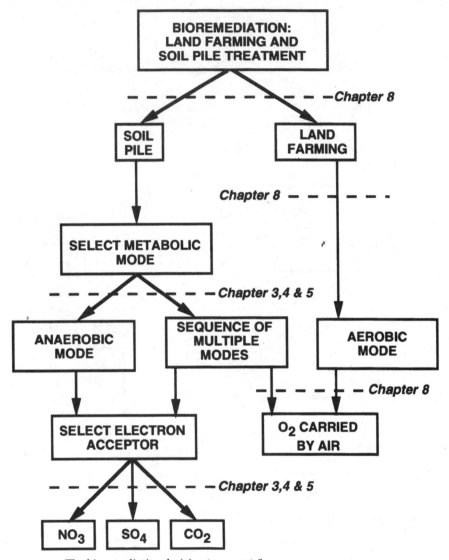

Figure 2.7 The bioremediation decision tree, part 6.

tant part of the health and safety program, and the method for process control and documenting the success of the bioremediation program.

Community relations plan

The public's desire to be involved in environmental decisions is only logical. We live in a society in which environmental regulations and

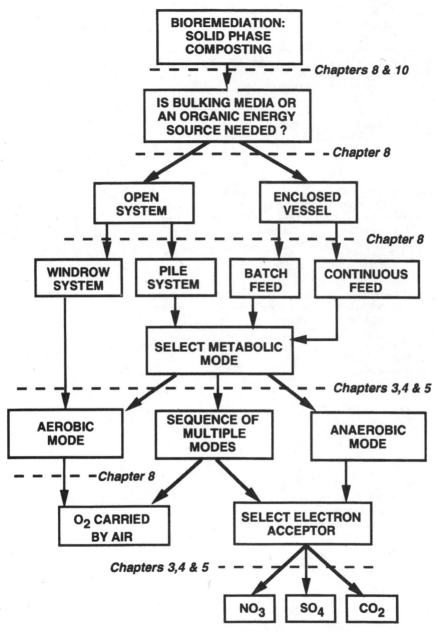

Figure 2.8 The bioremediation decision tree, part 7.

**TABLE 2.5   Supporting Elements of a
Remediation Project**

Develop a health and safety plan
Develop a spill and contingency plan
Develop a monitoring plan
Develop a community relations plan
Develop a residuals management plan
Formulate site utility requirements
Formulate permit requirements
Formulate and schedule reporting requirements

**TABLE 2.6   Consideration for Developing Health and Safety
Programs and Contingency Plans**

1. Toxicity of waste site chemicals:
   Acute toxicity
   Chronic toxicity
   Dermatology effects
   Epidemiological studies
   Other possible health effects
2. Material safety data sheets for major waste site chemicals
3. Potential routes of waste chemical exposure:
   Skin contact
   Inhalation
   Injection
4. Respiratory protection
5. Protective clothing and equipment
6. Classification of site work areas:
   Exclusion zone
   Contamination reduction zone
   Noncontaminated or clean zone
7. Industrial hygiene and safety requirements:
   Equipment use and protective guards
   Electrical hazards
   Facility design
8. Changing room requirements
9. Fire fighting techniques
10. Medical monitoring requirements
11. Medical training:
    First aid
    CPR
    Recognize medical symptoms
12. Standard operating procedures
13. Decontamination procedures
14. Waste-handling procedures
15. Personal hygiene and cleanliness
16. Training:
    Off-site hands-on practice
    On-site dry runs prior to start-up

SOURCE: Adapted from Sawyer, 1994.

their enforcement are driven by the public as based on their information. Public information is all too frequently limited to the news media and catchy phrases. Everyone is well aware of the potential horror of environmental contamination. The grisly scenes of dead alewives in Lake Michigan (Fig. 2.9), of crude oil–covered birds of the west coast (Fig. 2.10), or of cattle ill from drinking contaminated stream water (Fig. 2.11) are all too well known. Couple these visual

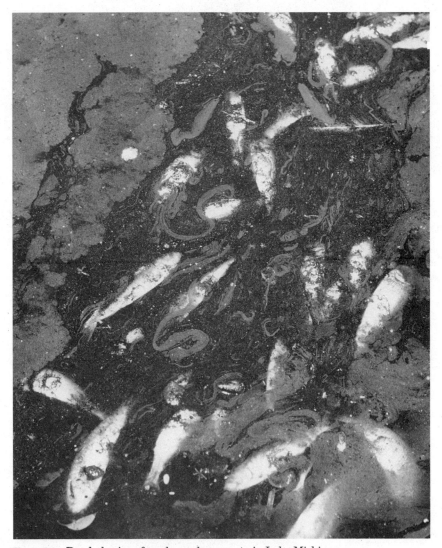

Figure 2.9 Dead alewives from hazardous waste in Lake Michigan.

Figure 2.10 One of the thousands of wildlife species impacted by petroleum spills, West Coast, United States.

impacts with the publicized correlations between cancer and hazardous chemicals, and one has the basis for understanding citizen interaction. For projects under the Superfund law, the client may have no more influence over the technology selection than other affected parties. This can include affected communities.

Sensitive projects require coordination of information flow (Table 2.7). Trust and communication are the key tools for effective commu-

Figure 2.11 Ill shorthorn afflicted by partial paralysis from hazardous chemicals in their watering source.

TABLE 2.7   Designing a Communications Program

Prepare a strategic goal:
  A communications program
  A public involvement project
Employee communication program:
  Informal meetings
  Technical summary sheets
Public communication program:
  Audience analysis
  Message design
Media relations program
Community relations projects:
  Meeting with community groups
  Fact sheets and comment solicitation
Agency relations program:
  Information meetings
  Technical fact sheets
  Reports
  Joint planning sessions
Conflict approach program:
  Analyze causes of conflict
  Establish criteria for solution
  Select a problem-solving course of action

nity relations. Providing clear presentations on data that support remediation decisions is an important part of program management. Equally important is recognizing potential concerns of either the community or regulatory agencies and addressing these. It is not enough to satisfy the needs of the client. More than once a successful on-time project has been transformed into a financial failure by either poor regulatory communications or poor community relations.

Residuals management

Most remediation actions generate some residuals that must be disposed of. These include treated water, solids, or contaminated equipment and waste supplies. A program for storage, transportation, and disposal is necessary. These actions frequently require various federal and local permits.

Site utilities

Remediation actions require site utilities in the form of electric, water, telephone, and waste discharge. Each need must be assessed for capacity and appropriate permits.

Permit requirements

Remediation activities require many permits that must be obtained on schedule to prevent costly field delays. The required federal permits are usually obvious and clear from interactions with the principal players in the remediation project. These frequently include permits for water discharges, air emissions, and hazardous waste treatment. The volume of potential local and state permits makes permitting a program effort of its own. After the potential permitting requirements are established, a matrix of permits, dates required, and agencies should be developed. Some permits require coordination and approval from several agencies. Possible permitting needs are tabulated in Table 2.8.

Treatability Studies

Treatability studies consist of laboratory evaluations, pilot studies, and field demonstrations. Laboratory studies yield information for selecting the best treatment program. Pilot studies yield information on the expected performance, process control parameters, and design criteria. Treatability studies are often a prerequisite for many bioremediation projects. Few, if any, contaminated sites are identical, and identical responses are unlikely. Bioremediation performance is a function not only of the contaminated media and the contaminants but also of the genetic variability of microbial species.

The design of treatability studies is discussed in Chap. 10. The first step is to establish the goals and the treatability budget. One goal is to determine if the bioremediation scheme will succeed in meeting cleanup limits at a specific location. A second goal is the development of data for design and operation. After the biodegradability of the contaminants is established, treatability studies can be used to evaluate the environmental conditions necessary for efficient degradation of the target compounds. This provides information for enhancing the degradation and determining cost-effective modifications to the microbial environment. Treatability studies allow one to refine projections on the following:

Assess the ability to meet the cleanup level.

Estimate the reduction in contaminant toxicity, volume, and mobility.

Develop cost and time estimates for the remediation program.

Estimate long-term operation, maintenance, and monitoring requirements.

TABLE 2.8   Potential Permits for Remediation Activities

Air navigation obstruction
Air noise zone permit
Air quality permit
Blasting permit
Building permit
Certificate of potability
Coastal zone permit
Corps of Engineers permit
County or city access permit
Electrical permit
Environmental impact statement
Erosion control permit
Fence permit
Fish & Wildlife permit
Grading permit
Groundwater appropriation permit
Hazardous waste management permit
Highway access permit
Landfill permit
Mechanical permit
Modular structure permit
Noise control variance
NPDES permit
Oil operations permit
Public service and utility permits
Radioactive materials license
Railroad approvals
Site plan approvals
Soil percolation test permit
Special exception (zoning opinion)
Surface water appropriation
Tank permit
Trailer permit
Tree cutting permit
Wastewater discharge permit
Water quality permit
Water and sewer connection permit
Waterway construction permit
Wetland permit
Zoning variance (zoning opinion)

Compare the performance with results of alternative treatment studies.

Identify contaminant fate and the removal due to biological and nonbiological mechanisms.

Determine the type and concentration of residual contaminants and by-products.

Determine the optimal conditions for biodegradation.

Evaluate the need for pretreatment and other treatment modes.

Produce the design information for final process design.

Analytical expertise is an important part of any bioremediation project. The first need for this expertise is during site characterization. The second need is for support in the treatability study. Finally, laboratory support is used to monitor the progress of bioremediation projects and confirm that cleanup levels are achieved.

Revising Process Design

Design revision incorporates the findings from treatability studies to yield the best treatment concept. Inputs from all participating parties are considered and the remediation program revised as necessary. After designs are updated to reflect new data, the revised program is submitted for approval by the client. Frequently regulatory agencies also will require the next to final design for review.

Finalize Design

Finally, design and specifications are completed. These are based on final discussions with appropriate regulatory agencies and the final approval of the client. This phase includes the finalization of the health and safety plan, spill and contingency plan, and a program to monitor the effectiveness of bioremediation.

Remediation Field Activities

On-site activities must be under a single person, the chief of field operations, whose day-to-day responsibilities include:

- Planning and scheduling daily activities
- Monitoring and directing site activities
- Managing resources for:
  - Workers
  - Materials
  - Equipment
- Line of communications to:
  - Field staff

- On-site agency coordinators
- Safety and health officers

■ Training program

The chief of field operations is in charge of and responsible for all on-site actions, including ceasing activities until corrective actions are taken when any unsafe procedures or disobedience occurs. The on-site agency coordinators and safety and health officers as well as other consultant/contractor staff members follow the chief's directives when on site.

One of the most important on-site activities is training. All site workers must be fully trained and informed. Each shift must be initiated with a clear set of objectives for the day's actions. Correct procedures and potential safety hazards should be reviewed for the day's activities. Classroom education before each work shift on the day's activities is an effective means to minimize the need for implementation of a contingency action.

3

Microbial Systems of Bioremediation

Introduction

This chapter discusses the basic microbial systems and requirements for successful bioremediation. It covers the principles of microbial metabolism, enzyme catalyzed reactions, and requirements for biological processes important to bioremediation. It discusses the nature of biochemical reactions and the mechanisms by which microorganisms mediate degradation of hazardous compounds.

Microbial Energy

The biological destruction of hazardous chemicals is based on the principles that support all ecosystems. These principles involve the circulation, transformation, and accumulation of energy and matter. The implementation of bioremediation requires an understanding of the interrelationships of these microbial functions. The design of a bioremediation process involves the optimization and control of select portions of biochemical cycles. The biochemical transformation of energy and matter requires catalysts. Microorganisms are the catalyst generator, and enzymes are the catalysts. These enzymes result in degradative (catabolic) reactions to provide energy and material for synthesis of additional microbial cells.

Process optimization for degradation of hazardous chemicals requires an understanding of the microorganisms involved, their nutrient needs, the biochemistry of their mediated reactions, and why they promote the reaction. This last point is the basic driving force of all biological reactions: Microorganisms must obtain energy. The specific reaction by which the organisms gain this energy is determined by the amount of energy yield. Thus thermodynamics can be used to predict the specific biochemical reaction. Microorganisms have several modes of metabolism to obtain this energy, which manifests itself through oxidation-reduction reactions. The mechanisms by which microorganisms obtain energy are the basis for their major classification subdivisions.

Organisms are subdivided into major categories according to their metabolic capabilities. These metabolic characteristics are a function of the energy source, the carbon source, and the electron donor and electron acceptor of the mediated oxidization-reduction reaction. The first major classification subdivision is based on the energy source to support biosynthesis and growth. One energy source is sunlight that is directly used by *phototrophs*. The second energy source is chemical energy that is used by *chemotrophs*. How microorganisms obtain their principal carbon source, disregarding requirements for specific growth factors, is the second parameter for classification. Organisms that use CO_2 as the principal carbon source are defined as *autotrophic*. Organisms that use organic compounds as the principal carbon source are defined as *heterotrophic*. The final category is based on the source of electrons that the organisms used for oxidization-reduction reactions. If the electron donor is an organic compound, the organisms are called *organotrophs*. If an inorganic compound is the electron source, they are called *lithotrophs*.

This classification scheme ignores the important fact that microorganisms are nutritionally versatile, so that a given organism may belong to more than one category. Thus it is impossible to assign many organisms to a single metabolic classification. Most sites contaminated with hazardous organic chemicals support chemoorganotrophic and chemoheterotrophic organisms. Chemoorganotrophs are the most important class of microorganisms for degrading hazardous organic compounds. However, chemolitrophic and chemoautotrophic organisms can be actively involved in the complete mineralization of hazardous chemicals. The chemoorganotrophs obtain their energy from such compounds as ammonia, nitrite, molecular hydrogen, hydrogen sulfide, and sulfur. Any bioremediation effort is not an open cycle but is closed at some point by the interaction of heterotrophs and autotrophs.

Microorganisms obtain energy by a complex sequence of redox reactions, oxidation-reduction reactions. Oxidation is the removal of electrons from an atom or molecule. Reduction is the addition of electrons. The oxidation of iron is an example of a simple oxidation reaction, in which a single electron is removed:

$$Fe^{2+} \rightarrow Fe^{3+} + e^-$$

Several bacteria, *Gallionella* and *Thiobacillus ferrooxidans*, can oxidize ferrous iron (Fe^{2+}) to ferric iron (Fe^{3+}) and deposit it as insoluble ferric hydroxide. Oxidation can occur only with reduction, and the electron must be accepted by another atom or molecule. This electron flow generates energy through a sequence known as the electron transport chain. This chain consists of molecules that undergo repeated oxidation and reduction and transfer the electron from one to another (Fig. 3.1). The electrons are transported in the microbial cell system by such compounds as nicotinamide adenine dinucleotide phosphate (NADP) (Fig. 3.2).

The pyridine nucleotides can readily undergo reversible oxidation and reduction due to the nature of the nicotinamide group (Fig. 3.2). The transition in oxidization state of these and similar compounds results in the storage of energy within the microbial cell in the form of energy-rich chemical bonds. It is the exploitation of these energy needs that the bioremediation specialist must engineer.

The chemoheterotrophic organisms get their energy by oxidation-reduction reactions that employ an organic compound as the carbon

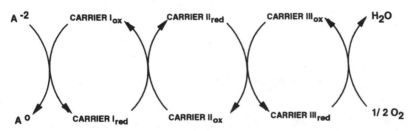

Figure 3.1 Respiratory electron transport systems, linking an oxidizable substrate A^{-2} with the reduction of molecular oxygen to water.

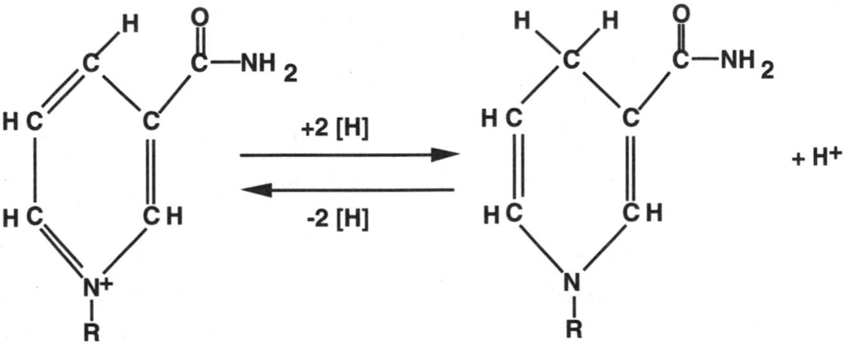

Figure 3.2 Oxidized and reduced forms of nicotinamide group of pyridine nucleotides.

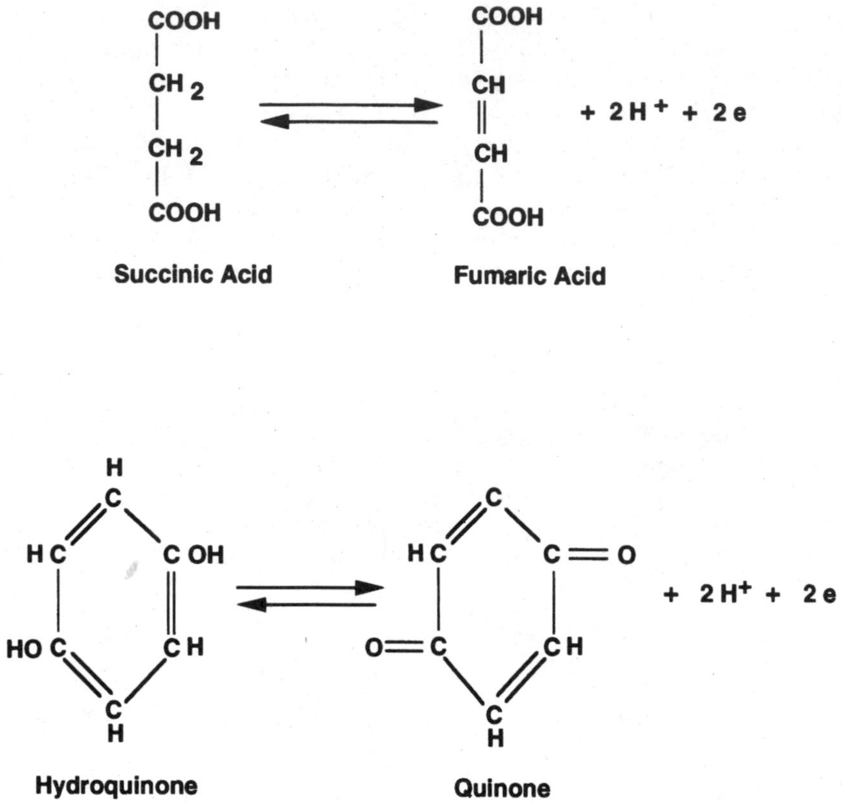

$$\text{Succinic Acid} \rightleftharpoons \text{Fumaric Acid} + 2H^+ + 2e$$

$$\text{Hydroquinone} \rightleftharpoons \text{Quinone} + 2H^+ + 2e$$

Figure 3.3 Oxidation of organic compounds: succinic acid and hydroquinone.

source. Each step in the oxidation of an organic compound involves the removal of two electrons and the simultaneous loss of two protons. This is equivalent to the removal of two hydrogen atoms and is called *dehydrogenation*. Reduction of an organic compound involves the addition of two electrons and two protons and can be called *hydrogenation*. The oxidation of succinic acid to fumaric acid and hydroquinone to quinone is an example (Fig. 3.3).

An electron acceptor must be present to accept the released electrons and is thereby reduced. The nature of the electron acceptor establishes the mode of energy-yielding metabolism.

Microbial metabolism

Bioremediation engineering results in the controlled manipulation of the environment to generate the proper enzymes for catalyzing the desired reactions. As discussed in Chap. 1, bioremediation is the application of chemistry, but more intricate because all reactions are catalyzed by a special enzyme that can only be brought to the system by specific microorganisms. Thus bioremediation starts with knowledge on the basic biological reactions and how these reactions are influenced by the environment. From this information one determines the environmental changes needed and designs the necessary process controls.

To design the bioremediation process, one must first determine the desired degradation reactions to which the target compounds will be subjected. This is the same as designating the metabolism mode that will occur in the process, since the metabolism mode is defined by the nature of the redox reaction. The metabolism modes are broadly classified as aerobic and anaerobic. Aerobic microorganisms and aerobic reactions occur in the presence of molecular oxygen with molecular oxygen serving as the electron acceptor. This form of metabolism is respiration. Anaerobic reactions occur only in the absence of molecular oxygen. The anaerobic reactions are subdivided under anaerobic respiration, fermentation, and methane fermentation (Table 3.1).

Bacteria have developed a wide variety of respiration systems. These can be characterized by the nature of the reductant and of the oxidant. In all cases of aerobic respiration, the electron acceptor is molecular oxygen (Table 3.2). Anaerobic respiration uses an oxidized inorganic or organic compound other than oxygen as the electron acceptor (Table 3.2). The respiration of organic substrates by bacteria is, in most respects, very similar. The substrates are oxidized to CO_2, with the successive removal of pairs of H^+ and electrons.

Fermentation is the simplest of the three principal modes of energy-yielding metabolism. During fermentation organic compounds serve as both electron donors and electron acceptors. The process maintains a strict oxidation-reduction balance. The average oxidation level of the end products is identical with that of the substrate fer-

TABLE 3.1 Metabolism Modes

Type	Electron acceptor
Aerobic respiration	Oxygen (O_2)
Anaerobic respiration:	
Denitrification	Nitrate (NO_3^-)
Nitrate reduction	Nitrate (NO_3^-)
Sulfate reduction	Sulfate (SO_4^{2-})
Ferric iron reduction	Iron (Fe^{3+})
Obligate proton reduction	Protons (H^+)
Sulfur reduction	Sulfur (S)
Organic respiration	Fumarate
Organic reduction	Trimethylamine oxide
Fermentation	Organic compound
Methane fermentation (methanogenesis)	Carbon dioxide (CO_2)

TABLE 3.2   Reductants and Oxidants in Bacterial Respiration

Reductant electron donor	Oxidant electron acceptor	Products	Example organism or group
Aerobic respiration:			
H_2	O_2	H_2O	Hydrogen bacteria
Organic	O_2	$CO_2 + H_2O$	Many bacteria and animals
NH_3	O_2	$NO_2^- + H_2O$	Nitrifying bacteria *Nitrosomonas*
NO_2^-	O_2	$NO_3^- + H_2O$	Nitrifying bacteria *Nitrobacter*
Fe^{2+}	O_2	Fe^{3+}	*Ferrobacillus*
S^{2-}	O_2	$SO_4^{2-} + H_2O$	*Thiobacillus*
Anaerobic respiration:			
Organic	NO_3^-	$N_2 + CO_2$	Denitrifying bacteria
Organic	SO_4^{2-}	$H_2O + S^{2-}$	Sulfate reducers
H_2	SO_4^{2-}	$H_2O + S^{2-}$	Sulfate reducer *Desulfovibrio*
H_2	CO_2	CH_4, H_2O	Methanogens
Organic	NADP	CH_4, CO_2	Methanogens
Organic	H^+	H_2	Acetogens
Organic	Organic	Organic	Many fermentation bacteria

TABLE 3.3   Major End Products of
Bacterial Fermentation

Propionic acid	Ethanol
Formic acid	Butanol
Acetic acid	Isopropanol
Butyric acid	2,3-Butanediol
	Acetone
	Carbon dioxide
	Hydrogen
	Methane

mented. Thus the substrate yields a mixture of end products, some more oxidized than the substrate and others more reduced. The end products depend on the type of microorganisms but usually include a number of acids, alcohols, ketones, and gases (Table 3.3).

Fermentation can proceed only under strictly anaerobic conditions, in the absence of molecular oxygen. For some organisms, molecular oxygen is a toxic substance. These organisms are called *strict anaerobes*. Other organisms can perform fermentation in the absence of oxygen, and in the presence of oxygen they change their mode of metabolism to respiration. These organisms are called *facultative anaerobes*.

Other important metabolism modes for bioremediation include the use of nitrate as an electron acceptor (performed by denitrifiers and nitrate reducers), the use of sulfate, thiosulfate, or sulfur as an electron acceptor (performed by sulfate reducers), the use of protons as an electron acceptor (performed by proton-reducing acetogens), the use of carbon dioxide as an electron acceptor (performed by methanogens), and the use of chlorinated organic compounds as electron acceptors. The use of chlorinated organic compounds as an electron acceptor for energy-yielding reactions is a recent finding (Dolfing and Tiedje, 1987).

Another important metabolism concept in bioremediation is cometabolism. In a true sense, cometabolism is not metabolism (energy-yielding) but fortuitous transformation of a compound. Traditionally, it was believed that organisms must obtain energy from an organic compound to biodegrade it. However, enzymes generated by an organism growing at the expense of one substrate also can transform a different substrate that is not associated with that organism's energy production, carbon assimilation, or any other growth process. This mode of activity is called cometabolism. Cometabolism is defined as the degradation of a compound only in the presence of other organic material that serves as the primary energy source (McCarty, 1987).

Hazardous chemicals that lend themselves to bioremediation by becoming a secondary substrate through cometabolism are only partially transformed. This transformation may or may not result in reducing toxicity. If all toxicity properties of a hazardous compound are removed, this is referred to as detoxification. Detoxification does not imply mineralization. Fortunately, the metabolites or transformation products from cometabolism by one organism can typically be used as an energy source by another.

Both the hazardous compound and the primary substrate will compete for enzymes. This competition for enzymes has been termed substrate competitive inhibition. Thus the rate of cometabolism is usually enhanced by reducing the overall substrate competition for enzymes. This is achieved by maintaining a low or near depleted primary substrate concentration. In some cases, cometabolism can continue for short periods from stored cellular energy even after the primary substrate is depleted. For a cometabolism metabolic mode, the degradation rate of the target compound is dependent on the electron flow from the primary substrate. This concept applies to the methanotrophs that are used for oxidation of halogenated aliphatic compounds as well as the reductive dehalogenation in anaerobic systems.

The term gratuitous metabolism has been used to describe the ability of a microorganism to break down a second compound that it does not use for an energy source. The distinction that has been made between cometabolism and gratuitous metabolism is that with gratuitous metabolism a compound can be acted upon by enzymes without the need for energy or reducing power. An example of this is a bioreactor in which a pregrown culture of microbial cells is brought in contact with a hazardous compound. This compound is degraded as a result of the enzymes present in the culture, and the degradation ceases since the culture does not continue to grow. With cometabolism, energy or reducing power must be present to transform the hazardous compound. In cometabolism, no transformation of the hazardous compound would occur since the pregrown cells could not extract energy to drive the energy-requiring step of cometabolism.

Cometabolism is important for many transformations, including some polynuclear aromatic hydrocarbons, halogenated aliphatic and aromatic hydrocarbons, and pesticides. Many compounds are subject to bioremediation by cometabolism (Table 3.4). Many microbial species exhibit the phenomenon of cometabolism. Some are listed in Table 3.5 (Horvath, 1972).

A final aspect of microbial metabolism is the recognition of preferential substrate degradation. Preferential degradation results in a sequence of attack. Typically the highest-energy-yielding compounds are degraded first. Preferential substrate metabolism is a function of the bacteria growth rate as supported by the individual compounds.

TABLE 3.4   Organic Compounds Subject to Cometabolism and Accumulated Products

Substrate	Product
Ethane	Acetic acid
Propane	Propionic acid, acetone
Butane	Butanoic acid, methyl ethyl ketone
m-Chlorobenzoate	4-Chlorocatechol, 3-chlorocatechol
o-Fluorobenzoate	3-Fluorocatechol, fluoroactate
2-Fluoro-4-nitrobenzoate	2-Fluoroprotocatechuic acid
4-Chlorocatechol	2-Hydroxy-4-chloro-muconic semialdehyde
3,5-Dichlorocatechol	2-Hydroxy-3,5-dichloro-muconic semialdehyde
3-Methylcatechol	2-Hydroxy-3-methyl-muconic semialdehyde
o-Xylene	o-Toluic acid
p-Xylene	p-Toluic acid, 2,3-dihydroxy-p-toluic acid
Pyrrolidone	Glutamic acid
Cicerone	Cierolone
n-Butylbenzene	Phenylacetic acid
Ethylbenzene	Phenylacetic acid
n-Propylbenzene	Cinnamic acid
p-Isopropyltoluene	p-Isopropylbenzoate
n-Butyl-cyclohexane	Cyclohexaneacetic acid
2,3,6-Trichlorobenzoate	3,5-Dichlorocatechol
2,4,5-Trichlorophenoxy acetic acid	3,5-Dichlorocatechol
p-p[pr]-Dichlorodiphenyl methane	p-Chlorophenylacetate
1,1-Diphenyl-2,2,2-trichloroethane	2-Phenyl-3,3,3-trichloropropionic acid
1,1,1-Trichloroethene	1,2-Dichloroethene

SOURCE: Horvath, 1972; Vogel, 1987.

TABLE 3.5   Microorganisms Exhibiting the Phenomenon of Cometabolism

Microorganism
Achromobacter sp.
Arthrobacter sp.
Aspergillus niger
Azotobacter chroococcum
Azotobacter vinelandii
Bacillus megaterium
Bacillus sp.
Brevibacterium sp.
Flavobacterium sp.
Hydrogenomonas sp.
Methanotrophs
Microbacterium sp.
Micrococcus cerificans
Micrococcus sp.
Nocardia erythropolis
Nocardia sp.
Pseudomonas sp.
P. fluorescens
P. methanica
P. putida
Streptomyces aureofaciens
Trichoderma viride
Vibrio sp.
Xanthomonas sp.

SOURCE: Horvath, 1972; Vogel, 1987.

Those compounds yielding the fastest growth rate will be degraded first (Arvin, 1988). In a mixed hydrocarbon spill benzene will degrade at a faster rate than naphthalene, and naphthalene at a faster rate than pyrene. If the energy source compounds are degraded before those that require cometabolism, such as some PAHs, the degradation of these PAHs may stop. Preferential substrate degradation also causes hindered degradation of many hazardous compounds. Most hazardous compounds yield significantly slower growth rates than energy sources like methanol or glucose. Thus care must be used when adding an energy source to supplement an energy-deficient system.

Other significant reactions

During the application of bioremediation, other microbial reactions can occur in conjunction with the transformation of the target compound. Those reactions that have potential impact on operational needs are briefly presented below so their importance can be considered.

The use of reduced inorganic compounds as substrates for respiratory metabolism is characteristic of a physiological group, known collectively as the chemoautotrophs. The substances that can serve as energy sources are H_2, CO, NH_3, NO_2, Fe^{2+}, and reduced sulfur compounds (H_2S, S, $S_2O_3^{2-}$). The cellular carbon is derived from CO_2.

The oxidation of ammonia to nitrate, nitrification, is an important aspect of the nitrogen cycle in nature. The nitrification process occurs in two steps, each carried out by a very specialized group of bacteria. The first step is the oxidation of ammonia to nitrite by bacteria of the genus *Nitrosomonas*. These bacteria are incapable of any other mode of growth. They depend on the enzyme ammonia monooxygenase, and ammonia must be oxidized during the degradation of organic compounds (Vanelli, 1991).

$$NH_3 + 1\tfrac{1}{2}O_2 \rightarrow NO_2^- + H^+ + H_2O + \text{energy}$$

Nitrosomonas is capable of degrading many organic compounds, including halogenated aliphatic compounds, except tetrachloroethylene and carbon tetrachloride (Table 3.6) (Vanelli, 1991).

The second step is the oxidation of nitrite to nitrate by the genus *Nitrobacter;* the oxidation of nitrite is the only source of energy these bacteria can use.

$$NO_2^- + \tfrac{1}{2}O_2 \rightarrow NO_3^- + \text{energy}$$

These nitrification reactions are important since they can create a significant oxygen demand. This oxygen need must be considered for any site remediation that contains appreciable ammonia, either

TABLE 3.6   Compounds Degraded by Nitrosomonas

Haloalkanes	Haloalkenes
Bromomethane	Chloroethylene
Chloromethane	cis-1,2-Dibromoethylene
Dibromomethane	1,1-Dichloroethylene
Dichloromethane	cis-1,2-Dichloroethylene
Trichloromethane	Trichloroethylene
Bromoethane	1,3-Dibromopropene
Chloroethane	2,3-Dichloropropene
Fluoroethane	1,1,3-Trichloropropene
Iodoethane	Nitrapyrin
1,2-Dibromoethane	
1,1,1-Trichloroethane	
Chloropropane	
1,2,3-Trichloropropane	
Chlorobutane	

SOURCE: Vanelli, 1991.

added as a nutrient source or as an existing compound in the site. Urea can also be used as a nitrogen source, yielding ammonia:

$$CO\,(NH_2)_2 + H_2O \rightarrow CO_2 + NH_3$$

Another group of chemoautotrophs of significance is the sulfur bacteria. The oxidation of reduced sulfur compounds is the energy source for *Thiobacillus thiooxidans,* which oxidizes elemental sulfur to sulfuric acid:

$$S^0 + H_2O + 1\tfrac{1}{2}O_2 \rightarrow H_2SO_4 + energy$$

This organism can grow in extremely acid environments.

Another group of sulfur-oxidizing bacteria is *Beggiatoa,* which oxidizes hydrogen sulfide to elemental sulfur:

$$H_2S + \tfrac{1}{2}O_2 \rightarrow S^0 + H_2O + energy$$

Beggiatoa has a long filament structure that becomes stuffed with minute granules of sulfur. This gives it a very characteristic appearance under the microscope.

A few aerobic bacteria (principally *Pseudomonas* and *Bacillus* species) can use nitrate as the electron acceptor by reducing it to molecular nitrogen. The formation of N_2 from nitrate is termed denitrification. Denitrification requires special enzymes and the normal aerobic respirator electron transport chain. It is therefore always an alternative mode of respiratory energy-yielding metabolism. The special enzymes needed for denitrification are not synthesized in the presence of air even when nitrate is present.

$$2NO_3^- + 10e^- + 12H^+ \rightarrow N_2 + 6H_2O$$

Denitrification can be coupled with the oxidation of inorganic compounds under anaerobic conditions. For example, certain chemoautotrophs of the genus *Thiobacillus* can reduce nitrate to N_2 and oxidize sulfur anaerobically.

$$5\,S + 6\,NO_3^- + 2H_2O \rightarrow 5SO_4^{2-} + 4H^+ + 3N_2$$

A small group of strictly anaerobic bacteria, the methanogens, use the mode of methane fermentation. A simple example is the oxidation of molecular hydrogen:

$$4H_2 + CO_2 \rightarrow CH_4 + 2H_2O$$

In this case carbon dioxide is reduced, resulting in an equilibrium shift of the bicarbonate-carbonate system.

Electron acceptors

All energy-yielding reactions are oxidation-reduction reactions. The reduction reaction, that is, the reaction involving the electron acceptor, establishes the metabolism mode. When a site contains more than one potential electron acceptor, the one that serves as the acceptor is determined by the potential energy yield of the reaction. Microorganisms have the innate ability to select the type of redox reaction that will yield the greatest energy. Thus the amount of free energy that a microorganism can obtain from coupling an oxidation reaction with potential reduction reactions establishes the preferred electron acceptor. This free energy is determined by thermodynamics of the chemical system.

The amount of free energy from a reaction depends on the Gibbs free energies for the substrates and products. The relationship is given by

$$\Delta G^\circ = \Sigma \Delta G_f^\circ \text{ (products)} - \Sigma \Delta G_f^\circ \text{ (substrates)}$$

where ΔG° is the increment in free energy for the reaction under standard conditions (25°C and 1 atm). For aqueous systems the standard condition of all solutes is 1 mol/kg activity.

A useful concept for solution-based redox reactions is electron activity (pE). Electrical activity pE is a measure of the availability of electrons in much the same manner that pH measures the availability of protons. Electron activity is related to Gibbs free energy by

$$\text{pE} = \frac{-\Delta G^\circ}{2.3} \, nRT$$

where R = gas constant
T = absolute temperature
n = number of electrons involved in the reaction

The amount of free energy that can be obtained by microorganisms from redox reactions is directly proportional to the electrical activity (pE) of the redox system. The pE values of energy-producing reactions of microbial significance are provided in Table 3.7.

A comparison of reduction and oxidation half-reaction potentials for microbially mediated reactions is presented in Fig. 3.4. The top half of Fig. 3.4 presents the reduction half reactions. The bottom half presents the oxidation half reactions. The complete oxidation-reduction reaction results in the coupling of the two half reactions. The arrow in Fig. 3.4 points in the favorable reaction direction and the arrow's base shows the potential of the half reaction. The greater the change in pE (that is, the greater the horizontal separation of an oxidation and reduction arrow's base), the greater the free energy of the reaction.

The oxidation of an organic compound (located at the far left of the

TABLE 3.7   Energy-Producing Levels of Electron Acceptance Reactions

Type	pE	Name of reaction
Aerobic:		
$O_{2(g)} + 4H^+ + 4_e \rightarrow 2H_2O$	+20.8	Aerobic respiration
Anaerobic:		
$2NO_3^- + 12H^+ + 10e \rightarrow N_{2(g)} + 6H_2O$	+21.0	Denitrification
$NO_3^- + 10H^+ + 8e \rightarrow NH_4^+ + 3H_2O$	+14.9	Nitrate reduction
$CH_2O + 2H^+ + 2e \rightarrow CH_3OH$	+3.99	Fermentation
Formaldehyde Methanol		
$SO_2^4 + 9H^+ + 8e \rightarrow HS^- + 4H_2O$	+4.13	Sulfate reduction
$CO_{2(g)} + 8H^+ + 8H^+ + 8e \rightarrow CH_{4(g)} + 2H_2O$	+2.87	Methane fermentation

SOURCE: Snoeyink, V. L., and Jenkins, D., *Water Chemistry,* John Wiley & Sons, New York, 1980.

oxidation portion of Fig. 3.4) will supply electrons to the electron acceptor located the greatest distance to the right of the reduction portion. Therefore, if molecular oxygen is present, the electron acceptor will be O_2. If molecular oxygen is not available, the electron acceptor successive levels are NO_3, organic compounds, SO_4, and CO_2. These thermodynamic concepts establish the sequence of chemical reactions that results in an ecological succession of microorganisms as each acceptor becomes depleted. For an organic substrate in the presence of oxygen, aerobic heterotrophs will be the first active microbial community. As oxygen is depleted, this succession is followed by denitrifiers, fermenters, sulfate reducers, and methanogens. Bioremediation should enhance and work with this ecological succession when at all possible. For the degradation of some hazardous compounds, a specific metabolism mode is necessary. Controlling the metabolism mode is one key principal to successful bioremediation. This is partially achieved by control of the available electron acceptor.

Enzyme activity

The microbial redox reactions are motivated by catalysts called enzymes. Enzymes are complex organic proteins generated by the microbial cell. A catalyst increases the velocity of the overall reaction and is both a reactant and a product of the reaction. Catalysts have the following properties:

- They are effective in extremely small amounts.
- They are unchanged in the reaction.
- They do not change the equilibrium of the chemical reaction.
- They are usually very specific in their ability to accelerate a chemical reaction.

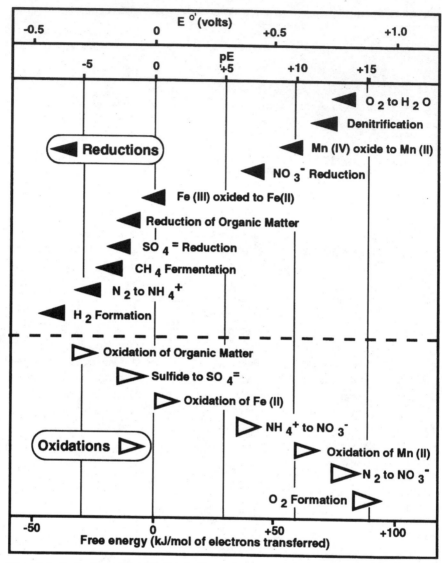

Figure 3.4 Reduction and oxidation half-reaction potentials for some microbially medi-ated redox reactions. Bases of arrows align with the potentials of the half reaction shown in volts. (*Stumm and Morgan, 1981.*)

Biotic reactions (biological mediated) will occur naturally owing to the reaction's thermodynamic equilibrium state. These nonbiological reactions are called abiotic reactions. Examples important to bioreme-diation are transformation pathways for the halogenated aliphatic compounds such as 1,1,1-trichloroethane. Another is the initial incor-poration of oxygen (in the absence of molecular oxygen) into the structure of nonoxygenated aromatic compounds under anaerobic conditions. Without this step, nonoxygenated aromatic compounds cannot be degraded under anaerobic conditions. These biochemical pathways are discussed in Chap. 4.

Under nonbiological conditions these reactions can take an indefinite time to reach equilibrium. Thus insignificant chemical degradation occurs over years. However, enzymes accelerate these reactions, resulting in rapid degradation of many hazardous contaminants. Enzymes do not change the free energy of these reactions but influence the reaction rate by lowering the activation energy for the reaction (Fig. 3.5).

Enzymes are usually very specific in their ability to lower activation energy of oxidation-reduction reactions. There are some exceptions such as the monooxygenase enzyme of methanotrophs and the extracellular enzymes (ligninases) of *P. chrysosporium*. Enzymes also may require cofactors. A cofactor is a nonprotein compound that combines with an inactive protein to give a catalytically active complex. This complex is frequently referred to as the enzyme. These cofactors may be metal ions which include Fe^{2+} or Fe^{3+}, Co^{2+}, Cu^{2+}, Mg^{2+}, Mn^{2+}, Ca^{2+}, and Zn^{2+}. In addition to metals they include organic compounds. One common organic cofactor is coenzyme A.

The first requirement for designing a bioremediation system is knowledge that enzymes exist that can degrade the target compound. The second requirement is having the enzyme generator, the required microorganisms, present in the contamination area. Third, the microorganism must be induced to generate the enzyme as illustrated in Fig. 1.2. The inducement is an energy and carbon source that is readily available. The production of enzymes is triggered by this energy-yielding source, which may or may not be the hazardous compound (target compound). Recall that the target compound is not the energy source or enzyme inducer with cometabolism.

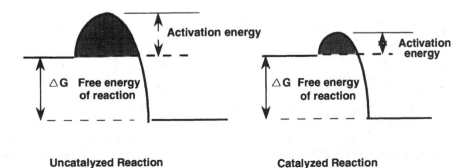

Figure 3.5   A catalyst lowers the activation energy.

For a microorganism to obtain energy, the compound must pass through the microorganism's cell membrane so the organism's electron transport system can be used for energy storage. Some hazardous compounds are too large to pass through the cell membrane, and some that are highly charged are rejected. If the compound is too large but there is still a net energy gain potential, the microorganism will generate exoenzymes. Exoenzymes are secreted through the cell wall to perform extracellular digestion. Exoenzyme production is triggered by the hydrolytic products of macromolecular substrates. These hydrolytic products may be initiated by abiotic reactions or through cometabolism. Although hydrolyzed reactions occur without microorganisms, enzymes significantly increase the rate of these reactions. Since exoenzymes are produced at a low rate, macromolecular compounds require longer acclimation periods and are degraded at a slow rate.

Enzyme production can be hindered by catabolite repression. Catabolic repression occurs when metabolites from degradation inhibit enzyme production or hinder the overall rate of reaction by maintaining their complex with the enzyme. This has been reported in both laboratory and pilot field studies for the aerobic dehalogenation of halogenated aliphatic compounds. Another type of repression (competitive inhibition) occurs when a more suitable or highly metabolized organic energy source becomes available.

The potential for enzyme repression must be considered with operating bioremediation systems. An example is the effect of trichloroethylene, a metabolite formed during dehalogenation of higher chlorinated structures. A mechanism must be operational to reduce the metabolite buildup. This can be done with other microorganisms that degrade the metabolites or by water flux rates that maintain a limit on metabolite concentration.

Microorganisms are capable of catalyzing a variety of reactions. Those of specific importance are illustrated in Table 3.8. Hydrolysis is frequently conducted outside the microbial cell by exoenzymes. Hydrolysis is simply a cleavage of an organic molecule with the addition of water. Examples are given for the hydrolysis of protein (Ia), ester (Ib), organic thiol (Ic), and a glycoside (Id) (Table 3.8). The symbols R and R' represent different forms of a carbon-hydrogen structure.

Cleaving a carbon-carbon bond is another important reaction. Cleavage is illustrated by reactions IIa and IIb (Table 3.8). An organic compound is split or a terminal carbon is cleaved off an organic chain.

Organic compounds can undergo substitution reactions in water without the presence of an enzyme catalyst. Substitution reactions involve replacing one atom with another. An example is the substitution of a halogenated organic compound through hydrolysis as illus-

TABLE 3.8   Important Reactions Mediated by Enzymes

I. Hydrolysis

(Ia) $RCO{-}NHR^1 + H_2O \longrightarrow RCOOH + R^1NH_2$

(Ib) $RCO{-}OR^1 + H_2O \longrightarrow RCOOH + 4R^1OH$

(Ic) $RCO{-}SR^1 + H_2O \longrightarrow RCOOH + R^1SH$

(Id) $R{-}CH{-}OR^1 + H_2O \longrightarrow RH + HO{-}CH{-}OR^1$

II. Cleavage

(IIa) $RCOOH \longrightarrow RH + CO_2$

(IIb) $HOCH{-}CH{-}OH \longrightarrow RCH_2OH + R^1CHO$

$\quad\quad\quad R \quad R^1$

III. Oxidation reduction

(IIIa) $AH_2 + B \longrightarrow A + BH_2$

(IIIb) $AH_2 + O_2 \longrightarrow A + H_2O_2$

IV. Dehydrogenation

(IVa) $R_2C\overset{H}{\underset{R}{-}}C\overset{H}{-}OH \longrightarrow R_2C{=}\overset{H}{\underset{R}{C}} + H_2O$

(IVb) $R_2C\overset{H}{\underset{R}{-}}C\overset{H}{-}NH_2 \longrightarrow R_2C{=}\overset{H}{\underset{R}{C}} + NH_3$

V. Dehydrohalogenation

(Va) $RC\underset{Cl}{-}CR^1\underset{H}{} \longrightarrow RC{=}CR^1 + HCL$

VI. Substitution

(VIa) $RCBr + H_2O \longrightarrow RCOH + Br$

(VIb) $RCH_2\underset{Br}{} + HS^- \longrightarrow RCH_2SH + Br$

trated by reaction VIa (Table 3.8). Sulfides also react by substitution to produce mercaptans. In general, substitution reactions proceed slowly with half-lives as long as centuries. However, enzymes acceler-ate these reaction rates.

Other important reactions include dehydrogenation, dehydrohalo-genation, and oxidation reduction. As illustrated in Table 3.8 (reac-tions IVa and IVb), dehydrogenation is an oxidation-reduction reac-tion that results in the loss of two electrons and two protons (Fig. 3.3). The net result is the loss of two hydrogen atoms.

Dehydrohalogenation, or dehalogenation, is the removal of halo-gens such as chloride, bromide, and fluoride from an organic com-pound. An example is illustrated by reaction Va (Table 3.8), which shows the release of both a hydrogen ion and a chlorine ion resulting in hydrochloric acid production.

Microbiology of Bioremediation

Microbial divisions

The structures of a plant, a mammal, and a bacterium may appear to have little in common, but the living world yields a fundamental unity. The cell has three basic features that are common to all biological systems. First is a common chemical composition, as shown by the presence of three types of complex macromolecules:

1. Protein

2. Deoxyribonucleic acid (DNA)

3. Ribonucleic acid (RNA)

Deoxyribonucleic acid carries the genetic information that is translated, through the intermediacy of RNA, into protein synthesis. The specific pattern of protein synthesis determines an organism's gross properties. Proteins also serve as the catalysts or enzymes responsible for the second common feature, metabolism. Finally, organisms have a common physical structure, being composed of subunits known as cells.

The two major components of the cell consist of the nucleus and the cytoplasm. The nucleus contains genetic information, DNA, and accordingly serves as the information center for cellular synthesis. The cytoplasm surrounds the nucleus and contains most of the RNA and protein of the cell. It is separated from the external environment by the cell membrane that is composed of proteins and lipids. All materials must pass through the cell membrane to enter the cell. The membrane is semipermeable; that is, it allows the passage of some substances into the cell and excludes others. It also allows the exit of the waste products of cellular metabolism. In many microorganisms the cell membrane is the only bounding structure.

The simplest organisms, such as bacteria, protozoa, some algae, and a few fungi, are unicellular. The more complex organisms are multicellular. In the mature state they consist of many cells, permanently attached to one another in a characteristic way, which determines the organism's external form. Many bacteria and algae are multicellular, but they contain a relatively small number of cells.

Two kinds of cells are recognized, the procaryotic and eucaryotic cell. Organisms have been divided as based on the structural differences of these two cell types (Table 3.9). Bacteria and blue-green algae are procaryotic. All other organisms are eucaryotic. The blue-green algae are also referred to as blue-green bacteria and cyanobacteria.

TABLE 3.9 Microbial Divisions
According to Cell Type

Procaryotic cell	Eucaryotic cell
Bacteria	Plants
Blue-green bacteria	Animals
or cyanobacteria	Crustaceans
	Rotifers
	Protozoa
	Fungi
	Most algae

The procaryotic organisms are the most important to bioremediation. Eucaryotes have not generally been recognized as important degradative organisms except fungi used in lignocellulose degradation.

The most important groups to bioremediation are bacteria and fungi. The bacteria are a large and very varied group of procaryotic organisms. It is convenient to recognize three principal subgroups among them, eubacteria, mycobacteria, and spirochetes. All can obtain energy by chemoheterotrophic processes. The principal characteristics of the three groups are given in Table 3.10. The eubacteria, the mycobacteria, and the spirochetes are distinguished by the differences in mode of movement and structure of cell walls. Eubacteria and mycobacteria have been identified as active microorganisms in

TABLE 3.10 Principal Characteristics of Bacteria

	Mycobacteria	Spirochetes	Eubacteria
Mechanism of cellular movement	Gliding	Axial filament	Flagella or immotile
Cell wall Structure:	Thin, flexible	Thin, flexible	Thick, rigid
Unicellular	+	+	+
Filamentous	−	−	+
Mycelial	−	−	+
Cell shape:			
Rods	+	−	+
Spheres	−	−	+
Spirals	−	+	+
Nutritional categories:			
Photoautotrophic	−	−	+
Photoheterotrophic	−	−	+
Chemoautotrophic	−	−	+
Chemoheterotrophic	+	+	+

+ = yes, − = no.

bioremediation. The eubacteria are frequently reported as the prevalent group for many sites.

Many mycobacteria are soil inhabitants, although some are found in water. They are important in the breakdown of complex polysaccharides. The spirochetes are unicellular organisms with a distinctive form. The cell length is much greater than its width and helicoidal in shape. A longitudinal filament is spirally wound about the cell. The spirochetes are chemoheterotrophs and most are anaerobic. They are found in anaerobic waters and mud.

Eubacteria. The eubacteria are unicellular organisms, multiply by binary transverse fission, and constitute the largest and most diversified group. Their significance to environmental engineering requires consideration, but their microbiology cannot be adequately discussed in a few pages. The following describes their general properties.

Eubacteria have three principal cell shapes: cocci, with spherical or ovoid cells; rods, with cylindrical cells; and spirilla, with helical cells. Since cell division always takes place at right angles to the long axis of the cell, the only possible type of aggregate in rod-shaped and helical cells is a chain of cells. Chain formation is common.

In cocci, cell divisions take place in the same plane, yielding a chain of cells. In some cocci, successive divisions occur in two planes and at right angles to one another to produce four-cell tetrads.

The outer structure of the bacteria may consist of a capsule or slime layer. This capsule may be 100 to 300 Å thick. It consists of a single polysaccharide or a polypeptide, which may protect the bacteria from engulfment by various protozoa. The capsule has significant importance to environmental engineers, since the ability of bacteria to form a gelatinous floc depends on these natural polymers. Formation of a good settling floc is necessary to remove the bacteria from the effluent of biological treatment reactors. This polymer production is a function of available nutrients and growth stages of the organism.

The cell wall is chemically very complex and responsible for the shape of the cell. The bacterial cell membrane regulates the entry of nutrients to the cell and plays an important role in metabolism and biosynthesis. The membrane is the site of different enzymatic activities.

The size of individual cells ranges from about 0.3 to 3.0 μm. The average rod-shaped bacteria is 0.5 to 1.0 μm wide to 1.5 to 3.0 μm long. The average sphere is 0.5 to 1.0 μm in diameter.

Bacterial cells are often motile. Their means of locomotion results from a whiplike appendage called a flagellum. The flagella are distributed over the surface of the cell in characteristic fashion. They may be restricted to one or both ends of the cell (polar flagella).

Some bacteria contain endospores. Bacterial spores result when cells find themselves in a slowly changing adverse environment. The spore develops when the nucleus becomes surrounded by a very tough polysaccharide coating. The spore can withstand extreme conditions, both chemical and physical. Some endospores can germinate even after several hours of immersion in boiling water. Only one spore is formed in a cell. It may lie in the center or toward one end of the cell. The spore can have a greater diameter than the cell, which gives a spindle or racket shape to the organism.

Fungi. The fungi are heterotrophic and have developed a highly distinctive biological organization. Soil is their most common habitat, but many primitive fungal groups are aquatic. Many species can grow over a wide pH range. Some are active at pH values as low as 2 and others at pH values as high as 10. Most fungi are composed of long, fine filaments of cells called hyphae, which together constitute the mycelium. The mycelium consists of a multinucleate mass of cytoplasm enclosed within a rigid, much-branched system of tubes.

Fungi are classified according to their mode of reproduction. The classes are:

Phycomycetes

Ascomycetes

Fungi Imperfecti

Basidiomycetes

Most fungi can be placed in the first three classes. Aquatic Phycomycetes occur on the surface of decaying plant or animal materials and in streams. The Phycomycetes are the largest group of freshwater fungi. However, the significance of fungi in ecosystems is not understood. They do not have the typical mycelial structure. The mature vegetative structure consists of a sac about 100 μm in diameter that is anchored to the solid substrate by fine, branched threads known as rhizoids. The aquatic Phycomycetes are a very varied group with respect to reproduction mechanism. Some terrestrial Phycomycetes play a role in the degradation of organic compounds in compost.

The Ascomycetes and Basidiomycetes are higher forms of fungi. The Ascomycetes are frequently seen on decaying fruit and bread. The Basidiomycetes are found on decaying trees and include mushrooms. These fungi have little significance in aquatic environments but are very active in soil containing organic energy sources. Several Basidiomycetes are effective in the degradation of hazardous chemicals.

Most fungi belong to the class Fungi Imperfecti. These molds produce spores only asexually. The mycelium is septate and the asexual spore is borne on specialized structures known as conidiophores (Alexander, 1977). This class of fungi is probably the most likely to be cultivated on agar media during laboratory studies.

Bioremediation organisms

It is unlikely that one specific organism is important to successful cleanup of contaminated sites because of the highly complex environments existing at these sites. Laboratory studies and microbial characterization of pilot and field bioremediation projects have identified many microorganisms responsible for degradation of hazardous chemicals (Tables 3.5 and 3.11). There are several important groups that require specific attention. However, it is unlikely that one, two, or three specific microorganisms result in successful bioremediation. For most

TABLE 3.11   Additional Microorganisms Found Active in Contaminated Soil and/or Groundwater Cleanup

Achromobacter xyloxidans
Acinetobacter sp.
Alcaligenes denitrifricans
Berijerinckia sp.
Desulfomonile tiedjei
Flavobacterium sp.
Hyphomicrobium sp.
Inonotus circinatus
Methanosarcina mazei
Methanosarcina sp.
Methanobacteriaceae
Mycobacterium sp.
Mycobacterium vaccae
Nitrosomonas eurupaca
Nocardia corallina
Phanerochaete laevis
Phanerochaete sanguinea
Phanerochaete filamentosa
Phanerochaete chrysorhiza
Pseudomonas aeruginosa
Pseudomonas sp.
Pseudomonas stutzeri
Pseudomonas vesicularis
Pseudomonas mendocina
Pseudomonas paucimobilis
Serratia marcescens
Trametes hirsuta
Xanthobacter autotrophicus

SOURCE: Rainwater, 1991; Lamar, 1990.

projects a high diversity of microorganisms must be present for successful bioremediation, and their specific identification is unimportant.

Microbial consortiums. The world is filled with diversity when considering the types of microorganisms and the diversity of their energy sources. This diversity makes it possible to break down thousands and thousands of different organic chemicals. However, it also results in complicated and extraordinary diverse needs of microorganisms. Microorganisms apply oxidation and reduction reactions in a variety of specialized mechanisms that are frequently specific to an organism or group of organisms. This fact leads to the need for microbial diversity when degrading complex organic compounds or performing bioremediation of a site contaminated with mixed organic compounds.

Microorganisms separately cannot mineralize most hazardous compounds. Complete mineralization results from a consortium of microorganisms (Table 3.12). A degradation sequence occurs where a second organism degrades the metabolic products of the first, and a third, etc., to yield complete mineralization of an organic compound. A mixed microbial consortium also is needed for microbial synergism and cometabolism. Both principals are vital to the mineralization of some hazardous chemicals. A few examples are discussed below.

In some cases, the mixed consortium of microorganisms can accomplish more than the individual sums of each microorganism. This principal is called synergism. One member of a microbial community may be unable to synthesize a particular requirement for the reaction but will degrade the compound when a second organism synthesizes the needed component. This is illustrated by the work of Jensen (1957) using soil organisms to degrade trichloroacetic acid (Fig. 3.6). An unidentified soil organism can dehalogenate trichloroacetic acid. This organism, however, cannot perform this dehalogenation when a second organism, *Streptomyces* sp., is absent. The *Streptomyces* produces a specific vitamin necessary for the unidentified organism that apparently acts as a coenzyme. Individually, neither can dechlorinate trichloracetic acid, but as a consortium one is capable of the dehalogenation.

TABLE 3.12   Multiple Roles of Microbial Synergism

Stepwide or sequential degradation
Synthesis of a necessary component
Removal of toxic metabolites
Enhancing overall rate of degradation
Microbial web, the need of complex associations
Favorable thermodynamics

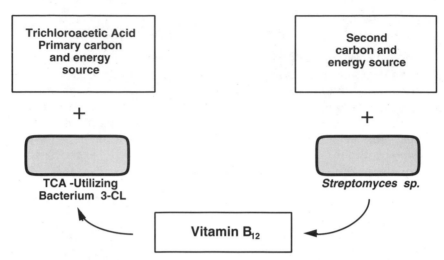

Figure 3.6 Microbial synergism—synthesis of a necessary component for a second organism to mediate transformation. (*Jensen, 1957; Slater, 1984.*)

Another example is illustrated by the work of Stirling (1976). A consortium of two organisms can oxidize cyclohexane. However, growth does not occur when one organism is absent. Although the unknown factor has not been isolated, one organism produces a component that is necessary for the second to undertake the oxidation of cyclohexane. Similar studies also have been reported on microbial relationships for the breakdown of hexadecane.

Microorganisms produce metabolites that may be self-inhibitory or inhibit the growth of other organisms. Maintaining the enzyme reaction depends on the activity of a second organism to remove compounds excreted by the first organism. This principal has been recognized for years in operating anaerobic digesters.

In anaerobic digestion, complex organic compounds are degraded to volatile acids by hydrolytic-fermentative bacteria. A second group of bacteria, the H_2-producing acetogens, must be present to remove the volatile acids so a neutral pH is maintained. These organisms produce hydrogen gas. The accumulation of hydrogen makes the overall degradation reaction thermodynamically unfavorable. A third group of organisms, the methanogens, must be active in concert with the other organisms to remove the hydrogen. Methanogens obtain energy from the oxidation of hydrogen, using carbon dioxide as the electron acceptor to generate methane. Furthermore, the methanogens require the removal of volatile acids by the acetogens to oxidize hydrogen. This consortium is illustrated in Fig. 3.7. This principal is important for the degradation of a variety of aromatic compounds. For example,

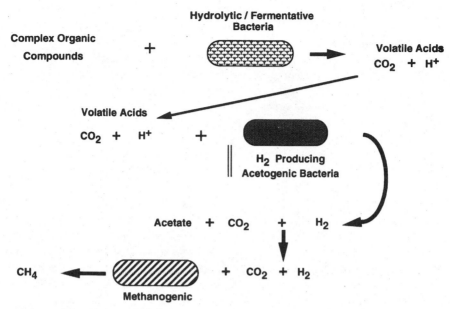

Figure 3.7 Microbial consortium responsible for balanced metabolic activity in anaerobic digesters.

methanogenic fermentation of benzoate, in the absence of nitrate and sulfate, is carried out by such a mixed microbial community.

Another interaction supporting the need for mixed cultures has been demonstrated during studies on 1,2-dichloroethane (DCE) degradation. *Xanthobacter autotrophius* is capable of using DCE as a sole carbon and energy source. However, in the absence of biotin, toxic compounds accumulate from DCE degradation, retarding the growth on DCE. The presence of *Pseudomonas* sp. results in the restoration of DCE degradation by *X. autotrophius* (vanderWiyngaard, 1993).

Synergism is also important in enhancing the overall rate of degradation of hydrocarbons in mixed cultures. This phenomenon has been documented for the degradation of polychlorinated biphenyls. The increased rate of degradation is probably a result of combined metabolic attack at different sites on an organic contaminant, increasing overall degradation rates.

There are other known degradation pathways that illustrate the importance of microorganisms altering the thermodynamics of a reaction. The interaction of *Methanospirillum* is necessary for one pathway in the oxidation of benzoate. This organism's only substrate is formate, hydrogen gas, and carbon dioxide. Its action on hydrogen reduces the partial pressure of the gas, making the oxidation of benzoate thermodynamically favorable for other organisms (Tiedje and Stevens, 1987).

Synergistic metabolic activity is illustrated by the degradation of Diazinon (*O, O*-diethyl *O*-(6-methyl-2-(1-methylethyl) 4-pyrimidinyl phosphorothioate) in soil (Gunner and Zuckerman, 1968). Two microorganisms are necessary for the degradation of this insecticide. Individually, neither of the two organisms can grow on Diazinon as the only carbon source, but they can as a microbial community.

The relationship between microorganisms is sometimes so important that the coupled associations have been called a "microbial web." These associations have resulted in the incorrect classification of two organisms as a single species. Three microorganisms have been reported for one degradation method of 3-chlorobenzoate (Tiedje and Stevens, 1987). The consortium consists of a dechlorinating bacteria designated as *Desulfomonile tiedjei*. This strain is an anaerobic bacterium that carries out reductive dechlorination. However, it does not grow on benzoate or chlorobenzoate. The second strain is a benzoate oxidizer designated strain BZ-2. Benzoate is the only substrate known to support its growth. The third strain is a methanogen, *Methanospirillum,* designated strain PM-1. The consortium of these three organisms will degrade 3-chlorobenzoate, producing methane, acetate, and chloride. Under microscopic examination, the benzoate-oxidizing bacterium is seen attached to *D. tiedjei*. This intimate contact may facilitate transfer of some intermediate in the 3-chlorobenzoate mineralization. The interaction of this consortium is as follows (Fig. 3.8):

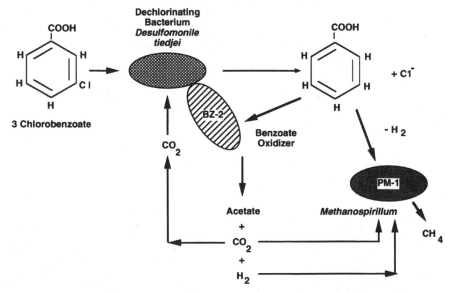

Figure 3.8 Consortium of microorganisms: a microbial web for degradation of chlorinated aromatic compounds. (*Tiedje, 1987.*)

1. *Desulfomonile tiedjei* is the organism capable of dechlorinating chlorobenzoate. Carbon dioxide is the major source of carbon for this organism.
2. BZ-2 is the strain responsible for benzoate oxidation, but the catalyzed reaction does not proceed because of an unfavorable thermodynamic situation.
3. Strain PM-1 uses only formate or hydrogen gas for its substrate. By reducing the hydrogen gas partial pressure this organism makes it thermodynamically favorable for BZ-2 to carry out benzoate oxidation.

Microbial diversity for bioremediation cannot be stressed strongly enough. A mixed microbial community is necessary for bioremediation of mixed organic contaminates, and frequently the only method for mineralization of specific hazardous compounds. The probability of successful bioremediation is dramatically increased by having microbial diversity.

Dominating organisms.   As discussed above, there is no one species that is responsible for bioremediation of a contaminated site. Hazardous compounds result in the selection of a mixed microbial population with improved abilities to tolerate and extract energy from the contaminants. Identification of significant populations consistently finds some as active participants in the microbial consortiums as shown in Table 3.13.

TABLE 3.13   Bioremediation Microorganisms Frequently Identified as Active Members of Microbial Consortiums

Identified in soil and groundwater systems:

Alcaligenes denitrificans	*Nocardia* sp.
Arthrobacter globiforms	*Pseudomonas aeruginosa*
Arthrobacter sp.	*Pseudomonas cepacia*
Bacillus megaterium	*Pseudomonas fluorescens*
Bacillus sp.	*Pseudomonas glatheri*
Berijerinckia sp.	*Pseudomonas mendocina*
Flavobacterium sp.	*Pseudomonas methanic*
*Methanobacteriaceae**	*Pseudomonas paucimobilis*
Mycobacterium sp.	*Pseudomonas putida*
Mycobacterium vaccae	*Pseudomonas* sp.
Nitrosomonas eurupaca	*Pseudomonas testosteroni*
Nocardia corallina	*Pseudomonas vesicularis*
Nocardia erythropolis	

*Genera have not been identified in field operations.

The group found with the highest frequency consists of those belonging to the genus *Pseudomonas*. *Pseudomonas* can degrade many chemicals. These include cresol, benzene sulfonate, phthalates, furans, benzoate, and anthranilate. *Pseudomonas vesicularis* can utilize fluorene as the sole source of carbon and energy (Grifoll, 1992). *Pseudomonas paucimobilis* can degrade fluoranthene.

The genus *Pseudomonas* consist of gram-negative, aerobic chemoheterotrophic organisms. About 30 species are recognized, and they are common soil and water microbes. Each species is known to be capable of utilizing 60 to 100 different organic compounds as sole energy and carbon source (Stanier, 1986). Thus it is not surprising to find them as a predominant group in contaminated soil and groundwater.

Another dominating group is the mycelial bacteria, Actinomycetes. *Nocardia* and *Mycobacterium* are typical representations. *Nocardia* is capable of oxidizing many aromatic hydrocarbons. The genus *Mycobacterium* is found in environments containing aliphatic and aromatic hydrocarbons. *Mycobacterium* are common soil organisms. *Mycobacterium* sp. is capable of degrading n-butane, 2-butanone, naphthalene, 3-naphthalene, phenanthrene, fluoroanthene, pyrene, 3-methylcholanthrene, 1-nitropyrene, and 6-nitrochrysene. In pure culture, *Mycobacterium* sp. mineralized pyrene through cometabolism (Heitkemp, 1988). Actinomycetes also have been demonstrated to transform p-nitrophenol, using it as sole carbon and nitrogen source (Herman, 1993).

Methanogens.   The group of bacteria known as methanogens are very significant to bioremediation because of their ecological relationship within bioremediation consortiums. Degradation products from fermentation are frequently altered in the presence of methanogens. These organisms make certain reactions possible by providing an improved thermodynamic environment. Some compounds cannot be dechlorinated unless methanogenic activity is present.

These methane-producing bacteria use hydrogen as an energy source and carbon dioxide as the electron acceptor. They are strict anaerobes and responsible for significant methane generation in the biosphere. Organic compounds such as acetic acid, formic acid, and butyric acid stimulate their growth. Acetate and formate may be used by methanogenic bacteria as a carbon source. The methanogens produce methane as shown:

$$CH_3COOH \rightarrow CH_4 + CO_2$$

$$CO_2 + 4H_2 \rightarrow CH_4 + 2H_2O$$

The hydrogen methanogenesis is a faster reaction than the acetate methanogenesis. In addition to coupling hydrogen oxidation to carbon dioxide reduction, they can use elemental sulfur as an electron acceptor.

Methanotrophs.   The methanotrophs, like the methanogens, bring a specific tool to bioremediation. These organisms, under aerobic conditions, are capable of dehalogenating hazardous compounds. They can transform a range of halogenated methanes, ethanes, and ethylenes. The term "methanotroph" designates bacteria that use methane for carbon and energy. The methanotrophs are procaryotic bacteria. They are obligate aerobes and exist as rods, vibrios, and cocci (Haber, 1983). However, many grow more rapidly under reduced oxygen tension (Stanier, 1986), and there is evidence that hydrogen peroxide may inhibit methanotrophic activity (Semprini, 1992). Alcohols are also reported to inhibit methanotrophs (Janssen, 1991). Methanotrophs are ubiquitous and are an important member of the global carbon cycle. High methanothrophic activity is found in the boundary regions between anaerobic habitats and aerobic zones. Methane produced from anaerobic decomposition of organic matter mixing with atmospheric oxygen provides for optimum growth.

Most methanotrophs are obligate methylotrophs; that is, they only use carbon-carbon bonds as energy and carbon sources. Many are able to utilize methanol and formaldehyde. A few can utilize a wider range of organic compounds. Facultative methylotrophs can grow on both one-carbon and multicarbon compounds. Most use ammonia or nitrate as their nitrogen source. They are inhibited by some amino acids. The methylotrophic bacteria are capable of oxidizing a wide range of compounds (Table 3.14). This capability results from the lack of specificity of the enzyme, monooxygenase. Thus cometabolism is responsible for the dehalogenation of hazardous compounds. Methanotrophic organisms can initiate the transformation of halogenated compounds that are completely mineralized by heterotrophic communities. Methanotrophs can catalyze the following oxidative reactions:

$$CH_4 + O_2 + 2H^+ \rightarrow CH_3OH + H_2O$$

$$CH_3OH \rightarrow HCHO + 2H^+$$

$$HCOH + \tfrac{1}{2} O_2 \rightarrow HCOOH$$

$$HCOOH \rightarrow CO_2 + 2H^+$$

The hydrogen ions are cycled by the nicotinamide group (NADH).

TABLE 3.14   Chemicals Oxidized by
Methanotrophs

Benzene	1-Methylnaphthalene
Bromomethane	2-Methylnaphthalene
Chloromethane	1,2-Dichloroethylene
Dichloromethane	1,1-Dichloroethylene
Chloroform	1,1,1-Trichloroethane
Naphthalene	1,1,2-Trichloroethane
Styrene	Ethane
Toluene	Propane
Vinylidene	Butane
Vinyl chloride	Isobutane
m-Cresol	Pentane
o-Cresol	Hexane
m-Chlorotoluene	Heptane
	Octane
	Hexadecane
	Methane
	Ethene
	Propene
	1-Butene
	trans-2-butene
	cis-2-Butene
	Butadiene
	Isoprene
	2-Propanol
	2-Butanol

SOURCE: Strand, 1991; Haber, 1983; Semprini,
1989.

The cell growth equation for methanotrophs, using a typical cell composition of $C_5H_7O_2N$, has been reported as (Strand, 1990)

$$CH_4 + 1.66\ O_2 + 0.0482\ H^+ + 0.0482\ NO_3^-$$

$$= 1.855\ H_2O + 0.759\ CO_2 + 0.0482\ C_5H_7O_2N$$

This yields 0.51 g of cells per gram of methane oxidized.

When using monooxygenase to cometabolize a halogenated aliphatic, reducing power must be provided by NADH (Segar, 1992). The reaction between NADH and the halogenated aliphatic can be written as

$$R-H + O_2 + NADH + H^+ \rightarrow R-OH + H_2O + NAD^+$$

where R—H is either the primary substrate that yields microbial energy and carbon, or the secondary substrate such as trichloroethylene. If R—H is the primary substrate, the transformation product (R—OH) is further metabolized to regenerate the reducing power of NADH. If R—H is trichloroethylene, an unstable epoxide intermediate can be formed, but no energy is yielded for the microorganism (Henry and Grbić-Galić, 1991). In the absence of the primary substrate the transformation of trichloroethylene will cease. Some methanotrophs, however, will continue to transform a halogenated compound such as trichloroethylene for extended periods (1 to 2 days). Others lose this transformation capacity (Tc) within several hours (Alvarez-Cohen and McCarty, 1991a). Transformation capacity is measured in units of mass of halogenated hydrocarbon transformed per unit of biomass. Typical units are g/g of volatile suspend solids. The addition of an electron donor, formate, has been shown to extend the transformation capacity for some methanotrophs. Formate increases the dehalogenation rate of trichloroethylene when the primary substrate is exhausted.

Transformation capacity is also decreased as a result of toxic intermediates during the transformation of halogenated compounds (Stensel, 1992). Trichloroethylene is one such toxin. Trichloroethylene produced from the transformation of higher halogenated compounds binds to bacterial protein, killing methanotrophic cells. This decrease in transformation capacity is a function of the halogenated hydrocarbon concentration. Typical values for Tc are discussed in Chap. 9 under the design of cometabolism bioreactors.

Carbon monoxide also inhibits the dehalogenation by competing with methane for monooxygenase. Carbon monoxide is one of the degradation products of trichloroethylene transformation and can be oxidized by monooxygenase (Henry, 1991). Carbon monoxide is a much stronger inhibitor to dehalogenation than methane (Henry, 1991a).

Two types of monooxygenases appear to exist: a membrane-bound and a soluble enzyme (Semprini, 1992). The soluble monooxygenases are produced under conditions of copper limitation and higher oxygen tension. This soluble form may have a broader substrate specificity. The presence of copper has been demonstrated to interfere with the degradation of trichloroethylene. The addition of a metal chelator such as EDTA can change trichloroethylene degradation rate by an order of magnitude (Semprini, 1992).

Wood-rotting fungi. Fungi have not been investigated as extensively as bacteria for waste degradation, but they have significant capabilities. Wood-rotting fungi have the ability to degrade lignin, a complex polymer composed of phenylpropane units. Because of this ability, the wood-rotting fungi are potential candidates for degrading hazardous waste.

There are 1600 different wood-rotting fungi known (Glaser, 1988). A great diversity exists among these organisms, and only a few have been studied. Their response is individual, requiring separate optimizing conditions for each species. Some are reported to degrade pentachlorophenol (PCP) in contaminated soils exceeding 500 ppm (Lamar, 1990a). PCP mineralization has been documented for *Phanerochaete chrysosporium, P. sordida,* and a fungus designated RAB-83-12. *P. chrysosporium* and *P. sordida* also are capable of mineralization of the pentachloroanisole (PCA), a transformation product formed during PCP degradation.

Fungi are significantly affected by soil moisture content. Slight changes in soil water suction, particularly the matric suction, act as a selective factor for life-cycle stages (Griffin, 1980). Soil water content is usually expressed as gravimetric soil moisture, as percent moisture, or as field capacity (see Chap. 6). Evaluating moisture by these parameters seriously limits the information for microbial optimization. A better measurement is water suction. Soil water suction is the sum of matric suction and osmotic suction. For evaluating microbial-soil systems, matric suction is the most important. The selective preference on microorganisms is related to interactions between matric suction and soil pore size distribution.

At matric suction less than 800 to 1000 kPa (100 kPa = 1 bar), mycelial growth may not occur, and for some fungi the tubes lyse (Griffin, 1980). Bacteria metabolism decreases at matric suctions above 50 to 300 kPa. As matric suction increases, the relative competitive advantage between bacteria and fungi moves progressively toward fungi. Fungal species can contribute significantly to the microbial composition of soil. Fresh soil samples have been measured to contain 655 mg of carbon biomass per kg of soil, representing 130 fungal species (Anderson, 1984).

The wood-rotting fungi are divided into three major categories: soft rot, white rot, and brown rot fungi. The color classification is a result of the residue left after fungus growth. The white rot fungi are the primary wood degraders in nature. One white rot fungus, *Phanerochaete chrysosporium,* has been found to degrade some complex hazardous waste constituents. *P. chrysosporium* degrades chlorocarbons, polycyclic aromatics, polychlorinated biphenyls, polychlorinated dibenzo(p)dioxins, pesticides such as DDT and lindane, and some azo dyes (Lewandowski, 1990; Dhawale, 1992). This ability is attributed to its extracellular peroxidase enzymes. Two groups of peroxidase enzymes are produced, the lignin peroxidases and Mn-peroxidases. The Mn-peroxidases require manganese for activation (Lamar, 1990). These enzymes are capable of transforming a wide variety of complex compounds. The lignin peroxidase is nonspecific because of its ability to oxidize the heterogenic lignin molecule.

P. chrysosporium is a filamentous fungus typed to be a member of the Hymenomycetes subclass of Basidiomycetes (Glaser, 1988).

Hymenomycetes fungi develop many structures during their vegetative, sexual, and asexual reproductive phases. Growth occurs in all directions rather than from an apical point. The thallus, the basic vegetative body, absorbs, assimilates, and accumulates food. The thallus serves as the base for development of reproductive bodies and becomes a reproductive structure itself (Glaser, 1988). The fungal mycelium consists of interwoven filamentous hyphae. The mycelium has three growth stages. The vegetative phase is the dominant growth phase. During this phase the fungus secretes the most extracellular enzymes. Asexual reproduction occurs during the vegetative stage as well as the other stages of its life cycle.

P. chrysosporium grows over a wide temperature range. No growth is observed below 10°C and growth rate appears to change insignificantly between 30 and 39°C. Optimum growth conditions for P. chrysosporium have been reported at a temperature of 39°C, pH range of 4.0 to 4.5, and high oxygen content. Lignin degradation is reported to be 2 to 3 times greater using pure oxygen rather than air (Lewandowski, 1990). Growth increases significantly with increasing water potentials from 1.5 to 0.03 MPa. Growth in soil increases directly with nitrogen content (Glaser, 1988).

During growth, P. chrysosporium requires a source of fixed nitrogen and a primary substrate, such as glucose. Adequate carbohydrates must be available for energy production or the rate and extent of degradation are diminished. The hazardous components are secondary substrates. The enzyme production of lignin peroxidases occurs under nitrogen starvation. Thus nitrogen deficiency is used to induce enzyme production. Studies confirm that many PAHs are degraded under nitrogen-limiting conditions. Nitrogen levels one-tenth of the required level have been used (Lewandowski, 1990).

The importance of limiting nitrogen, however, has been questioned for some organic compounds. Some studies indicate that extracellular ligninases are not responsible for the degradation of DDT (Dhawale, 1992). Pentachlorophenol, azo, and heterocyclic dyes and crystal violet are degraded in the presence of adequate nitrogen.

P. chrysosporium degrades lignin by first fragmenting the lignin extracellularly. They are degraded intracellularly after the fragments pass through the cell wall. P. chrysosporium cleaves lignin alkyl side chains by a hemoprotein ligninase. The fungus produces at least 10 extracellular hemoproteins. Hydrogen peroxide is consumed in the reaction (Glaser, 1988). The one-electron oxidative mechanism is the primary extracellular pathway. In addition to alkyl side chains, ring demethylation and ring cleavage are performed by P. chrysosporium.

A major limitation on P. chrysosporium applications is that it is sensitive to biological process operations. The fungus does not grow well in suspended cell systems, mixing action appears detrimental to enzyme induction, and the fungus does not attach well to fixed media.

Bioprocess Control Parameters

Control parameters can be measured and manipulated by the process operator. In many cases, the actual parameter of importance cannot be easily measured and another parameter is used that is known to vary in some relationship with the prime parameter. For example, the amount of enzyme present is of key importance for operating biological reactors. For waste treatment systems, a practical means to measure the enzyme of significance is not available. However, the amount of enzyme present in a reactor is related to the mass of microorganisms. Thus the weight of biomass present in a given volume of liquid is used as an operational parameter for suspended growth reactors. This measurement is called mixed liquor volatile suspended solids (MLVSS) (see Chap. 9). For the bioremediation of hazardous compounds, direct microbial counts are frequently used, but these are inadequate for process control.

In addition to the amount of microorganisms present, other process control parameters must be measured and controlled for optimization. Proper environmental conditions must be maintained for growth of the organisms, enzyme generation, and occurrence of the catalyzed reactions. The environmental requirements are discussed below and integrated into process control in those sections addressing process design.

Biological process requirements

The above review on microbial concepts provides the foundation on which the basic biological requirements and therefore the process controls are built. These basic biological requirements are:

1. The presence of organisms with the capability to degrade the target compound or compounds

2. The presence of a recognizable substrate that can be used for an energy and carbon source

3. The presence of an inducer to cause synthesis of specific enzymes for the target compounds

4. The presence of an appropriate electron acceptor-donor system

5. Environmental conditions that are adequate for the enzymatically catalyzed reactions with emphasis on adequate moisture and pH

6. Nutrients necessary to support the microbial cell growth and enzyme production

7. A temperature range that supports microbial activity and catalyzed reactions

8. The absence of materials, or substances toxic to the desired microorganism

9. Presence of organisms to degrade the metabolic products

10. Presence of organisms to prevent transit buildup of toxic intermediates

11. Environmental conditions that minimize competitive organisms to those conducting the desired reactions

Of first and prime importance is knowledge that the target compounds are biodegradable. Chapters 4 and 5 review the degradation of the most important contaminants found in soil and groundwater. Knowledge of the metabolic mode necessary for degradation and the nature of the responsible microorganisms is important for establishing the necessary environmental conditions. A large diversity of microorganisms exists and many species have been documented as suitable enzyme generators for bioremediation. The desired microorganisms must be present and located within the contaminant–cell membrane transfer zone. The more available and imminent the contact, such as dissolved contaminants versus adsorbed, the faster the rate of degradation.

Contamination at most sites has existed for long periods. Old spills provide adequate time for the natural development of seed organisms that respond to the available energy of the spill. For most bioremediation projects the indigenous microorganisms are adequate. In some cases, the addition or bioaugmentation with specific seed microorganisms may prove advantageous. Special requirements for biodegradation are discussed in Chap. 4. The engineering challenge is to control the environmental conditions for an optimized degradation rate and complete mineralization of the target compounds.

Substrate

Given that the proper microorganisms are present, the second major consideration is an available substrate for energy and a carbon source. The ability of hazardous compounds to serve as energy sources is discussed in Chaps. 4 and 5. Microorganisms must obtain all the carbon necessary for cellular growth and reproduction from the environment. Most hazardous chemicals are degraded by chemoheterotrophic activity. This class of microorganism requires some form of organic compound as the energy source, and it is this factor that leads to the ability to degrade complex organic compounds. The need of this energy for growth is exploited for bioremediation purposes.

Many organic contaminants will serve as a substrate. However, for some hazardous chemicals, another substrate is necessary. The compound targeted for degradation may be at concentrations below that necessary for good biological response for bioremediation. The addition of a primary substrate supplies additional energy. When another energy and carbon source is made available to stimulate biological growth, it becomes the primary substrate and the target compound to be degraded is the secondary substrate. The primary substrate can be a single compound or several compounds. Frequently sites contaminated with mixed wastes provide both primary and secondary substrates.

There are hazardous compounds that must be degraded as a secondary substrate regardless of their concentration. These compounds are degraded through a mechanism referred to as cometabolism as discussed earlier. Under some conditions the primary compound must be a simpler chemical structure of the more complex target compound. These primary substrates are called analog enrichment. The analog substrate induces enzyme production. The induced enzyme indiscriminately reacts with both the analog and target compound. An example is the need for naphthalene or another analog substrate to achieve a reasonable rate of degradation of pyrene. Studies indicate that some mixed cultures are unable to degrade pyrene and 1,2-benzanthracene without analog enrichment.

If the target compound is too large to pass through the cell membrane, exoenzymes must be produced and secreted to initiate degradation of the compound. However, if the net energy to be gained is less than that required to generate the exoenzymes, enzyme production is not induced. Thus the enzyme induction step does not occur and one has unsuccessful bioremediation unless a primary substrate is added.

Electron acceptors

All biological reactions that yield energy are redox reactions. Thus adequate electron acceptors are an important parameter to control for bioremediation. Without an adequate supply or type, bioremediation response will fail. The electron acceptors of importance are:

Oxygen	O_2
Nitrate	NO_3
Sulfate	SO_4
Carbon dioxide	CO_2
Organic compounds	Various

The type of electron acceptor establishes the metabolism mode and therefore the specific degradation reactions. The microbial requirements for an energy source and electron acceptor are more difficult to control. It is difficult to deliver these requirements to a contaminated location. The engineering of the delivery systems and their control provides far more challenge than understanding the biochemical process.

Required environmental conditions

The environment must be adequate to support microbial life and for enzyme reactions to proceed. The engineering design must provide for adequate control of environmental conditions during the complete bioremediation program.

Moisture. A moisture content of about 80 percent field capacity, or about 15 percent water on a weight basis, has been reported as optimum for bioremediation of soil (English, 1991). Inadequate moisture, less than 40 percent, is reported to significantly reduce the rate of bioremediation. A higher moisture content of 50 percent appears necessary for the bioremediation of petroleum oils and fuels. Soil moisture above 70 percent hinders gas transfer for oxygen and significantly reduces aerobic activity (Cookson, 1990).

Moisture is a very important variable relative to bioremediation. Moisture content of soil affects the availability of contaminants, the transfer of gases, the effective toxicity level of contaminants, the movement and growth stage of microorganisms, and species distribution. Low moisture levels restrict bacteria activity by limiting cellular movement and metabolism reactions. The competitive advantage between bacteria and fungi in soil changes markedly with moisture levels (Griffin, 1980). Typical moisture levels for enhanced growth of the fungus *P. chrysosporium* in field soil pile studies has been 20 percent (Lamar, 1993).

Soil moisture is frequently measured as a gravimetric percentage or reported as field capacity (Chap. 6). Evaluating moisture by these methods provides little information on the availability of moisture at low percentages. The interactions between moisture and soil spore size distribution are far more important than a simple percent measurement. A better parameter for evaluating moisture is total water suction or matric suction. Matric suction provides information on the availability of moisture for plant or microbial metabolism.

Total water suction is the sum of matric suction and osmotic suction. For most conditions, matric suction is an adequate measurement since it is the major component of total water suction. The capillary phenomenon that causes water to rise in a small tube creates this matric suction. The smaller the tube or the soil's pore size, the greater the capillary rise height. This capillary water has a negative

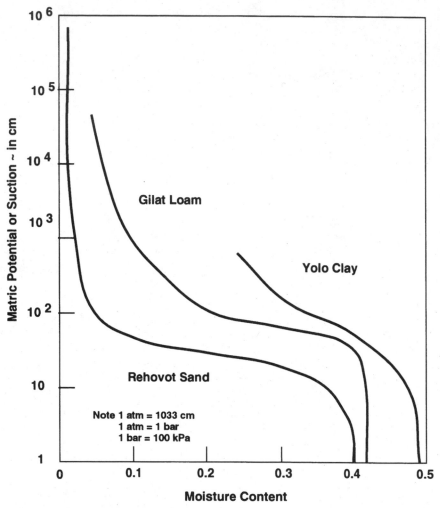

Figure 3.9 Characteristic curves of moisture content as a function of matric potential for three different soils. (*Fredlund, 1993.*)

pressure with respect to air pressure. Soils of different pore size distributions will have significantly different moisture percentages at equal matric suction or matric potential (Fig. 3.9). The relationship between water content of soil and the matric suction is the soil-water characteristic curve (see Chap. 6). At a soil-water suction of 15 bars (100 kPa = 1 bar = 1 atm), most plants cannot overcome this suction. A matric suction of 15 bars is also the point at which bacteria metabolism is negligible (Griffin, 1980). Thus wilting point may be as important to bioremediation as it is to agriculture.

The ability of microorganisms to move and metabolize in soil can be related to the matric suction. For sample, bacteria movement is negli-

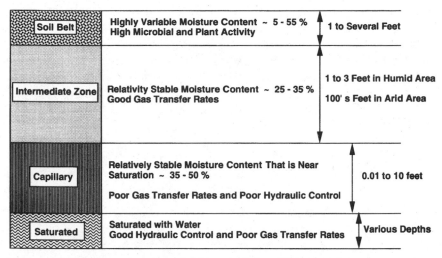

Soil Belt	Highly Variable Moisture Content ~ 5 - 55 % High Microbial and Plant Activity	1 to Several Feet
Intermediate Zone	Relativity Stable Moisture Content ~ 25 - 35 % Good Gas Transfer Rates	1 to 3 Feet in Humid Area 100' s Feet in Arid Area
Capillary	Relatively Stable Moisture Content That is Near Saturation ~ 35 - 50 % Poor Gas Transfer Rates and Poor Hydraulic Control	0.01 to 10 feet
Saturated	Saturated with Water Good Hydraulic Control and Poor Gas Transfer Rates	Various Depths

Figure 3.10 Moisture and gas transfer characteristics of the subsurface zones.

gible in soils drained between 0.2 and 1 bar, and bacteria metabolism decreases sharply as matric suctions rise between 0.5 and 3 bars (Griffin, 1980). Most bacterial metabolism occurs at values of less than 3 bars. Total microbial respiration above 8 bars is mainly a result of fungi. Many fungal mycelia will lyse or stop growth at matric suctions greater then 10 bars.

The moisture content of the subsurface can be divided into four zones. These zones correspond with the soil belt, intermediate belt, capillary fringe, and saturated zone (Fig. 3.10). The saturated zone is below the groundwater table and has all soil pores occupied by water. Moisture content on a weight basis varies from 40 to 55 percent. The capillary fringe zone has a moisture content that is relatively stable and high, near saturation. The height of this zone is a function of the specific soil. Distribution of oxygen as a gas and other gas transfers are hindered in this zone. Unfortunately, this zone frequently contains the major mass of contaminants with specific gravities less than water.

The intermediate belt has the most stable moisture content. It typically has adequate moisture for bioremediation, except in very arid climates. The soil belt has the most variable water content. It is a function of climate, infiltration rates, organic content, evaporation rates, and plant uptake.

pH. The pH affects both the microorganism's ability to conduct cellular functions, cell membrane transport, and the equilibrium of catalyzed reactions. Most bacteria grow best at neutral to slightly alkaline pH. Growth is poor at pH 5 or lower. Generally, pH should be maintained near 7 and within the range of 4 to 10. For nitrogen oxidation and methane fermentation this range is limited to pH 6 to 8. The conversion of ammonia to nitrate, nitrification, is limited at pH values below 6.0 and stopped at a pH below 5.0 (Lee, 1988). *Nitrobacter* has an optimum pH growth near 8.0 (Gaudy, 1988). Hydrocarbon degradation has been reported faster at pH values above 7 compared with values of 5. Fungi desire pH levels below neutral. For example, *Phanerochaete chrysosporium* is active at pH 4.5 to 5.5 (Dhawale, 1992; Brodkorb, 1992).

Metabolic activities of microorganisms produce organic acids and HCl from the dehalogenation of organic contaminants. When high concentrations of organic compounds are present in soil or groundwater with low alkalinity, addition of lime or other alkalinity may be necessary.

Temperature.   Temperature has a marked influence on the rate of bioremediation. The rate of degradation is a direct function of temperature and obeys the relationships originally described by Arrhenius. The logarithm of the velocity of the reaction is a function of 1 over the temperature, where temperature is expressed in degrees Kelvin. A rise of 10° in temperature will approximately double the speed of reaction.

Most bacteria stop metabolic activities at temperatures just above the freezing point of water. Each microorganism, however, has a minimum temperature below which growth no longer occurs. Some microorganisms have the ability to adapt to temperature changes. *Micrococcus cryophilus* can alter its membrane fluidity by changing the ratio of fatty acids when subjected to temperature changes (Sontakke, 1988). Microorganisms commonly found effective in bioremediation perform over a temperature range of 40°C to levels below 10°C (Fig. 3.11). Bacteria can operate at 0°C and slightly below as a result of solute concentrations that have reduced the freezing point of water.

A frequently used expression for estimating changes in degradation rates according to temperature is

$$k_2 = k_1 \, \theta(T_2 - T_1)$$

where k_1 = first-order degradation rate at temperature T_1, °C
k_2 = first-order degradation rate at temperature T_2, °C
θ = temperature coefficient

For hydrocarbon degradation in contaminated soils, θ equal to 1.088 has been experimentally measured over a temperature range of 15 to 42°C (Troy, 1993). As temperature increases, organisms reach a maximum level of activity that decreases with increasing temperature. Maximum microbial metabolism has been measured in compost processes at 50 to 55°C (Cookson, 1990). At temperatures above 60°C, the microbial diversity drops significantly, causing a drop in degradation rates.

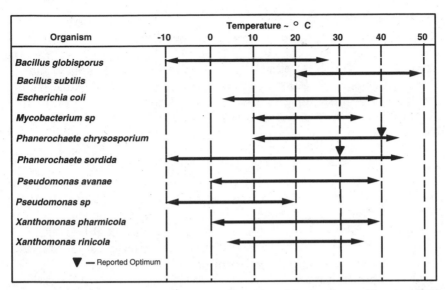

Figure 3.11 Temperature growth range of certain microorganisms. (*Stanier, 1986; Heitkemp, 1988; Lamar, 1990.*)

For many regions of the country, groundwater does vary significantly with seasons. The temperatures of the upper 10 m of the subsurface vary seasonably, while depths between 10 and 18 m are about the mean air temperature for the region (Lee, 1988).

Osmotic level. For most contaminated sites, the groundwater salinity does not vary significantly and the subsurface microorganisms have had adequate time to become adapted. Nutrients should be added as required throughout the bioremediation program rather than in one nutrient loading. Any single-dose application to soil should be limited to 3 lb of nutrients per cubic yard to prevent osmotic shock (Hicks, 1993). Time-release nutrients can be applied as a single dose. Microorganisms can withstand a range of salt concentrations when these changes are introduced slowly. This range can vary as a result of organisms' ability to adjust their internal osmolarity so it always exceeds that of their environment. A period of adaptation with controlled salinity change may be necessary for seed organism use for groundwater with salt intrusion. Organisms have the ability to adapt to stressful or changing environmental conditions. In some cases they acquire new gene functions. Many protect their cells by changes in the cell membrane. For example, *Methanosarcina thermophia* has a heteropolysaccharide layer that it loses when grown in a high-salt-level nutrient-deficient environment. This also requires the organism to metabolize compounds with a net higher energy yield.

Growth nutrients

Although microorganisms obtain energy by different metabolism modes, they all have the same function: the generation of energy for biosynthesis. This biosynthesis results in an immense range of nutritional requirements. Some need only sunlight, carbon dioxide, water, and inorganic nitrogen to grow. Others cannot synthesize all their building units and must be supplied amino acids and vitamins. Needed vitamins are called growth factors.

Carbon is one basic element of all living material. It forms the skeletons upon which most compounds found in cells are built. In addition to carbon, cells are mainly composed of the elements hydrogen, oxygen, and nitrogen. These four chemical elements constitute about 95 percent by weight of living cells. Two other elements, phosphorus and calcium, contribute about 70 percent of the remainder. Other essential elements are usually found in surface water, groundwater, soil, and sludges in the trace concentrations needed by microbial life.

Microorganisms must get all their nutrients necessary for growth from the environment. The microbial requirements for nutrients are approximately the same as the composition of their cells (Table 3.15). The chemical structure of bacteria can be expressed as $C_5H_7O_2N$ with only minor, but important, traces of other atoms. Carbon is usually supplied by organic carbon but in a few cases may be provided by inorganic carbon such as carbonates and bicarbonates. Hydrogen and

TABLE 3.15 Composition of the Microbial Cell

Element	Percentage of dry weight
Carbon	50
Oxygen	20
Nitrogen	14
Hydrogen	8
Phosphorus	3
Sulfur	1
Potassium	1
Sodium	1
Calcium	0.5
Magnesium	0.5
Chlorine	0.5
Iron	0.2
All others	0.3

SOURCE: Stanier, R. Y., Ingraham, J. L., Wheelis, M. L., and Painter, P. R., *The Microbial World,* 5th Edition, Prentice-Hall, Englewood Cliffs, N.J., 1986.

TABLE 3.16 Nitrogen and Phosphorus Supplement Sources

Ammonium sulfate
Ammonium phosphate
Disodium phosphate
Monosodium phosphate
Urea
Potassium monophosphate
Orthophosphoric salts
Phosphoric acids
Polyphosphate salts:
  Pyrophosphate
  Tripolyphosphate
  Trimetaphosphate

oxygen are supplied by water. Nitrogen, phosphorus, and sulfur are provided through both inorganic and organic sources (Table 3.16).

Design of a successful bioremediation project considers the availability of carbon, nitrogen, and phosphorus. Carbon is usually provided by the substrate. Nitrogen and phosphorus are naturally found in soil and groundwater, but for highly contaminated sites, their concentrations may be too low. The available nitrogen and phosphorus at a site are compared with the microbial need for mineralization of the available organic compounds.

The remaining elements, potassium, manganese, calcium, iron, cobalt, copper, and zinc, are all provided in the form of inorganic salts. These are present in adequate concentrations in most soil and aquifer systems and usually need no further attention in the design of a bioremediation process.

When one of these essential elements is not present in adequate quantities, microbial growth is limited. This condition is stated by Liebig's law of the minimum: The essential constituent that is present in the smallest quantity relative to the nutritional requirements of organisms will become the limiting factor for growth. This law can be expanded to include the electron acceptor, temperature, etc.

Present laboratory and field data on the importance of adding nutrients for bioremediation yields mixed conclusions. Many laboratory studies show significant improvement on the rate of degradation with nutrient supplements. Field data often demonstrate little effect. This could be a problem of inadequate delivery and distribution of nutrients in the field, or it may result from other parameters limiting degradation. If oxygen delivery rate is the limiting factor, adequate nutrient cycling may occur to support bioremediation under conditions that would be nutrient-limiting with an adequate oxygen supply rate.

Chapter

4

Optimizing Microbial Transformation of Hazardous Chemicals

Introduction

The feasibility of bioremediation for the cleanup of contaminated sites starts with an assessment of the degradation potential of the contaminants. This assessment includes the ease or difficulty of degradation, the ability to achieve total mineralization, as well as the environmental conditions necessary for mineralization (Table 4.1). Regulatory officials and citizens want information on the potential degradation compounds and their potential hazard.

TABLE 4.1   Assessing the Potential for Successful Biodegradation of Organic Compounds

1. Is the compound biodegradable?
2. Is the compound easily degraded or difficult?
3. Is the compound mineralized?
4. What type of environment and electron acceptor is necessary for the degradation?
5. Are there any specific microbial, substrate, or other conditions that must be met to achieve degradation?
6. What environmental conditions inhibit or enhance the degradation?
7. What are the metabolic intermediates that should be monitored for gauging the degree of degradation?
8. Are there any potential intermediates that are of greater health concern than the parent compound?

Years of research on establishing the degradation pathways for specific chemicals provides many of these answers. A review of pertinent degradation pathways provides information on intermediates and insight on where the reaction sequence may be blocked. Knowledge of these intermediates is necessary to address health concerns and monitor the degree of bioremediation. Monitoring intermediate compounds during transformation can provide insight on rate-limiting factors during the mineralization process.

Microbial transformations of organic compounds are frequently described by the terms detoxification, degradation, and mineralization. Detoxification is the transformation of the compound to some intermediate form that is less toxic. Degradation means that the parent compound no longer exists. Mineralization refers to the complete conversion of the organic structure to inorganic forms. Thus detoxification and degradation are necessary consequences of mineralization.

Providing microbial transformations for each possible hazardous compound is not the intent of this chapter. First, these pathways have not been delineated for many compounds. Second, the goal of this chapter is to provide information on the general nature of the most needed transformations in bioremediation. Fortunately microorganisms frequently use similar degradation pathways for chemicals of similar structure.

The chapter is organized to support engineering design and applications according to chemical categories that are frequently found in contaminated sites. For this reason, all metabolism modes and the responding transformations are presented for each hazardous chemical category.

Compounds that differ significantly from these generalized categories, that pose special environmental problems, or that have specific commercial applications are presented in Chap. 5. These include pesticides, polychlorinated biphenyls, and dyes.

Petroleum Aliphatic and Alicyclic Hydrocarbons

Chemical nature

Hydrocarbons are compounds containing carbon and hydrogen. Aliphatic hydrocarbons are straight or branched-chain hydrocarbons of various lengths. The aliphatic hydrocarbons are divided into the families alkanes, alkenes, alkynes, alcohols, aldehydes, ketones, acids, and cyclic analogs. The alicyclic hydrocarbons have diverse structures. They include common components of petroleum oils and complex substituted compounds manufactured for pesticides. These more complex halogenated alicyclic compounds are discussed in Chap. 5 under pesticides. Typical structures are illustrated in Figs. 4.1 and 4.2.

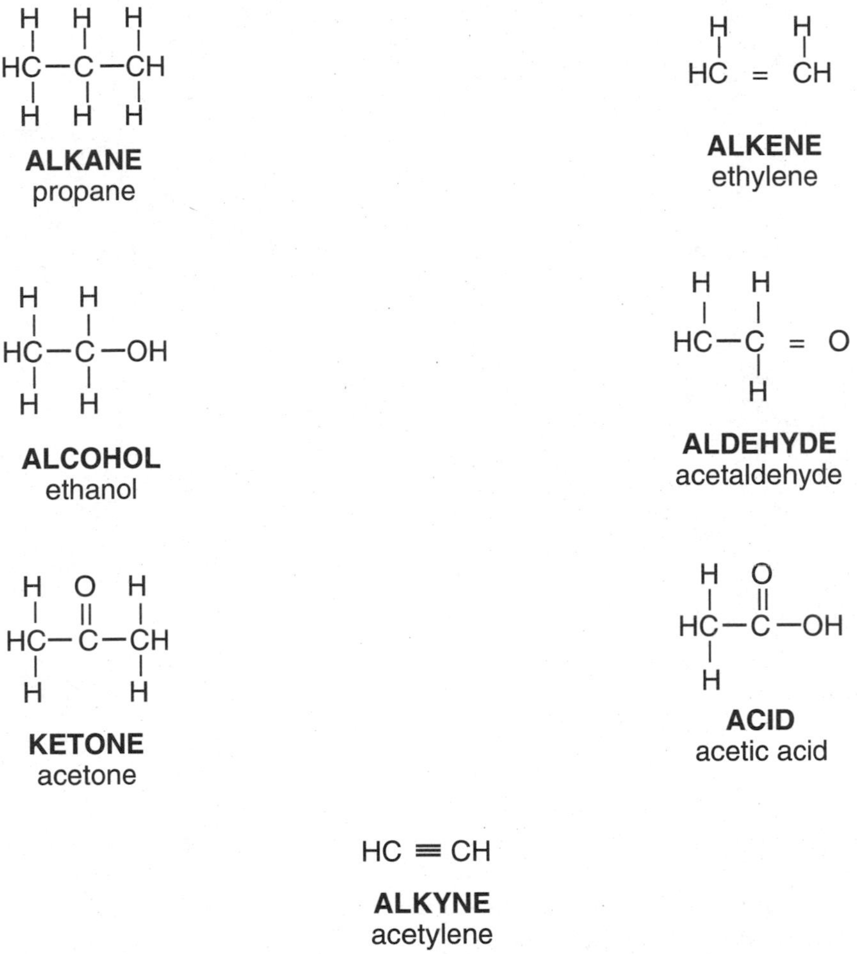

$$HC-\overset{\overset{\displaystyle H}{|}}{\underset{\underset{\displaystyle H}{|}}{C}}-CH$$

ALKANE
propane

$$HC = CH$$

ALKENE
ethylene

$$HC-\overset{\overset{\displaystyle H}{|}}{\underset{\underset{\displaystyle H}{|}}{C}}-OH$$

ALCOHOL
ethanol

$$HC-\overset{\overset{\displaystyle H}{|}}{\underset{\underset{\displaystyle H}{|}}{C}} = O$$

ALDEHYDE
acetaldehyde

$$HC-\overset{\overset{\displaystyle O}{||}}{C}-CH$$

KETONE
acetone

$$HC-\overset{\overset{\displaystyle O}{||}}{C}-OH$$

ACID
acetic acid

$$HC \equiv CH$$

ALKYNE
acetylene

Figure 4.1 Typical structures of aliphatic hydrocarbons.

Microbial metabolism

The most frequent application of bioremediation has been to hydrocarbon spills. The degradation potential of alkanes is a function of
the carbon chain length. Short chains are more difficult to degrade
than the longer chains. In fact, short chains, less than 10 carbons,
tend to be toxic as a result of their higher solubility. In some cases,
the toxicity of short-chain hydrocarbons has been reversed by adding
a less toxic aliphatic hydrocarbon. For example, inhibition of microbial growth on n-decane can be reversed by adding n-tetradecane
(Bitton, 1984). The longer-chain n-tetradecane appears to reverse
toxicity by partitioning the more toxic shorter-chain hydrocarbon n-
decane between the aqueous phase and the larger hydrocarbon.

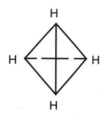

Cyclopentane

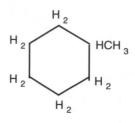

Methylcyclohexane

Tricyclobutane

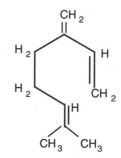

Terpene Hydrocarbon
Myrcene

Figure 4.2   Typical structures of cyclic aliphatic hydrocarbons.

Fortunately, alkanes shorter than 10 carbons are seldom found at most contaminant sites because of vaporization. The longer-chain aliphatic hydrocarbons are readily degraded by a wide variety of microorganisms under aerobic conditions. Soil contains significant populations of microbes that can use select hydrocarbons as sole sources of carbon and energy. Soil populations capable of degrading hydrocarbons have been reported as high as 20 percent of all soil microbes (Bitton, 1984). Over 160 genera of fungi alone have been demonstrated to grow on hydrocarbons. Filamentous fungi appear to be more versatile than yeast in degradation of short-chain hydrocarbons, but they follow the same pattern of utilizing long-chain alkanes more readily than short chains. Microbes capable of oxidizing hydrocarbons are also prevalent in the aquatic environment including seawater. A partial list of bacterial and yeast species capable of oxidizing hydrocarbons is presented in Table 4.2. The best inoculum for remediation of hydrocarbon contamination is the contaminated soil.

As with most chemicals, intimate contact between the microbial cell surface and the hydrocarbon appears necessary for high degra-

TABLE 4.2   Bacteria and Yeasts That
Have Been Demonstrated to Oxidize
Aliphatic Hydrocarbons

Bacteria	Yeasts
Achromobacter	Candida
Acinetobacter	Cryptococcus
Actinomyces	Debaryomyces
Aeromonas	Endomyces
Alcaligenes	Hansenula
Arthrobacter	Mycotorula
Bacillus	Pichia
Beneckea	Rhodotorula
Brevibacterium	Saccharomyces
Corynebacterium	Selenotila
Flavobacterium	Sporidiobolus
Methylobacter*	Sporobolomyces
Methylobacterium*	Torulopis
Methylococcus*	Trichosporon
Methylocystis*	
Methylomonas*	
Methylosinus*	
Micromonospora	
Mycobacterium	
Nocardia	
Pseudomonas	
Spirillum	
Vibrio	

SOURCE: Bitton, L. N., "Microbial Degradation of Aliphatic Hydrocarbons," *Microbial
Degradation of Organic Compounds,* D. T.
Gibson (Ed)., Marcel Dekker, New York,
1984.
*Methane is the only hydrocarbon used as a
sole carbon source.

dation rates. Bacteria and filamentous fungi are frequently attached to alkane droplets. Dispersing oil makes it more susceptible to microbial attack; however, the enhanced availability can cause toxicity problems.

Microbial cultures produce extracellular surfactants that aid in solubilizing hydrocarbons. Biological surfactants consist of complex mixtures of protein, lipid, and carbohydrates. A few identified surfactants are rhamnolipids, trehalose-containing glycolipids, phospholipids, and lipopolysaccharide (Bitton, 1984).

Nonbiological surfactants have been used to promote hydrocarbon dispersion. Solubility of hydrocarbons has been enhanced with nonionic alkylphenol ethoxylate surfactant (Triton X-100) (Edwards et al., 1991).

For hydrocarbon-contaminated soils, moisture levels below 50 percent appear to inhibit degradation. This finding may be related to the hydrophobic nature of these compounds. Hydrocarbon degradation rate also is limited at pH levels above 8.5.

Aerobic metabolism

Aliphatic. Bioremediation of aliphatic hydrocarbons should be performed as an aerobic process. Oxygen is usually the rate-limiting factor in most cases for contaminated soil and groundwater. The first step in the aerobic degradation of hydrocarbons is the incorporation of molecular oxygen in the hydrocarbon. This is performed by oxygenase enzymes. There are two groups of oxygenases, monooxygenases and dioxygenases (Wackett, 1989). Complete mineralization by aerobic respiration has been well documented in both laboratory and field applications.

The most common pathway of alkane degradation is oxidation at the terminal methyl group. The alkane is oxidized to an alcohol and then to the corresponding fatty acid. Once an organic compound has been oxidized to the acid form, resulting in the formation of a carboxyl group, the oxidation proceeds by successive removal of two carbon units (Fig. 4.3). The removal of two carbon units is universal to most living systems and is termed the "beta oxidation sequence." Under beta oxidation, the beta methylene group is oxidized to a ketone group followed by the removal of a two-carbon fragment from the compound. In addition to terminal oxidation, subterminal degradation occasionally occurs (Fig. 4.4).

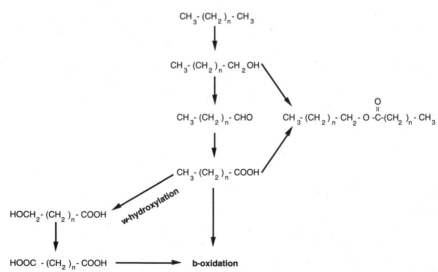

Figure 4.3 Oxidation of *n*-alkanes by attack on the terminal methyl group. (*Britton, 1984.*)

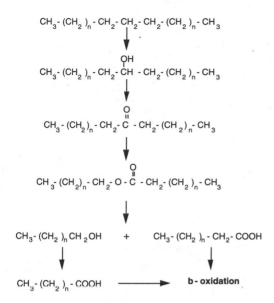

Figure 4.4 Aliphatic hydrocarbon oxidation by subterminal attack. (*Britton, 1984.*)

Short-chain hydrocarbons except methane are more difficult to degrade. Under aerobic conditions methane is readily used as the sole carbon source by methanotrophs. Successful degradation of the other short-chain hydrocarbons may require cometabolism. Cometabolism degradation of ethane, propane, and butane has been demonstrated (Horvath, 1972). *Pseudomonas methanic* will use methane as a primary substrate and oxidize a secondary substrate such as ethane, propane, and butane to the corresponding alcohols, aldehydes, and acids. None of these compounds serve as growth substrates for *Pseudomonas methanic*. For short-chain alkanes, cometabolism is an important degradation process. The cometabolism of propane and butane yields acetone, propionic acid, butanoic acid, 2-butanone, and methyl ethyl ketone as intermediates.

Alkene degradation is more varied since microbial attack can occur at either the methyl group or the double bond. Unsaturated straight-chain hydrocarbons are generally less readily degraded than saturated ones. Metabolism results in the formation of intermediates consisting of unsaturated alcohols and fatty acids, primary or secondary alcohols, methyl ketones, 1,2-epoxides, and 1,2-diols. Typical alkene degradation is illustrated in Fig. 4.5. The methyl group oxidation is considered the major degradation pathway.

Hydrocarbons with branch chains are much less susceptible to degradation. This fact can be recalled by the resistance of such detergents as alkyl benzene sulfonates to biological breakdown. Branched-chain compounds such as in quaternary carbon and β-alkyl-branched compounds are recalcitrant and accumulate in the biosphere. There

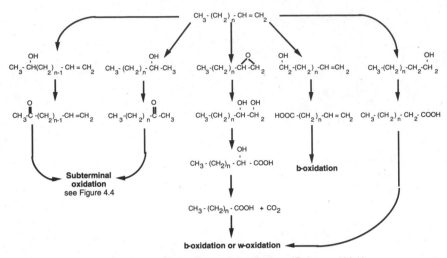

Figure 4.5 Potential pathways for 1-alkene degradation. (*Britton, 1984.*)

are, however, a few organisms capable of utilizing these alkyl-branched components as a sole carbon and energy source.

Even more resistant to degradation are the quaternary carbon compounds. For example, the compound 2,2-dimethylheptane is oxidized by microorganisms by attacking the unhindered terminal branch chain yielding 2,2-dimethylpropionic acid, apparently the end product. No organism has been isolated or found capable of degrading this compound. Recalcitrant compounds can be potentially remediated by combining chemical oxidative techniques with bioremediation.

The presence of one hydrocarbon substrate frequently influences the biodegradation of another. This can be a positive response as with cometabolism, or a negative response due to catabolite repression or toxicity. Oxidation of hydrocarbons may be subject to catabolite repression. An example is given by studies with *Pseudomonas aeruginosa* in which alkane oxidation was repressed by the presence of glucose and malate.

Alicyclic. Microorganisms capable of degrading alicyclic hydrocarbons are not as predominant in soils as those for the degradation of aliphatic alkane and alkene hydrocarbons.

The degradation of cycloalkanes is usually the oxidation of the terminal methyl unit yielding a primary alcohol. Many microorganisms capable of oxidizing noncyclic alkanes have a broad enough specificity to also hydrolyze cycloalkanes. Hydroxylation is vital to initiate the degradation of cycloalkanes. An example pathway for the degradation of cyclohexanol and cyclohexanone is illustrated for *Acinetobacter* in

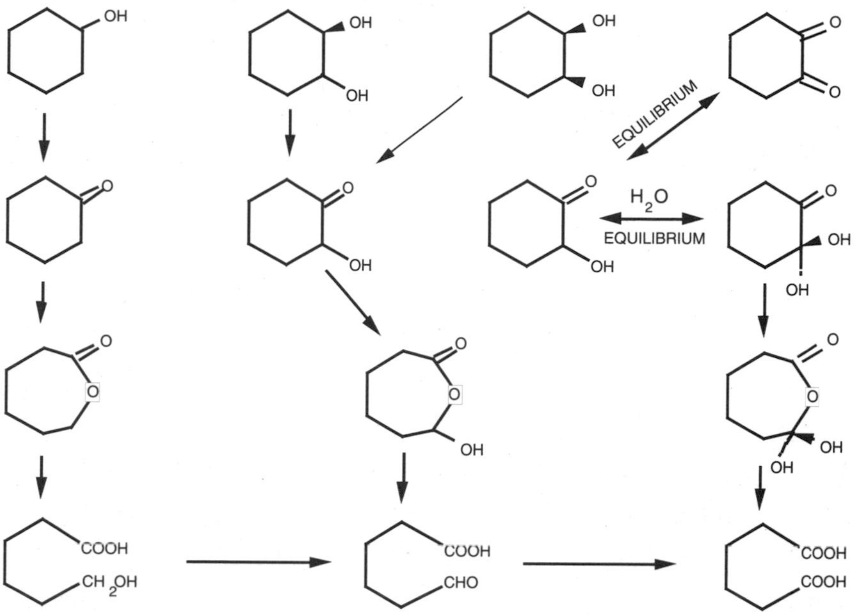

Figure 4.6 Example pathways for the degradation of cyclohexanols and cyclohexanones. (*Trudgill, 1984.*)

Fig. 4.6. Several alternate degradation pathways exist for microbial attack of the alicyclic structures.

The cycloalkanols and cycloalkanones are more readily metabolized by microorganisms than their parent hydrocarbons. Bacteria capable of using these as sole carbon sources abound in the environment and can be found in almost any source material.

The substituted cycloalkyl compounds vary significantly in their microbial stability. Those containing carboxylic acid groups are readily degraded. Those containing chlorine are very resistant and make the base formulation for a number of pesticides (Chap. 5).

Microorganisms capable of degrading cycloalkyl carboxylic acids such as cyclohexane carboxylic acid and cyclopentane carboxylic acid are numerous in the environment. A typical sequence for cyclohexane carboxylate degradation is illustrated in Fig. 4.7.

Cycle alkanes with substitutions of long carbon side chains are more resistant to degradation. The main contamination source of these compounds is from crude oil spills. The *n*-alkyl cycle alkanes with an odd number of carbon atoms in the side chain are susceptible to degradation by β-oxidization. The *n*-alkyl cycloalkanes with an even number of carbon atoms are attacked by methyl hydroxylation. This intermediate is oxidized to form the corresponding acid that is

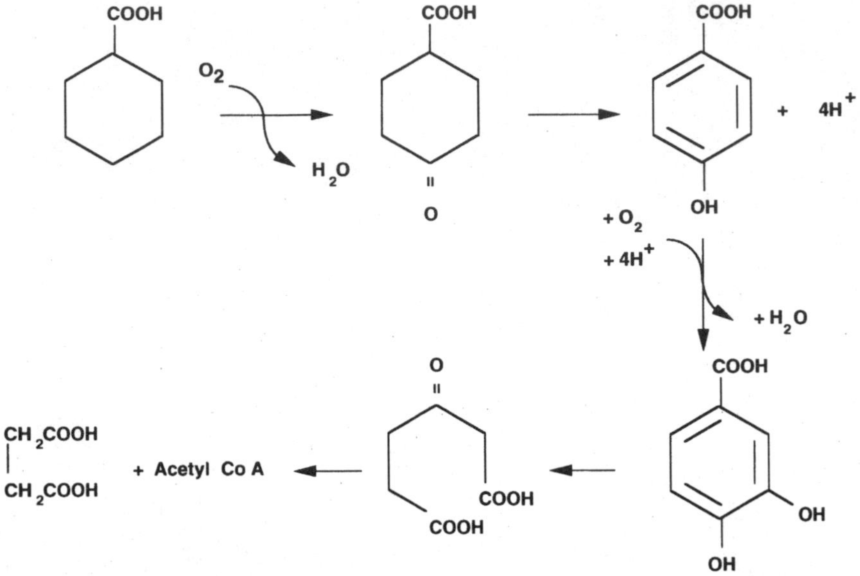

Figure 4.7 Example pathway for degradation of cyclohexane carboxylate. (*Modified from Trudgill, 1984.*)

then subjected to β-oxidization. A cyclohexanone is formed and degraded as presented in Fig. 4.7.

Many complex alicyclic compounds exist and numerous intermediates have been identified during their degradation. The degradation of these more complex structures can be illustrated by the degradation of camphor (Fig. 4.8). The initial step in the degradation is transformation at the ring-carrying keto group to form lactone and hydroxylation of the second ring. This is followed by ring cleavage (Trudgill, 1984).

Little is known about the degradation pathways of terpenes except that many intermediates are produced during metabolism. Limonin is a complex triterpenoid dilactone which contains a cyclic hexane ring, a cyclohexanone ring, a furan ring, and a tetrahydrofuran ring. During the degradation of limonene by *Pseudomonas strain L.*, many intermediates have been identified (Fig. 4.9). Once this compound is hydrolyzed it can serve as a sole carbon source for soil microorganisms.

Anaerobic metabolism. Conclusive evidence for anaerobic degradation of aliphatic hydrocarbons, although referenced, is uncertain and, at this stage, too ill defined for application. Bioremediation of hydrocarbons should be performed as an aerobic process. Oxygen is required.

OH
H
CO Succinyl CoA
COO⁻
CO Succinyl CoA
CO Succinyl CoA

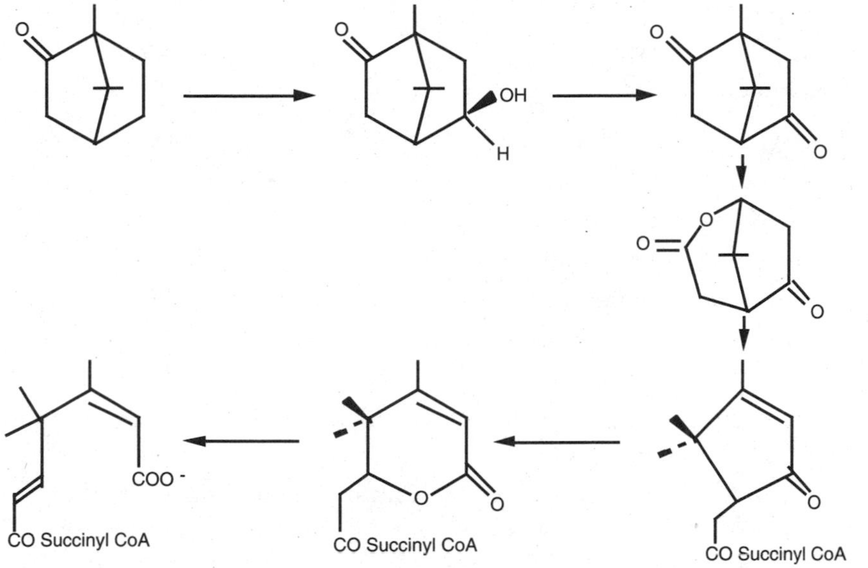

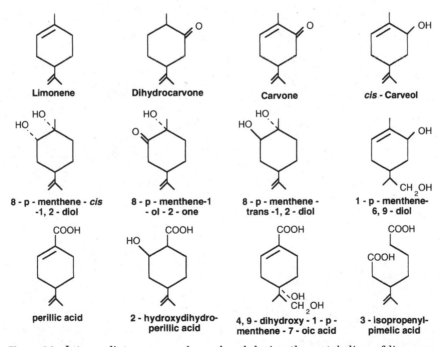

Figure 4.8 A pathway for the degradation of camphor by *Pseudomonas putida*. (*Modified from Trudgill, 1984.*)

Limonene
Dihydrocarvone
Carvone
cis - Carveol
8 - p - menthene - *cis* -1, 2 - diol
8 - p - menthene-1 - ol - 2 - one
8 - p - menthene - trans -1, 2 - diol
1 - p - menthene- 6, 9 - diol
perillic acid
2 - hydroxydihydro- perillic acid
4, 9 - dihydroxy - 1 - p - menthene - 7 - oic acid
3 - isopropenyl- pimelic acid

Figure 4.9 Intermediate compounds produced during the metabolism of limonene. (*Trudgill, 1984.*)

Summation

When bioremediating aliphatic hydrocarbon, the following should be considered:

1. Oxygen is required. Bioremediation is an aerobic process.
2. Many microorganisms will degrade aliphatic hydrocarbons.
3. Soil normally contains an adequate inoculum of natural organisms for bioremediation.
4. Long-chain alkanes are degraded more readily than short chains.
5. Saturated aliphatic hydrocarbons are degraded more readily than unsaturated ones.
6. Hydrocarbon with branched chains is degraded less readily than that with straight chains.
7. Hydrocarbons that are less than nine carbon atoms in length are significantly more difficult to bioremediate. For these hydrocarbons, cometabolism may be necessary.
8. For hydrocarbons, soil moisture requirements for maximum activity may be different from optimum moisture levels for bioremediation of hydrophilic substrates. Moisture levels below 50 percent appear to inhibit degradation.
9. Inhibition of hydrocarbon degradation may occur at pHs above 8.5.
10. The alicyclic hydrocarbons vary significantly in their degradation potential, depending on the type of substitutions and complexity of their structure.

Petroleum Aromatic Hydrocarbons

Petroleum fuels contain significant amounts of aromatic hydrocarbons (Tables 1.8 and 1.10). The aromatic hydrocarbons of benzene, ethylbenzene, toluene, xylene, trimethylbenzene, and other benzene forms make up over 40 percent of the gasoline's composition.

Benzene and other single-ring aromatic hydrocarbons are released from storage and transportation facilities, from industrial production facilities, and as an intermediate in the production of numerous chemicals. Some single-ring aromatic compounds are used as solvents in industrial processes. For example, ethylbenzene is a solvent used in the manufacture of styrene (Howard, 1989). Substituted forms of single-ring aromatic compounds are used in many industrial processes for the preparation of dyes, resins, antioxidants, polyurethane foams, fungicide stabilizers, coatings, insulation materials, fabrics, plastics, metals, leather, rubber, etc. Almost any industrial activity results in a potential source for these compounds.

The substituted aromatic compounds such as aromatic alcohols, aldehydes, ketones, and acids are all derivatives on the benzene structure. The degradation and chemistry of these compounds are very similar. They include benzoic acid, phenol, vanillin, phenyl, and benzyl alcohol, to mention a few. These compounds result from many industrial operations and include plasticizers, medicinals, cosmetics, preservatives, fibers, adhesives, fragment odors, and numerous other manufacturing operations.

Heterocyclic aromatic compounds have one other element in the ring in addition to carbon. Many are nitrogen or sulfur heterocycles such as pyrrole, indole, and piperidine. Generation sources include solvents, rubber curing, epoxy resins, and chemical synthesis operations.

Chemical nature

The aromatic hydrocarbons contain the benzene ring as the parent hydrocarbon. They are all ring compounds and have only one free valance bond. The benzene ring is represented by double bonds between alternate carbon atoms (Fig. 4.10). The single-ring structures consist of benzene, toluene, ethylbenzene, and the three isomeric forms of xylene. These are frequently referred to as BTEX compounds. The benzene series include various substitution benzene compounds. Typical structures are illustrated in Fig. 4.11.

Several benzene rings are joined at two or more ring carbons to form polynuclear aromatic hydrocarbons (PAH). The hydrogens in the aromatic hydrocarbon may or may not be substituted by a variety of

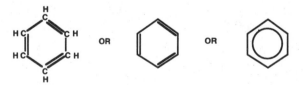

Figure 4.10   Benzene formula and simplified representations.

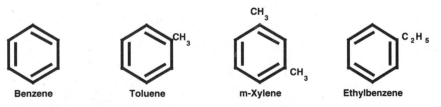

Benzene   Toluene   m-Xylene   Ethylbenzene

Figure 4.11   Structures of single-ring aromatic hydrocarbons.

groups. Some of the common substitutes are Cl, chlorol; Br, bromo; I, iodo; NO_2, nitro; NO, nitroso; and CN, cyano.

Microbial metabolism

The degradation of aromatic compounds by bacteria has been extensively studied, and significant information is being developed on degradation of aromatic compounds by fungi. Aromatic compounds can be readily degraded, are extremely resistant, or yield undesirable intermediates. These differences depend upon the number of rings in the structure, the number of substitutions, the type and position of substituted groups, and the nature of atoms in heterocyclic compounds. Aromatic chemical mixtures also can influence the degradation rates. For example, inhibition has been reported for mixtures of benzene and o-xylene (Arvin, 1988). The substituted structures of polynuclear aromatic hydrocarbons and aromatic pesticides are discussed in Chap. 5.

Aerobic metabolism.   Soil microorganisms capable of aerobically metabolizing single-ring aromatic hydrocarbons are ubiquitous. The degradation is achieved by two alternate modes of oxidation. The basic method of microbial attack on single-ring aromatic compounds involves the formation of a dihydrodiol as illustrated by the oxidation of benzene in Fig. 4.12. The formation of the dihydrodiol compound is typical for single-, double-, and triple-ring aromatic hydrocarbons. The next step in the metabolism is oxidative attack on the aromatic nucleus leading to the formation of alkyl catechol. Catechol, or benzene-1,2-diol, is a common intermediate during the oxidation of many aromatic compounds.

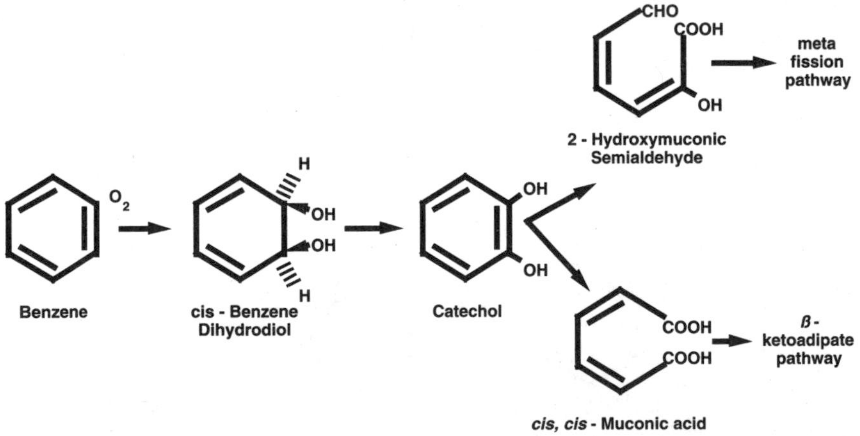

Figure 4.12   Aerobic degradation of benzene. (*Gibson and Subramanian, 1984.*)

These oxygenated benzene rings are related to alcohols and phenols and readily undergo oxidation. Oxidation results in ring fission, forming either an aldehyde or an acid. This step results in the destruction of the aromatic ring, leaving an oxidized aliphatic hydrocarbon. As discussed earlier, these compounds are readily oxidized, releasing hydrogen atoms. The acceptor for the hydrogen atoms is nicotinamide adenine dinucleotide (NAD). The oxidation of toluene and ethylbenzene by the organism *Pseudomonas putida* has been extensively studied. One proposed pathway for degradation is illustrated in Fig. 4.13.

Bacteria also can utilize *p*- and *m*-xylene as a sole carbon and energy source, but not *o*-xylene. Aerobic oxidation of these compounds is believed to be similar to the pathways presented in Fig. 4.14. There is no confirmed evidence that an organism can grow solely on *o*-xylene. However, cooxidation of *o*-xylene has been reported for several different strains of *Nocardia*. Organisms belonging to the genus *Nocardia* are well known for their ability to oxidize alkaline-substituted aromatic hydrocarbons.

The second mechanism for degradation of aromatic compounds is oxidation of any alkyl substitutes. Oxidization of alkyl substitutes results in the formation of aromatic carboxylic acids that are then oxidized to dihydroxylated ring fission substrates. The intermediate products of aldehyde and acid are then readily degraded by the well-known beta-ketoadipic and meta fission pathways. When microorganisms act upon alkyl substitutes of different chain length, they usually initiate oxidation at the smaller substitute. In general, aromatic compounds are less degradable the longer or more numerous the alkyl groups (Arvin, 1988).

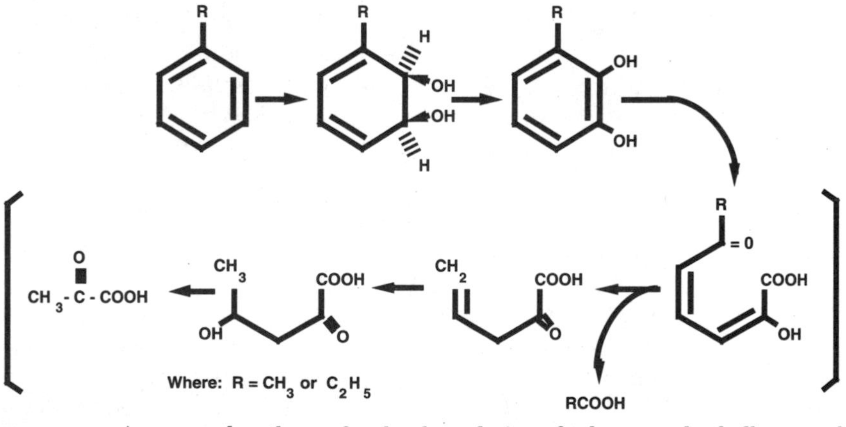

Figure 4.13 A proposed pathway for the degradation of toluene and ethylbenzene by *Pseudomonas putida*. (*Gibson and Subramanian, 1984.*)

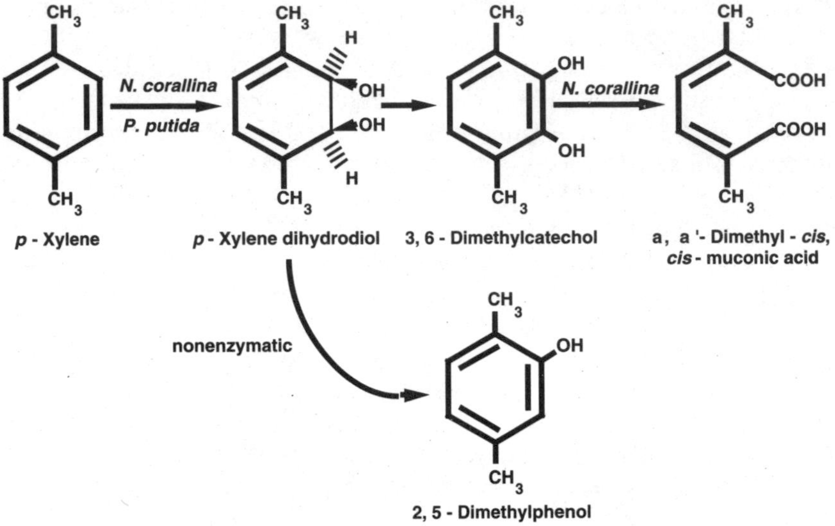

Figure 4.14 Proposed pathway for aerobic degradation of p-xylene by *Pseudomonas putida* and *Nocardia corallina*. (*Gibson and Subramanian, 1984.*)

Anaerobic metabolism. Confirmation on anaerobic degradation of aromatic compounds dates back to 1934, when Tarvin and Buswell reported on the anaerobic metabolism of benzoate, phenylacetate, phenylpropionate, and cinnamate. These compounds are mineralized to carbon dioxide and methane.

Aromatic hydrocarbons are transformed anaerobically under conditions of denitrifying, sulfate-reducing fermentation, and methanogenic conditions. Methanogenic transformation of complex organic compounds involves the interaction of several metabolic groups of bacteria, hydrolytic, acetogenic, homoacetogenic, and methanogenic (Kataoka, 1992). These interactions have been discussed in Chap. 3 (Fig. 3.7).

Denitrifying conditions have been observed in some field situations during the degradation of aromatic compounds in fuel, but little is documented on using nitrate as the electron acceptor. The rate of denitrifying reactions appears to be 50 percent less compared with oxygen as the electron acceptor (Hutchins, 1991). However, the cost of using nitrate as an electron acceptor is projected to be 50 percent less per cubic yard as compared with hydrogen peroxide.

The presence or absence of oxygen in the aromatic structure plays a significant role in the degradation mechanism and rate of degradation. During anaerobic degradation, the oxygenated aromatic compounds are reduced. However, for a nonoxygenated aromatic compound, reduction would cause the compound to become more stable

and less susceptible to further microbial attack. Thus the nonoxygenated aromatic compounds were considered recalcitrant to anaerobic microbial attack. This belief was supported by the presence of
large oil deposits in anaerobic subsurface regions. One of the first
indications that these compounds are broken down anaerobically
came from the *Amoco Cadiz* oil spill in 1979.

Nonoxygenated compounds. To understand the anaerobic transformation of nonoxygenated aromatic compounds, one must look at the role
of abiotic oxidative substitution reactions. There are several abiotic
oxidative reactions that occur with monoaromatic hydrocarbons. Metal
electron oxidants can be used as a catalyst. Electrophylic substitution
reactions can occur followed by a reductive elimination. A number of
transition metals participate in these reactions. Microorganisms also
use metal complexes to catalyze reactions.

The first step in the degradation of nonoxygenated aromatic compounds is the conversion to an oxygenated form. This can be achieved
by an abiotic reaction, or bacterial enzymes enhanced abiotic reactions. Under anaerobic metabolism oxygen is first incorporated from
water into the aromatic structure via hydroxylation (Vogel, 1986).

The transformation of compounds such as benzene and toluene can
be characterized as fermentation in which the substrates are partly
oxidized and partly reduced. The oxidation might include both methyl
group oxidation and ring oxidation. Reduction results in the formation of saturated cyclic rings (Grbić-Galić, 1987).

The degradation of benzene is initiated by ring oxidation, resulting
in the formation of phenol. With toluene, three pathways are possible
starting with ring oxidation to yield *p*-cresol or *o*-cresol and methyl
group oxidation to form a benzyl alcohol (Fig. 4.15) (Grbić-Galić, 1990).
Once the initial oxidation occurs, the degradation follows pathways
similar to that for anaerobic transformation of the oxygenated aromatics as discussed below.

Transformation of nonoxygenated aromatic compounds has been
demonstrated under denitrifying conditions as well as methanogenic
conditions. For example, in the degradation of benzene, toluene, and
xylene, they undergo oxidative microbial transformation under strictly anaerobic conditions with nitrate as an electron acceptor. They also
respond to degradation by fermentation metabolism with production
of methane. Under fermentative conditions, organic compounds are
used as the electron donor, carbon and energy source, and the electron acceptor.

Higher rates of degradation are reported under denitrifying conditions than under methanogenic conditions. This is expected when one

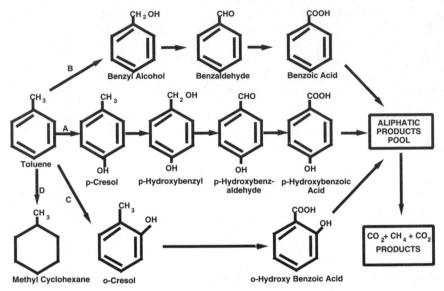

Figure 4.15 Anaerobic pathways for toluene degradation: (*a*) *p*-cresol pathway; (*b*) benzyl alcohol pathway; (*c*) *o*-cresol pathway; (*d*) reductive pathway. (*Grbić-Galić, 1990.*)

considers the thermodynamics of these reactions. The amount of energy obtainable from toluene with nitrate as an electron acceptor is 20 times higher than under the condition of methanogenic fermentation (Grbić-Galić, 1990).

Oxygenated compounds. During anaerobic metabolism, the oxygenated aromatic compounds such as the alcohols, aldehydes, acids, phenolics, and aromatic amino acids are transformed by a reductive mechanism. Reduction occurs, converting the aromatic ring to an alicyclic ring, followed by hydrolytic cleavage and mineralization (Young, 1984). The anaerobic metabolism of benzoate demonstrates this principle (Fig. 4.16). The reduction reaction occurs in contaminated aquifers under methanogenic, sulfate-reducing, and nitrate-reducing conditions. Thus the potential electron acceptors are:

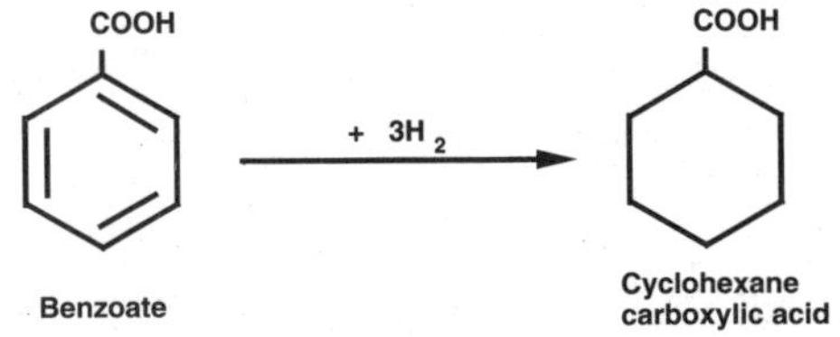

COOH   COOH

$\longrightarrow{} + 3H_2$

Benzoate   Cyclohexane
                                  carboxylic acid

Figure 4.16 Anaerobic reduction of benzoate to an alicyclic ring. (*Young, 1984.*)

- Organic compound
- Carbon dioxide
- Sulfate
- Nitrate

The ability to perform reductive fission of the benzene ring has been clearly demonstrated for three different physiological groups of microorganisms: purple nonsulfur bacteria, anaerobic nitrate reducer, and an anaerobic consortium of bacteria that includes methanogens. These anaerobic microbial transformations are well known. The anaerobic degradation of phenol, 2-methylphenol, 3-methylphenol, and 4-methylphenol has been observed not only in laboratory studies but in groundwater aquifers under methanogenic and sulfate-reducing conditions. In groundwater aquifers, phenol is preferentially utilized followed by 4-methylphenol and ultimately 3-methylphenol. The nitrate respiration of aromatic compounds is a reductive phase followed by the beta oxidation sequence and ring cleavage to aliphatic acids (Fig. 4.17). During the degradation of aromatic compounds in the presence of nitrate, methane is not produced (Young, 1984).

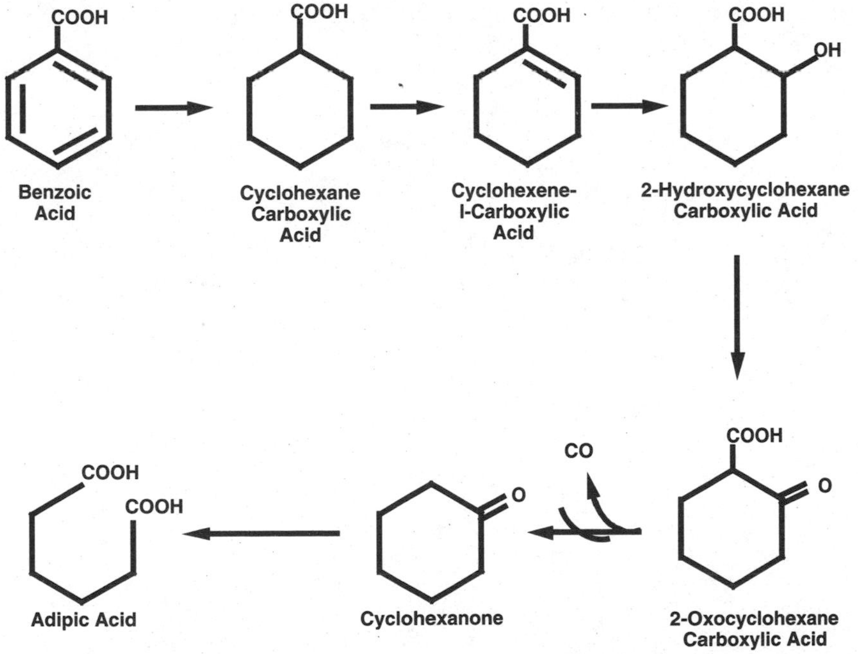

Figure 4.17   Anaerobic degradation of benzoate by nitrate respiration. (*Young, 1984.*)

Organisms that metabolize benzoate with nitrate as an electron acceptor also metabolize other aromatic compounds. These include such compounds as protocatechuate, phenylalanine, and p-hydroxybenzoate. A common soil organism, *Pseudomonas* has been demonstrated to metabolize either m- and p-hydroxybenzoate and protocatechuate. After 1 to 2 weeks of acclimation, this organism also metabolized phenylpropionate and o-hydroxybenzoate. However, no growth occurred when phenol, catechol, benzyl alcohol, or cyclohexane carboxylate were used as substrates. Benzoate is a central intermediate in the transformation of these aromatic compounds.

Under nitrate-reducing conditions the species *Moraxella* can metabolize a range of aromatic compounds including benzaldehyde, benzyl alcohol, m- and p-hydrobenzoate, protocatechuate, phenylacetate, cinnamate, p-hydroxycinnamate, caffeic acid, and phloroglucinol. Sulfate also can be used as an electron acceptor. Several sulfate-reducing bacteria have been demonstrated to metabolize a wide range of aromatic substrates.

For mixed cultures capable of methanogenic fermentation of aromatic compounds, the benzene nucleus is first reduced and then cleaved to aliphatic acid. A probable pathway for fermentation of benzoate and phenol is illustrated in Fig. 4.18. These reactions are not carried out by a single microorganism but require a consortium. If one organism of this consortium is inhibited, the activity of the other members is affected.

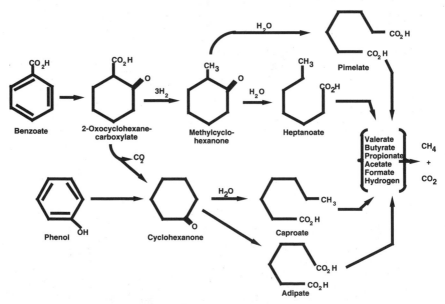

Figure 4.18 Methanogenic fermentation of benzoate and phenol. (*Young, 1984.*)

The methanogenic fermentation of aromatic compounds results in intermediates such as phenol, cyclohexanone, and the end products of carbon dioxide, hydrogen, methane, and organic acids. These organic acids include:

- Valerate
- Butyrate
- Propionate
- Acetate
- Formate

The monitoring of methane and organic acids can provide useful information for confirming the activity of methanogenic bacteria and the rate of treatment.

The presence of some of these organic acids is found to enhance the dehalogenation of chlorinated solvents under anaerobic metabolism modes. These principles are discussed below. The anaerobic degradation of mixed spills that contain aromatic hydrocarbons and halogenated aliphatic compounds has been successfully performed in field cleanup activities. The aromatic hydrocarbon serves as an electron donor and the halogenated aliphatic, CO_2, SO_4, or NO_3 serve as the electron acceptors.

The removal of ring substitutes from aromatic compounds has been demonstrated under anaerobic conditions. It is reported that none of the substituted monoaromatic compounds examined to date have proved to be resistant to anaerobic degradation (Young, 1984). Some of these aromatic compounds are listed in Table 4.3 with acclimation periods necessary before significant biodegradation occurs. Anaerobic organisms can conduct demethoxylation of the methoxylated compounds such as syringic, ferulic, and vanillic acids and isomers. For these compounds, ring fission occurs, resulting in potentially complete mineralization. Compounds of similar structure to these should be readily metabolized.

Ferulic acid has a number of substituted groups and is decomposed by a consortium of bacteria under methanogenic conditions. The degradation proceeds through benzoate as illustrated in Fig. 4.18. Since a consortium of organisms are responsible, several different pathways occur during degradation. Final products include organic acids, methane, and CO_2.

One interesting aspect from studies on substituted aromatic compounds is that specific species have been isolated that carry out various functions of the degradation. For example, in the breakdown of caffeic acid (3,4-dihydroxycinnamic acid) none of the microbes in the

TABLE 4.3   Anaerobic Degradation of Aromatic Compounds in
Methanogenic Enrichments

Substrate	Acclimation lag, days	Conversion of substrate carbon to gas, %
Vanillin	12	72
Vanillic acid	9	86
Ferulic acid	10	86
Cinnamic acid	13	87
Benzoic acid	8	91
Catechol	21	67
Protocatechuic acid	13	63
Phenol	14	70
p-Hydroxybenzoic acid	12	80
Syringic acid	2	80
Syringaldehyde	510	102

SOURCE: Young, L. L., "Anaerobic Degradation of Aromatic
Compounds," *Microbial Degradation of Organic Compounds,* Marcel
Dekker, New York, p. 487, 1984.

consortium isolated could carry out more than a single reaction in the
degradation sequence. From a practical standpoint, the best solution
for bioremediation is to promote a consortium of organisms.

Summation

When bioremediation is applied to the mineralization of single-ring
aromatic compounds, the following usually applies:

1. These benzene-ring-based compounds are mineralized under aerobic metabolism and degraded under anaerobic metabolism modes.

2. The ease of degradation for both aerobic and anaerobic modes is influenced by the number, type, and position of substitutions.

3. Organisms capable of aerobic degradation are ubiquitous in the soil.

4. The BTEX compounds are readily degraded in soil and groundwater systems.

5. There is no confirmed evidence that o-xylene can serve as a sole energy source.

6. o-Xylene has been demonstrated to degrade when cometabolism is applied.

7. A large number of fungi can oxidize aromatic hydrocarbons.

8. Successful degradation has been demonstrated for all the anaerobic modes: nitrate respiration, sulfate reducing, and methanogenic fermentation.

9. The anaerobic degradation of non-oxygen-containing aromatic compounds requires the incorporated of oxygen from water into the aromatic structure. This may be the rate-limiting step.

10. Once aromatic compounds are oxygenated, they are mineralized by anaerobic metabolism to carbon dioxide, water, and in some cases methane.

11. Aromatic compound degradation rates are usually higher under denitrifying conditions than methane fermentation.

Polynuclear Aromatic Hydrocarbons

Polynuclear aromatic hydrocarbons (PAHs) are ubiquitous pollutants. They are generated and released from incomplete combustion of organic materials, including automobile exhausts and municipal sources. They are also naturally formed from diterpenes, triterpenes, steroids, and plant quinone pigments that are reduced and aromatized over geological time from highly condensed aromatic hydrocarbons.

Many of the PAHs are associated with petroleum refining and coal-tar distillation. A large number of PAHs can be found in coal-tar-contaminated sites. Coal tar is a by-product from coal gasification and was frequently stored in unlined lagoons. Singer (1988) identified a large number of these PAHs in coal-tar waste (Table 4.4). Petroleum-related activities are reported to account for over 70 percent of the artificially generated source. Other industrial activities include the production of solvents, pesticides, plastics, paints, resins, and dyes (Young, 1984).

Chemical nature

The aromatic hydrocarbons include the benzene series discussed above and the polyring series. These are a wide variety of polyring aromatic hydrocarbons. Typical structure of the polyring aromatic hydrocarbons is illustrated in Figs. 4.19 and 4.20. They include the frequently found compounds of naphthalene and anthracene and the more complex compounds of pyrene and benzo[a]pyrene.

Microbial metabolism

The degradation of polynuclear aromatic hydrocarbons depends on the complexity of the PAH chemical structure and the extent of enzymatic adaptation. In general, PAHs containing two or three aromatic rings are readily degradable. PAHs containing four or more aromatic rings are significantly more difficult, and some may be recalcitrant. The ease of degradation of PAHs by either aerobic or anaerobic metabolism is a function of:

TABLE 4.4   PAH Extracted from Coal-Tar-Contaminated Soil

Compound name	Compound name
Naphthalene	Methylpyrene (2 isomers)
2-Ethylnaphthalene	Dimethylpyrene
5-Dimethylnaphthalene	Benzo[c]phenanthrene
Acenaphthene	1,2-Benzanthracene
Dibenzofuran	Chrysene
3-Trimethylnaphthalene	Triphenylene
Ethylmethylnaphthalene	Methylbenzo[c]phenanthrene
9H-Fluorene	Methylbenz[a]anthracene (3 isomers)
Phenanthrene	Benzo[b]fluoranthene
Anthracene	Benzo[e]pyrene
Methylphenanthrene	Benzo[a]pyrene
Methylanthracene	Perylene
Phenylnaphthalene	3-Methylbenzo[j]aceanthrylene
Dimethylphenanthrene (2 isomers)	Dibenzo[a,h]anthracene
Fluoranthene	Benzo[b]naphthacene
Pyrene	Indeno[1,2,3-cd]pyrene
11H-Benzo[b]fluorene	Benzo[ghi]perylene
11H-Benzo[a]fluorene	1,2,4,5-Dibenzopyrene

SOURCE: Singer, 1988.

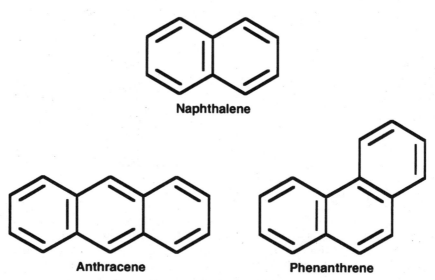

Figure 4.19 Structures of two- and three-ring aromatic hydrocarbons.

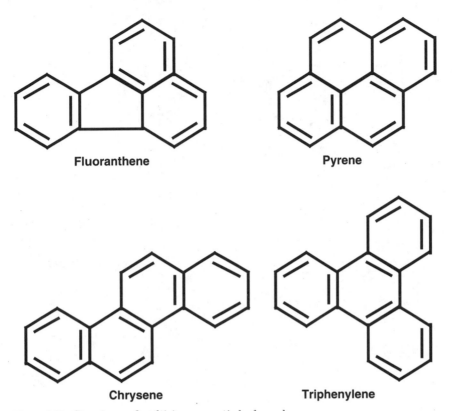

Fluoranthene   Pyrene

Chrysene   Triphenylene

Figure 4.20 Structures of multiring aromatic hydrocarbons.

- Solubility of the PAH
- Number of fused rings
- Number of substitutions
- Type of substitution
- Position of substitution
- Nature of atoms in heterocyclic compounds

PAHs can be readily degraded or extremely resistant or can yield intermediate products that are highly carcinogenic. The degree of degradation is influenced by chemical structure, solubility, and toxicity. It is difficult to develop generalized conclusions on PAH degradation potential; however, some basic findings appear to apply to most contaminated sites.

Under optimized metabolic conditions, the degradation rates of some PAHs are dependent on the soil-to-water ratio and the partition

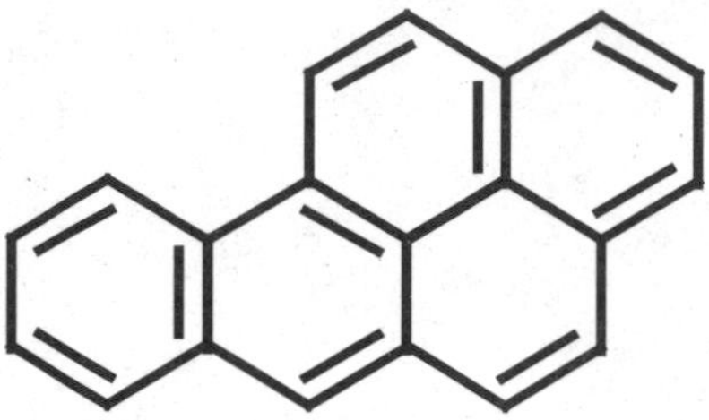

Benzo [a] pyrene

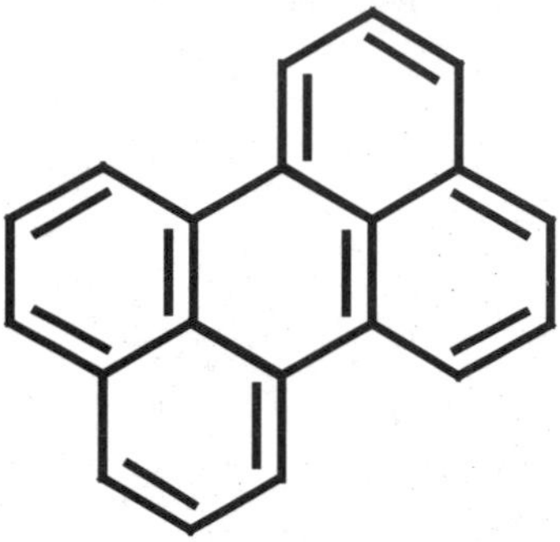

Perylene

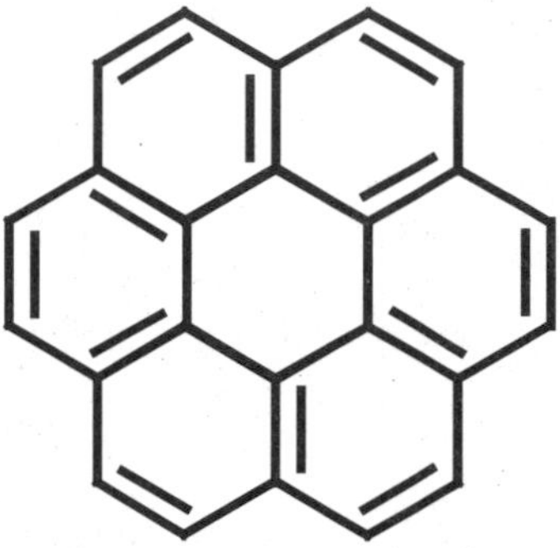

Coronene

Figure 4.20   (*Continued*).

coefficient. PAHs are strongly hydrophobic. Therefore, a significant portion of the PAHs is adsorbed on soil particles and possibly entrapped in intraparticle micropores. Microbial degradation reactions are significantly slower for sorbed-phase organic compounds than for dissolved compounds. For strongly sorbed compounds, the rate of degradation may be controlled by sorption-desorption kinetics. Enhancing PAH solubilization will improve the rate of degradation as long as the total organic solute concentrations do not become toxic. Toxicity increase has occurred in slurry bioreactors owing to enhanced desorption.

A coupled desorption–biological degradation model has been used to illustrate the degradation response of three PAHs for different soil-to-water ratios and PAH partition coefficients (Mihelcic, 1988). This model used biological metabolism constants for naphthalene degradation under denitrification conditions. For naphthalene 11 percent of the PAH would remain after 34 days of biological treatment. In contrast, 90 percent of the mass of anthracene and 99.9 percent of the mass of benzo(a)pyrene would remain after 1 year. Since the same microbial metabolism constants were used in these modeling studies, the persistence of anthracene and benzo(a)pyrene is a result of the rate-limiting desorption as discussed above. Benzo(a)-pyrene is more difficult to degrade microbially than the three-ring aromatics, making it even more persistent than illustrated above, since the more favorable microbial degradation constants of naphthalene were used.

Aerobic metabolism. Aerobic oxidation of the di- and tri-ring compounds is accomplished by a number of soil bacteria. The effect of the number of fused rings, substituted groups, and group position on oxidation of PAHs is clearly illustrated by the work of McKenna (1976). The relative degree of oxidation of various structures by two microorganisms has been evaluated as presented in Table 4.5.

The first obvious conclusion is that these two organisms respond differently to the PAH structures. It is very probable that this finding will hold true for the indigenous populations of contaminated sites. Degradation differences of microbial response emphasize the value of a treatability study to predict and optimize bioremediation.

As the number of fused rings increase in PAHs, the relative degree of degradation decreases. The influence of alkyl substituents is more difficult to predict. One methyl addition significantly decreases the degree of degradation, and its effect varies with the substituted position. The addition of three methyl groups causes severe retardation of degradation (Table 4.5). Increasing the degree of saturation of PAHs, adding hydrogen atoms by removing the double bond between carbons to provide another free valence bond, significantly reduces the relative degree of degradation. For phenanthrene, the addition of two hydrogens results in an 82 and 77 percent decrease of degradation for the two microorganisms studied (Table 4.5).

For the unsubstituted di- and tri-ring PAHs, degradation occurs readily in the presence of soil bacteria. Microbial populations isolated from freshwater and marine environments can also metabolize these compounds. The bacterial degradation rates for the three-ring PAHs (naphthalene, phenanthrene, and anthracene) appear to be related to the water solubilities of these compounds (Gibson, 1984). The lower water solubility of anthracene (75 μm/L) limits its availability. As with benzene, the oxidation of di- and tri-ring PAHs such as naphthalene, anthracene, and phenanthrene involves the formation of dihydrodiol intermediates (Fig. 4.21). Catechol is a principal intermediate product.

A considerable amount of information exists on the microbial metabolism of anthracene and phenanthrene. Phenanthrene degradation has been confirmed for bacteria genera of *Aeromonas, Beijerinckia, Flavobacterium, Nocardia, Pseudomonas,* and the yeast, *Cunninghamella.* (Hogan, 1988). *Pseudomonas putida* and

TABLE 4.5   Effect of Structure on the Oxidation of Polynuclear Aromatic
Hydrocarbons

Compound	Relative degree of degradation, %	
	Pseudomonas putida	*Flavobacterium* sp.
Effect of Number of Fused Rings		
Naphthalene	100	80
Anthracene	10	36
Phenanthrene	67	100
1,2-Benzanthracene	0	10
2,3-Benzanthracene	3	1
Chrysene	0	8
Pyrene	1	10
Triphenylene	0	0
Effect of Alkyl and Phenyl Substituents		
1-Methylnaphthalene	42	60
1-Ethylnaphthalene	39	36
1-Phenylnaphthalene	0	0
2-Methylnaphthalene	81	85
2-Ethylnaphthalene	70	45
2-Vinylnaphthalene	79	40
2-Phenylnaphthalene	1	9
1,3-Dimethylnaphthalene	39	0
1,4-Dimethylnaphthalene	9	20
1,5-Dimethylnaphthalene	9	0
1,6-Dimethylnaphthalene	1	14
2,3-Dimethylnaphthalene	88	84
2,6-Dimethylnaphthalene	16	63
2,6 (and 2,7)-di-tert-Butylnaphthalene	0	0
2,3,5-Trimethylnaphthalene	9	18
2,3,6-Trimethylnaphthalene	7	10
1,2,3,4-Tetraphenylnaphthalene	0	0
Effect of Position of Methyl Substituent		
Naphthalene	100	80
1-Methylnaphthalene	42	60
2-Methylnaphthalene	81	85
Phenanthrene	67	100
1-Methylphenanthrene	5	107
2-Methylphenanthrene	24	88
3-Methylphenanthrene	22	75
Fluorene	32	37
1-Methylfluorene	0	76
2-Methylfluorene	5	42

Beijerinckia sp. strain B-836 oxidized anthracene in the 1,2 positions
to cis-1-2-dihydroxy-1,2-dihydroanthracene (Cerniglia, 1984).
Intermediate products of cis-3,4-dihydrophenanthrene and cis-1,2-
dihydroxy-1,2-dihydrophenanthrene are produced by these organisms. Two pathways for phenanthrene are illustrated in Fig. 4.22.
Anthracene has been proposed to break down to 2-hydroxy-3-naph-

TABLE 4.5   Effect of Structure on the Oxidation of Polynuclear Aromatic Hydrocarbons (Continued)

Compound	Relative degree of degradation, %	
	Pseudomonas putida	*Flavobacterium* sp.
Effect of Saturation		
Naphthalene	100	80
1,2-Dihydronaphthalene	32	60
Tetralin	15	15
cis-Decalin	4	0
trans-Decalin	4	0
Phenanthrene	67	100
9,10-Dihydrophenanthrene	12	23
1,2,3,4,5,6,7,8-Octahydrophenanthrene	9	26
Perhydrophenanthrene	1	1
Indene	32	15
Indane	25	13
Hexahydroindane	7	1
Fluorene	32	37
Perhydrofluorene	0	3

SOURCE: McKenna, 1976.

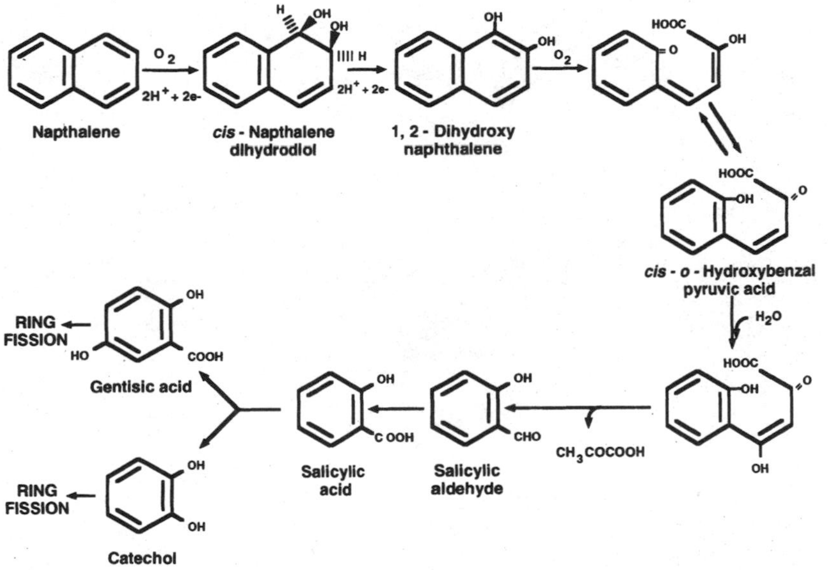

Figure 4.21   Proposed pathway for the degradation of naphthalene by bacteria. (*Gibson, 1984.*)

thoic acid as illustrated in Fig. 4.23. Fluorene can be used as the sole source of carbon and energy by *Pseudomonas vesicularis* and *Arthrobacter* sp. The degradation pathway for fluorene is illustrated in Fig. 4.24 (Grifoll, 1992).

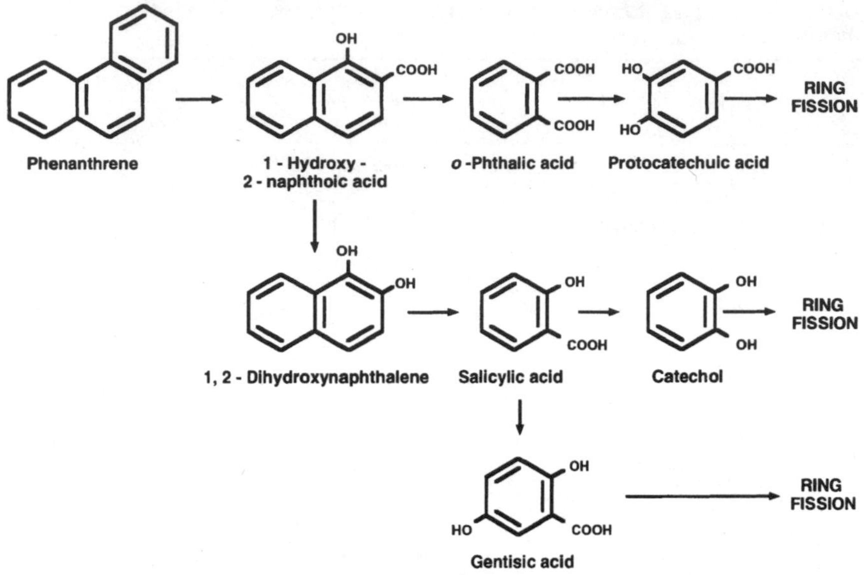

Figure 4.22 Divergent pathways for the degradation of phenanthrene by bacteria. (*Gibson, 1984.*)

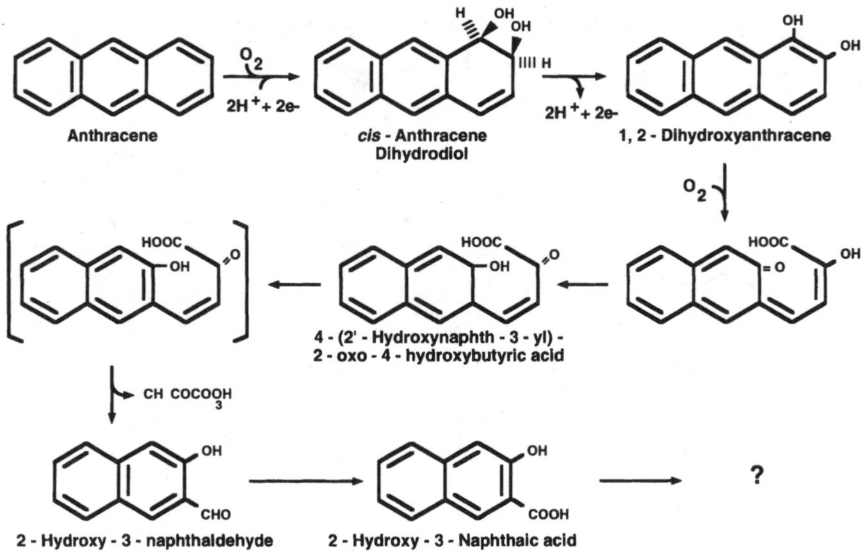

Figure 4.23 Proposed pathway for the bacterial oxidation of anthracene. (*Gibson, 1984.*)

Little information is available on the bacterial degradation of aromatic hydrocarbons that contain more than three rings. Confirmation on the ability of bacteria to use these multiple-ring compounds as growth substrates has not been documented. Microbial oxidation,

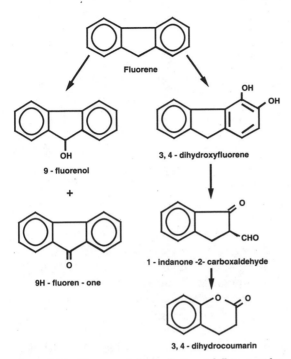

Figure 4.24 Proposed degradation of fluorene by *Arthrobacter* sp. (*Grifoll, 1992*).

however, of compounds such as fluorene, fluoranthene, pyrene, benzo[a]pyrene, benzo[a]anthracene, and dibenzo[a]anthracene does occur. The initial metabolite formed is the cis-dihydrodiol. Fluoranthene is mineralized by *Pseudomonas paucimobilis* and *Alcaligenes denitrificans* (Grifoll, 1992). *Mycobacterium* sp. can mineralize pyrene, but it does not use pyrene as a sole source of carbon and energy (Heitkamp, 1988; Mueller, 1990). When provided with a carbon energy source, *Mycobacterium* sp. is induced to mineralize several polycyclic aromatic hydrocarbons. PAHs in sediment-water samples were mineralized by the following percentages within 2 weeks after a 2- to 3-week enzyme induction and lag phase (Heitkemp, 1988):

Naphthalene	59.5
Phenanthrene	50.9
Fluoranthene	89.7
1-Nitropyrene	12.3
3-Methylcholanthrene	1.6
6-Nitrochrysene	2.0

Mycobacterium sp. has the exceptional ability to mineralize pyrene faster than the two- or three-ring PAHs. However, one cannot depend on the presence of these organisms throughout a contaminated site.

TABLE 4.6   Polynuclear Aromatic Hydrocarbons for Which Cometabolism May Be Required*

Pyrene
3,4-Benzopyrene
1,2-Benzanthracene
1,2,5,6-Dibenzanthracene
Benzo[a]anthracene

*Although a sole organism may be capable of mineralization, most studies confirm the need for cometabolism.

The importance of cometabolism for PAHs having four or more rings has been demonstrated by several investigations. In fact, cometabolism may be the only metabolism mode for degradation of the high-molecular-weight PAHs. Cometabolism coupled with analog substrate enrichment may be the only successful procedure for achieving cleanup of high-molecular-weight PAHs to regulatory limits.

The need for cometabolism has been demonstrated in laboratory studies for several high-molecular-weight PAHs (Table 4.6). Although a sole organism may be capable of mineralizing these compounds, most studies confirm the need for cometabolism. The number of high-molecular-weight PAHs that have been studied is very limited. It would be wise to assume that other PAHs will also fall into the same category as those in Table 4.6.

Cometabolism is the main mode for metabolism of benzo[a]anthracene, addition of the substrate biphenyl, m-xylene or salicylate-induced oxidation of benzo[a]anthracene to carbon dioxide, and a mixture of o-hydroxy polyaromatic compounds. The formation of 2-hydroxy-3-phenanthroic acid and 3-hydroxy-2-phenanthroic acid from degradation probably results from oxidative cleavage of catechol formed at the 10,11, and 8,9 positions of benzo[a]anthracene, respectively (Mahaffey, 1988). The proposed degradation pathway is given in Fig. 4.25. The structures shown in brackets are proposed intermediates and have not been characterized.

In addition to cometabolism, the addition of an analog substrate may be necessary for high-molecular-weight PAHs. As discussed in Chap. 3, enzyme induction occurs when a net energy gain can be achieved as a result of degradation. For many organisms, pyrene, benzo[a]anthracene, and other large PAHs may not induce enzyme production. The presence of an analog substrate such as naphthalene will induce the enzyme production that is then available to degrade

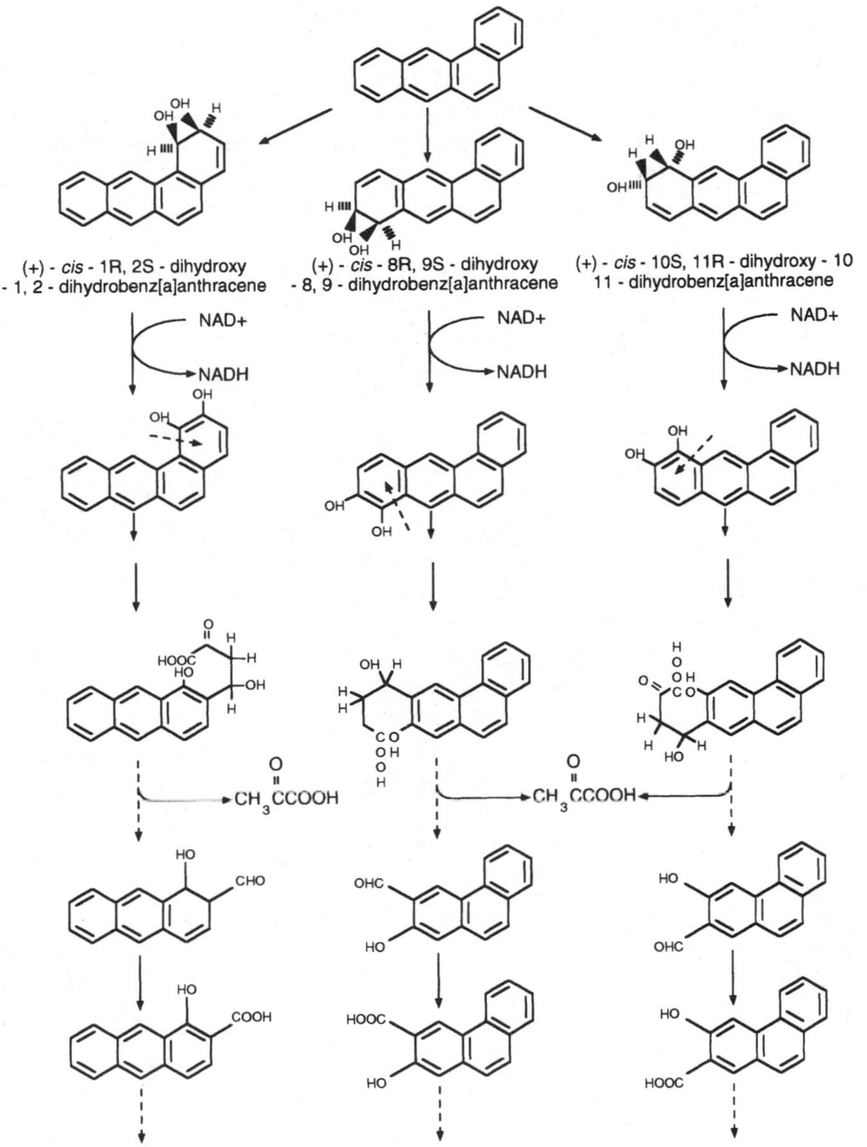

Figure 4.25 Proposed pathways for the metabolism of benzo[a]anthracene by *Beijerinckia* strain B-1. The major pathway on the left is initiated by oxidation at the 1.2 position. (*Mahaffey, 1988.*)

the higher-molecular-weight PAHs. Under this mode of metabolism the analog substrate is the primary substrate and the higher-molecular weight PAH is the secondary substrate. Contaminated sites that become exhausted of analog substrates, whether a spilled contaminant or an added substrate, may result in inadequate removal of high-molecular-weight PAHs.

Mycobacterium oxidative attack on pyrene results in the formation of dihydrol at the 4,5 position. Pyranol was detected as the initial ring oxidation product. The major metabolite produced by the *Mycobacterium* was 4-phenanthroic acid. Cinnamic acid and phthalic acid were also detected. The formation of both cis and trans-4,5-dihydrodiols of pyrene by *Mycobacterium* sp. suggests multiple degradation pathways.

The rate of degradation of higher-molecular-weight PAHs is enhanced when a lower-molecular-weight PAH is present as the primary substrate. In the presence of either naphthalene or phenol naphthalene, cometabolism of pyrene, 1,2-benzanthracene, 3,4-benzpyrene, and 1,2,5,6-dibenzanthracene by mixed culture of *Flavobacteria* and *Pseudomonas* has been demonstrated (Mueller, 1990).

The need for diversity in the microbial population at a site is highly advantageous for the degradation of aromatic compounds of more than three rings. Synergism has been demonstrated as important for several PAHs. For example, the organism *Pseudomonas putida* is unable to degrade pyrene, except when it is present with *Flavobacterium* sp. and its growth substrate, phenanthrene. Mixed cultures from a polluted stream were able to degrade both pyrene and 1,2-benzanthracene in the presence of either growth substrate, naphthalene, or phenanthrene by this mixed culture (McKenna, 1976).

Once PAH compounds have desorbed from soil, the rate-limiting step for the aerobic degradation of PAHs is the initial ring oxidation reaction, after which the degradation by bacterial communities proceeds quickly with little or no accumulation of intermediates. This theory is supported by the lack of these intermediates being found in PAH-contaminated sediments. Thus PAH-oxidizing bacteria inoculated into a PAH-contaminated site may enhance degradation by promoting the rate-limiting step of ring oxidation. Another possible approach is treatment with a strong oxidizing solution to initiate the oxidation of high-molecular-weight PAHs.

Studies on the aerobic oxidation of PAHs by bacteria have been the main research area. Unfortunately, one major microbial activity, that of fungus degradation of PAH, is poorly understood. The ability of fungal species to oxidize aromatic hydrocarbons does not appear as limited as bacteria. In one screening more than 85 species of fungi had the ability to oxidize naphthalene. The fungus, *Cunninghamella elegans,*

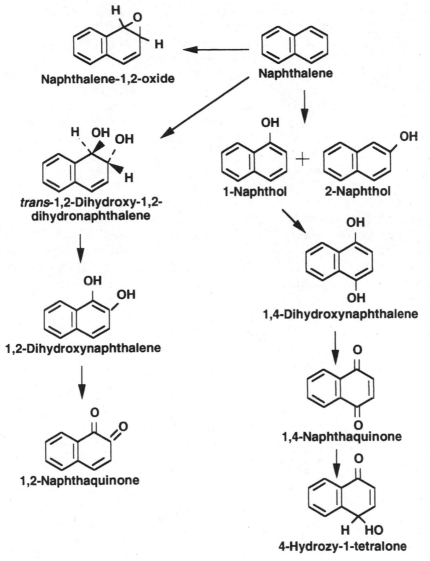

Figure 4.26 Proposed fungal degradation of naphthalene by *Cunninghamella elegans. (Gibson, 1984.)*

oxidizes anthracene, yielding trans-1,2-dihydroxy-1,2-dihydroanthracene and 1-anthryl sulfate. *C. elegans* oxidized phenanthrene at the 1,2 and 3,4 positions to form phenanthrene trans-1,2- and trans-3,4-dihydrodiols (Gibson, 1984; Sutherland, 1991). A proposed pathway for naphthalene degradation by fungal metabolism is presented in Fig. 4.26.

The fungus *Phanerochaete chrysosporium* degrades many PAHs including benzo[a]pyrene, pyrene, fluorene, phenanthrene, and more complex halogenated and heterocyclic aromatics (Chap. 5). The lignin peroxidases enzyme is believed responsible for these degradation reactions. Nitrogen-limiting conditions are used to achieve this degradation (Chap. 3). The optimum pH for *P. chrysosporium* degradation of phenanthrene is 4.5 (Dhawale, 1992; Brodkorb, 1992).

Phanerochaete sordida has been evaluated in field test plots for the ability to degrade PAHs. *P. sordida* was capable of decreasing the three- and four-ring PAHs by 91 and 45 percent, respectively. No detectable depletion of five- and six-ring PAHs occurred (Davis, 1993). However, only 3 percent of the total weight of PAHs was composed of ring structures of five or greater. Degradation of five and six rings by fungal treatment has been observed for aqueous systems.

Fungi catalyze the oxidation of aromatic compounds in a manner that differs from bacteria. The metabolism by fungi is much more closely aligned with that of mammals. The proposed pathway for initial oxidization and hydration of phenanthrene by *P. chrysosporium* is illustrated in Fig. 4.27 (Sutherland, 1991). The intermediates, phenanthrene-9, 10-oxide, and 9-phenanthrol are weakly mutagenic. When pyrene is the substrate, both pyrene-1,6-dione and pyrene-1,8-

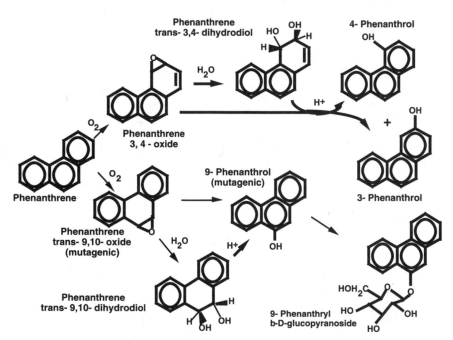

Figure 4.27 Proposed pathway for the generation of metabolites phenanthrene during transformation of phenanthrene by *phanerochaete chrysosporium*. (*Sutherland, 1991.*)

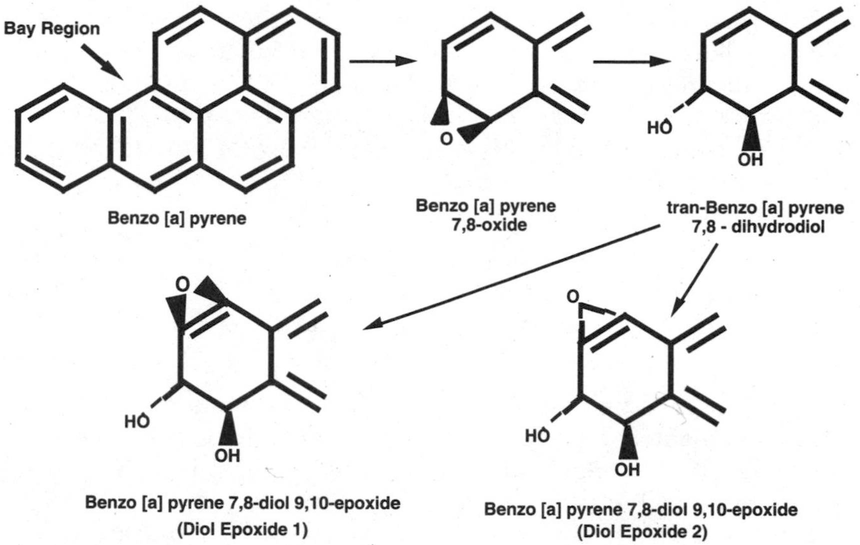

Figure 4.28 The formation of diastereomeric diol epoxides from benzo[a]pyrene. (*Gibson and Subramanian, 1984.*)

dione are the major products. In a similar fashion, anthracene is converted to anthraquinone, and benzo[a]anthracene yields 7,12-benzo[a]anthraquinone. Both the pyrene-1,6-dione, and pyrene-1,8-dione are mutagenic by the Ames test. However, these diones do not accumulate (Glaser, 1988).

The sometimes carcinogenic properties of aromatic hydrocarbons are related to oxidation products of the parent molecules. Although few aromatic hydrocarbons have been investigated in detail, the intermediates of greatest concern are those containing an epoxide group. These reactive electrophilic intermediates can react with genetic materials, yielding the carcinogenic property.

The key structural feature of carcinogenic metabolites is a result of the epoxide group on a saturated, angular benzo-ring. These are formed in part of what is called the Bay-Region that is the angular junction of three aromatic rings by analogy to a bay or harbor. Figure 4.28 shows the pathway for the formation of these epoxides from benzo[a]pyrene.

Anaerobic metabolism. The anaerobic metabolism of polynuclear aromatic hydrocarbons has not been extensively studied. Until recently, little has been known about the anaerobic degradation of PAHs. PAHs can be transformed anaerobically under conditions of denitrifying, sulfate-reducing, fermentation, and methanogenic conditions. Degradation of naphthalene and acenaphthene has been demonstrated under denitrifying conditions.

Under denitrification conditions, the degradation rate for naphthalene is independent of nitrate concentration but dependent on the soil-to-water ratio (Mihelcic, 1988). The rate of microbial degradation of anthracene is dominated by solute desorption. Under optimized biological conditions, desorption becomes the rate-controlling step. Similar results are obtained for benzo[a]pyrene. The use of surfactants can enhance the biodegradation rate.

The major benefit of anaerobic metabolism for PAHs has been the application of methanogenic metabolism to the dehalogenation of complex aromatic compounds. These applications are discussed in Chap. 5.

Degradation of naphthalene by methane fermentation has been clearly documented. The methanogenic degradation pathway for naphthalene follows a similar route to that of monoaromatic hydrocarbons, as illustrated in Fig. 4.29 (Grbić-Galić, 1990). As with the methanogenic fermentation of benzene, phenol appears to be a major intermediate in the fermentation of naphthalene.

Some information is available on the anaerobic degradation of PAHs that contain a nitrogen or sulfur atom in the ring structure. The presence of nitrogen or sulfur in a ring structure usually makes the chemical less stable than the homocycles. As a result, they are more readily degraded anaerobically. The heterocyclic aromatics that contain nitrogen degrade under denitrifying, sulfate-reducing, and fermentative conditions.

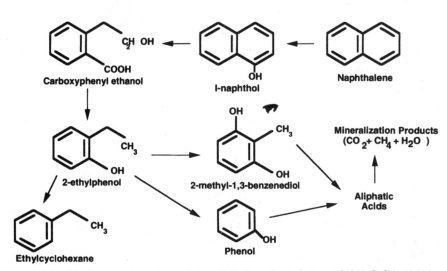

Figure 4.29 Methanogenic pathway for naphthalene degradation. (*Grbić-Galić, 1988.*)

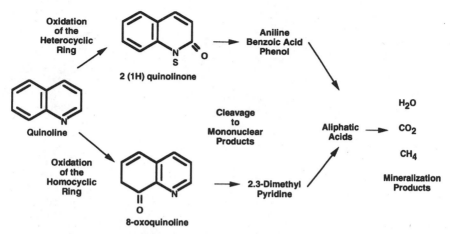

Figure 4.30 Methanogenic pathway for quinoline degradation. (*Grbić-Galić, 1988.*)

Microorganisms capable of anaerobically transforming the nitrogen heterocycles are widespread in the environment. A typical transformation pathway for quinoline is illustrated in Fig. 4.30 (Godsy, 1988; Grbić-Galić, 1988). As discussed above for anaerobic degradation of the nonoxygenated aromatics, the initial reaction step is the incorporation of oxygen from the hydroxyl group of water into the compound. Anaerobic degradation of nonoxygenated heterocycles has been demonstrated with:

- Indole
- Quinoline
- Isoquinoline
- 4-Methylquinoline
- Benzothiophene
- Dibenzothiophene
- Pyridine

Many of these have been transformed under nitrate-reducing conditions. Pyridine, indole, and quinoline transformation has been demonstrated under sulfate-reducing conditions. Under methanogenic conditions, indole has been fully mineralized to methane and CO_2 (Berry, 1987). The suggested pathway for degradation is given in Fig. 4.31.

Summation

The application of bioremediation to multiring aromatic compounds can be summarized as:

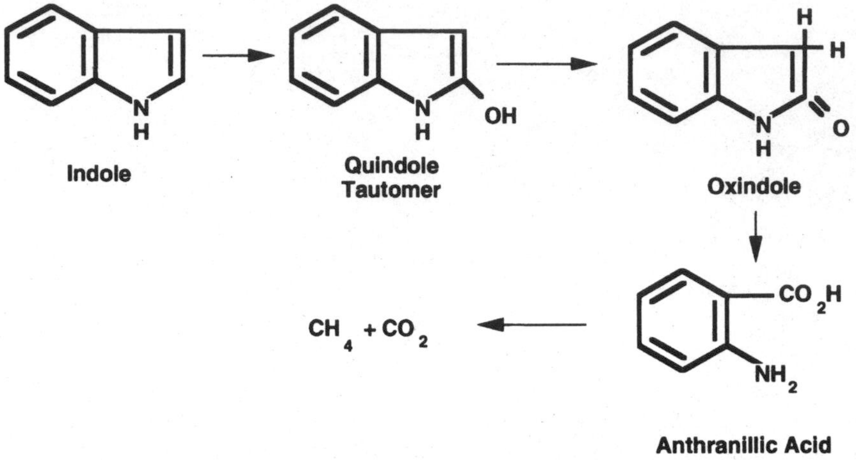

Figure 4.31 Proposed pathway for indole degradation under methanogenic conditions. (*Berry, 1987.*)

1. The two- and three-ring PAHs are readily degraded by soil bacteria and fungi.

2. Aromatic compounds of four and more rings are significantly more difficult to degrade, and some may be recalcitrant.

3. The ease of degradation, either aerobic or anaerobic, is a function of number of fused rings, substitution, position of substitution, and number and nature of heterocyclic atoms.

4. The degradation responses of specific microorganisms are significantly different for like PAH compounds.

5. The addition of three methyl groups to a ring structure severely decreases degradation potential.

6. Increasing the degree of saturation significantly reduces the relative degree of degradation.

7. Cometabolism and analog substrates appear important for the degradation of four-, five-, and more-ring PAHs.

8. Microbial synergism and diversity in microbial populations at a site are highly advantageous.

9. The rate-limiting step is initial ring oxidation, after which oxidation proceeds quickly for three and fewer ring structures.

10. PAH-oxidizing bacteria inoculated into a contaminated site may enhance the rate.

11. The degradation of polynuclear aromatic compounds under solely anaerobic conditions has not been extensively studied or applied as field systems.

12. Two- and three-ring aromatic compounds have been transformed under denitrifying, sulfate-reducing, fermentation, and methanogenic conditions.

Halogenated Aromatic Compounds

Chemical nature

The chemical nature of halogenated aromatic compounds covers a very broad spectrum. They include benzotrichloride, the chloronitrobenzenes, the chlorophenols, the chlorobenzenes, chloroaniline, polychlorinated biphenyl, and many pesticides. Their generation sources are equally broad. Degradation potential of the polychlorinated biphenyls and complexed halogenated aromatic pesticides is presented in Chap. 5.

Many of these compounds are produced directly for commercial use and as chemical intermediates during chemical synthesis operations. Potential releases are associated with the industrial operations dealing with dyes, pharmaceuticals, pesticide formulations, rubber, solvents, cleaners, deodorants, pigments, polyurethane, and numerous compounds.

Microbial metabolism

Aerobic metabolism. One requirement for microbial degradation of any compound is the need to induce enzyme production. Not all halogenated aromatic compounds will induce enzyme production even when their nonhalogenated form can (Reineke, 1984). For example, some organisms that obtain energy from diphenylmethane cannot grow on diphenylmethane when chlorine atoms are substituted in the para position of both benzene rings. If the nonhalogenated diphenylmethane is added as a primary substrate, the chlorinated substitute form is degraded by cometabolism.

Most halogenated aromatic compounds that are degraded under aerobic conditions are probably acted upon through cometabolism. It is also possible that a halogenated aromatic compound is transformed to a toxic product that prevents further aerobic degradation.

Many aerobic organisms are reported to cometabolize halogenated aromatic compounds. The relatively nonspecific nature of the enzymes that transform benzoate to catechol accounts for this (Fig. 4.12). However, in some cases this degradation is not complete in that

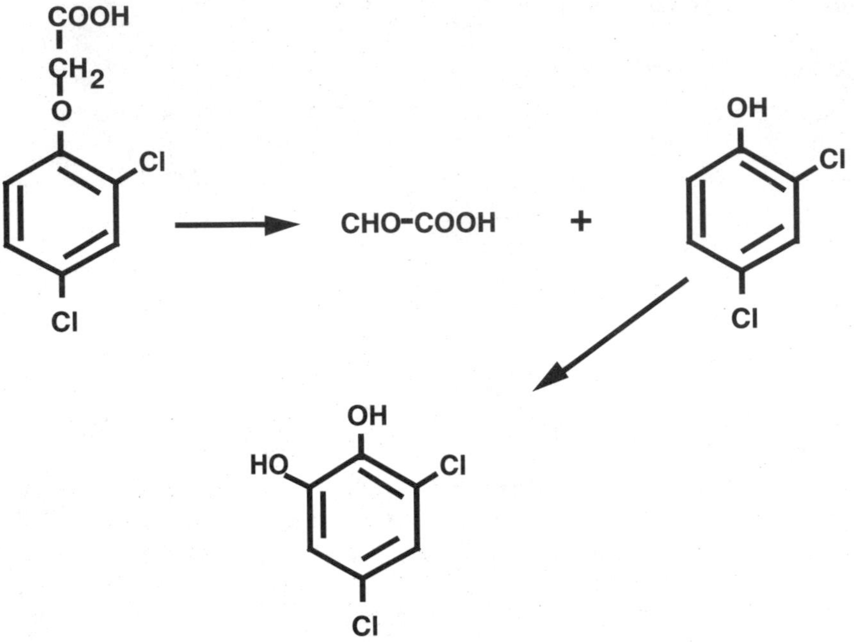

Figure 4.32 Formation of halocatechols from 4-chlorophenoxyacetate. (*Reineke, 1984.*)

chlorinated benzoates and catechols are the end result of oxidative degradation (Fig. 4.32) (Reineke, 1984).

Once a halogenated catechol or phenol is formed, there appear to be more organisms that can degrade these under aerobic conditions. An extensive list of microorganisms have been identified that can transform various chlorinated phenol compounds.

The most common transformation pathway is the elimination of the halide after ortho cleavage of the halocatechols. An example is the degradation of 3-methyl-5-chlorocatechol (Fig. 4.33) (Reineke, 1984).

The complete aerobic mineralization of halogenated aromatic compounds has been confirmed for several chemicals and microbial consortiums. However, the persistence of these compounds illustrates the ineffectiveness of many microorganisms. The nature, the number of substitutions, and their substitution position affect the ease and degree of degradation. The degradation of these halogen-substituted aromatic compounds frequently does not follow the reaction pathways of the unsubstituted parent compounds.

Chlorinated aromatic compounds can be used as sole energy sources. The use of 3-chlorobenzoate as a sole carbon source for *Pseudomonas* sp. is one example (Fig. 4.34). Once the chlorine has been removed, the compound is mineralized as typical of oxygenated aromatic compounds. The aerobic replacement of the halide with the hydroxyl group is typi-

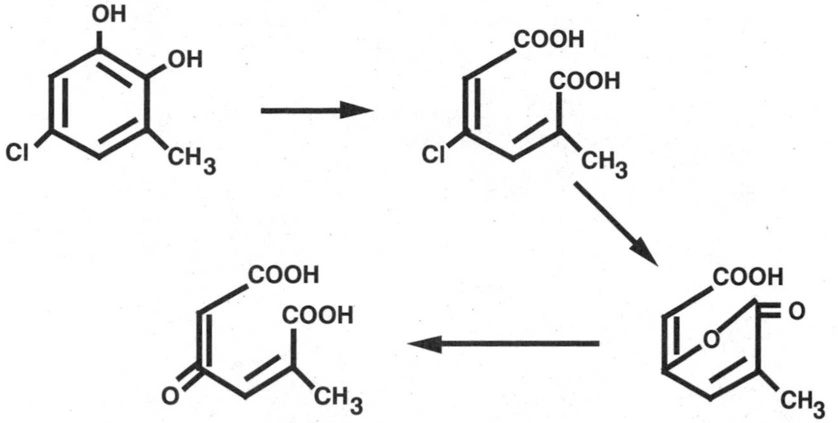

Figure 4.33 Aerobic degradation of a chlorinated catechol, 3-methyl-5-chlorocatechol. (*Reineke, 1984.*)

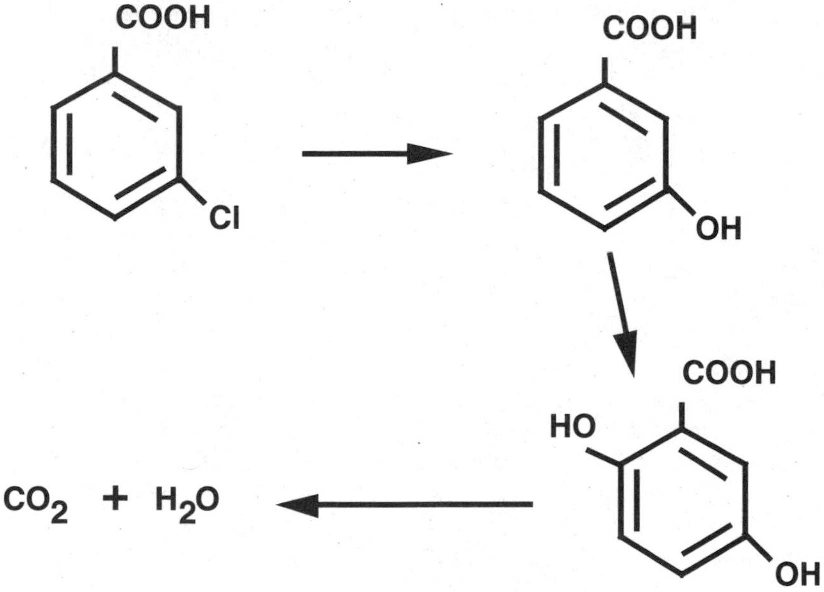

Figure 4.34 Transformation of 3-chlorobenzoate by *Pseudomonas* sp. (*Reineke, 1984.*)

cal of many halogenated aromatic transformations. However, if the substituted chlorine atom is changed to form 2-chloro- or 4-chloroben-zoate, the microorganism cannot use the compound as a sole energy source. Other microorganisms such as *Nocardia* sp. and *Arthrobacter globiformis* can utilize the 4-chlorobenzoate as a carbon and energy source. Present evidence indicates that the aerobic enzymes that provide energy and carbon source by elimination of halogens from aromatic compounds are highly specific. For most compounds, the aerobic rate of transformation decreases as the chlorine substitutions increase. This is the opposite for anaerobic transformations, as discussed below. When using aerobic systems for bioremediation of halogenated aromatic compounds, design must be based on adequate treatability studies.

Anaerobic metabolism. Chlorinated aromatic hydrocarbons that are refractory to aerobic systems are sometimes degraded by one or more reductive dehalogenations anaerobically. The halogenated organic compound serves as the electron acceptor. The electrons are supplied by the oxidation of a primary substrate. Since reaction rates are influenced by the availability of electron acceptors and donors, the rate of dehalogenation is linked to the rate of substrate oxidation.

Microbial populations can mediate the anaerobic dehalogenation of iodo-, bromo-, fluoro-, and chlorobenzoate compounds. Halogens at the ortho and para positions are more resistant to dehalogenation than those at the meta position. The number of halogen substitutions, except for PCBs, is not important. Once the halogen is removed, ring fission leads to methane and carbon dioxide (Young, 1984). Information presently available on anaerobic degradation pathways for halogenated aromatics is limited.

The main mechanism for transformation of halogenated aromatic compounds is by reductive dehalogenation. This reaction results in the replacement of the halogen by hydrolysis. Although the chlorine is usually replaced with hydrogen, it can be replaced with an amino, chloryl, hydroxyl, or methyl substitute (Dolfing, 1992). Reductive dehalogenation has been confirmed on a number of halogenated aromatic acids, mono- to hexa-chlorobenzenes, chlorophenols, chlorophenoxyacetate, herbicides, DDT, and PCBs. In reductive dehalogenation the halogenated organic compound serves as the electron acceptor.

Most anaerobic dehalogenation is probably a result of cometabolism. However, theoretical evidence from calculation of Gibbs free energy available from dechlorination indicates that microorganisms can benefit from the use of halogenated aromatic compounds as electron acceptors under anaerobic conditions (Dolfing, 1992). Experimental evidence comes from studies on the dechlorination of 3-chlorobenzoate to benzoate by *Desulfomonile tiedjei* (see Fig. 3.8) (Tiedje, 1987).

Several chlorinated benzoic acids are used as sole sources of carbon and energy by bacteria. Metabolism of 3- and 3,5-dichlorobenzoate proceeds by direct deoxygenation, followed by decarboxylation of the aromatic ring to give 3- and 3,5-dichlorocatechol, respectively, prior to ring fission and subsequent dehalogenation. In contrast, 4-chlorobenzoate is dehalogenated hydrolytically to 4-hydroxybenzoate prior to ring fission.

Methanogenic metabolism has successfully dehalogenated several chlorinated aromatic compounds. These include 2,4,5-trichlorophenoxyacetate, 3-chlorobenzoate, 2,4-dichlorophenol, 4-chlorophenol, 2,3,6-trichlorobenzoate, and 2,4-, 2,5-, 2,6-, 3,4-, and 3,5-dichlorobenzoates (Zhang, 1990; Young, 1984; Gibson, 1990). Methanogenic cultures show preferential removal of ortho-chlorines, with meta- or para-chlorines removed at slower rates (Häggblom, 1991). Relative rates of transformations from higher to lower are 2,4-, 2,6-, 2,3-, and 2,5-dichlorophenol, 2,3,6-trichlorophenol, and 2,4,6-trichlorophenol.

These anaerobic dehalogenation reactions and the final mineralization of the compounds are not simple degradation pathways but involve as many as six microorganisms, as demonstrated with 2,4-dichlorophenol (Fig. 4.35).

Under methanogenic conditions, 2,4-dichlorophenol is dehalogenated to 4-chlorophenol by an organism tentatively identified as a *Clostridium*. Reductive dehalogenation is performed by a second organism to yield phenol. A third microorganism carboxylates phenol to benzoate. Breaking of the benzoate ring and mineralization of acetate is then performed by two or three organisms including methanogens (Zhang, 1990).

Information on the sequential degradation of 2,4-dichlorophenol illustrates the importance of process operation. The transformation metabolite 4-chlorophenol is more toxic than the parent compound,

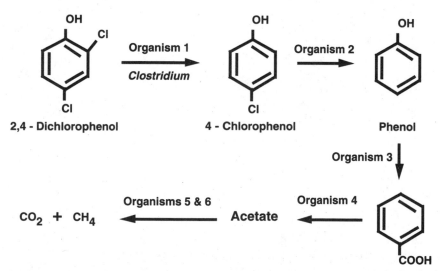

Figure 4.35 Degradation pathway of a microbial consortium for mineralization of 2,4-dichlorophenol. (*Zhang, 1990.*)

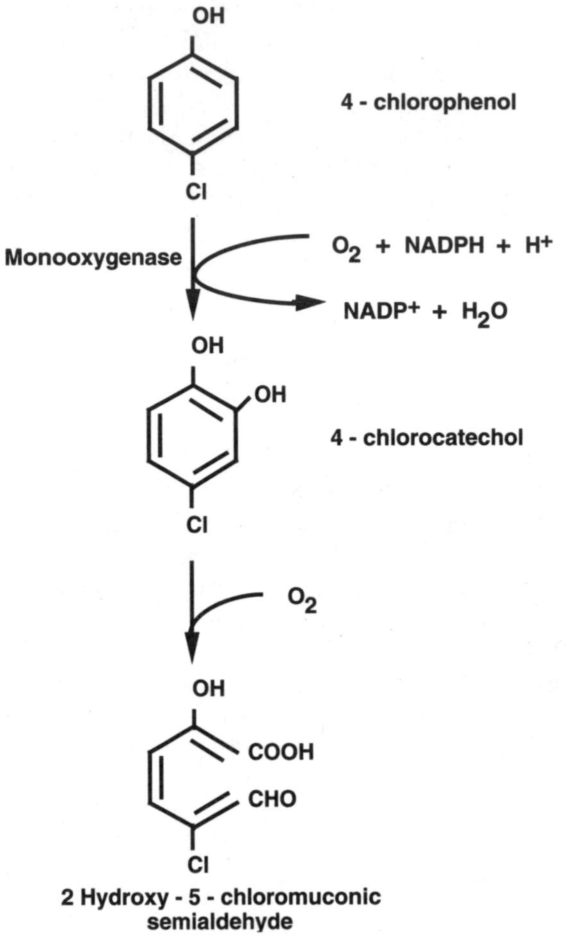

Figure 4.36 Proposed degradation of 4-chlorophenol by *Pseudomonas putida*. (*Saez, 1991.*)

and the conversion of 4-chlorophenol is the slowest step in this degradation sequence. An unbalanced microbial consortium can result in 4-chlorophenol buildup, which will strongly inhibit the other sequence of the degradation chain (Zhang, 1990). *Pseudomonas putida* is reported as one of the potential degraders of 4-chlorophenol through cometabolism (Fig. 4.36) (Saez, 1991). This work has confirmed that high concentrations of 4-chlorophenol inhibited the degradation by *P. putida*. Furthermore, as discussed in Chap. 3, an increase in hydrogen gas partial pressure makes the thermodynamics unfavorable for mineralization (see Fig. 3.7). A final potential problem is the effect of phenol, an intermediate on the degradation pathway. Phenol buildup can prevent the dechlorination of 4-chlorophenol.

The interactions of these microbial systems and the sensitivity of the overall mineralization of 2,4-dichlorophenol to methane and carbon dioxide can be achieved only by good engineering. The process must be controlled to maintain a balanced consortium so these inhibitory effects do not commence. Similar interactions may be necessary for mineralization of a number of halogenated aromatic compounds.

Anaerobic dehalogenation of aromatic compounds is stimulated by the addition of an electron donor. Addition of butyrate, propionate, ethanol, or acetate not only increase the rate of dehalogenation but increases the extent of degradation (Gibson, 1990). The addition of p-cresol and propionate enhances the methanogenic degradation of 2,4-dichlorophenol (Häggblom, 1991). The enhancement of dehalogenation by volatile fatty acids has also been found for halogenated aliphatic hydrocarbons. An adequate supply of these electron donors is necessary for the dehalogenation reaction.

Another important consideration is the influence that various electron acceptors have on the reaction. As discussed in Chap. 3, certain electron acceptors will block the desired reduction reaction. Sulfate has been reported to have an adverse effect on the dehalogenation of some aromatic compounds. Other researchers, however, report sulfate-stimulated anaerobic degradation of 4-chlorophenol in a rich organic sediment (Gibson, 1990). Information on the transformation of halogenated aromatic compounds under sulfate-reducing conditions is limited. Documentation has been provided for only five chlorinated phenols: 2-chlorophenol, 3-chlorophenol, 4-chlorophenol, 2,4-dichlorophenol, and 2,6-dichlorophenol (Colberg, 1991). Sulfate-reducing conditions yield relative rates of degradation of 4-chlorophenol > 3-chlorophenol > 2-chlorophenol, and 2,4-dichlorophenol (Häggblom, 1991).

Although the chlorophenols under sulfidogenic conditions are used as a source of carbon and energy, degradation is very specific and successful only after acclimation to each monochlorophenol. Sulfate is used according to the stoichiometric values as given by (Häggblom, 1991)

$$C_6H_5ClO + 3.25\ SO_4^{-2} + H_2O \rightarrow 6HCO_3^- + 3.25\ H_2S + Cl^- + 0.5H^+$$

In soil containing sulfate-reducing bacteria, sulfate may be the preferred electron acceptor, and dehalogenators and sulfate reducers may compete for suitable electron donors. During the dechlorination of 3-chlorobenzoate by *Desulfomonile tiedjei* hydrogen was present at partial pressures of 1 to 10 Pa and probably served as reducing equivalents needed for the reductase dechlorination. In the absence of hydrogen, other compounds such as acetate can serve as a source of reducing equivalents. The presence of high levels of organic electron donors may aid the dehalogenators under these sulfate-reducing conditions. For aliphatic hydrocarbons, it has been demonstrated that anaerobic dehalogenation is highly dependent on the availability of both electron donors and acceptors.

Summation

The application of bioremediation to halogenated aromatic compounds can be summarized as:

1. The halogenated aromatics cannot be expected to degrade as the unsubstituted parent compound.

2. Both aerobic and anaerobic microbial systems are possible, but either cannot be universally applied.

3. Microbial consortiums appear important for aerobic transformation of most halogenated aromatic compounds.

4. Under aerobic metabolism, the rate of transformation and potential for mineralization decreases with increasing degree of halogenation of the compound.

5. The substituted position of the halogen significantly affects the potential for degradation.

6. Anaerobic transformation is usually by reductive dehalogenation of the halogenated aromatic compound.

7. Greater success for removing the halogenated substitution can be expected under anaerobic systems versus aerobic systems.

8. Halogenated aromatic compounds that are refractory to aerobic systems are sometimes degraded by anaerobic systems.

9. For anaerobic systems, the number of halogen substitutions on the aromatic compounds is not as important as with aerobic systems.

10. Methanogenic metabolism has yielded the greatest success for degradation under anaerobic conditions.

11. Anaerobic dehalogenation requires a consortium of microorganisms that must be balanced in microbial activity to prevent the buildup of toxic conditions.

12. Organic fatty acids appear to be important as electron donors during methanogenic dehalogenation.

13. Complex halogenated compounds have been mineralized by combining metabolism under anaerobic conditions followed by an aerobic system.

14. Treatability studies are necessary for successful bioremediation of sites with mixed halogenated aromatic compounds.

Halogenated Aliphatic Compounds

Chemical nature

The halogenated aliphatic compounds made up a major portion of the EPA volatile chemicals list. They include vinyl chloride, carbon tetrachloride, chloroform, dichloromethane, trichloroethylene, 1,1,1-trichloroethane, and tetrachloroethane.

These chemicals are produced in large quantities and used as solvents for cleaning and chemical synthesis. Typical uses include metal manufacturing and finishing, paint and ink formulations, dry cleaning agents, synthetic rubber production, fluorocarbon production, fumigants, aerosol propellants, gasoline additives, paint and varnish removers, degreasers, pesticide solvent, adhesives, photographic supplies, pharmaceutical products, and many household and office supplies.

Several water and soil chemical properties will influence the stability of halogenated aliphatic compounds. These compounds undergo abiotic transformations in the environment (Table 4.7). The important abiotic transformations include substitution, dehydrohalogenation, and reduction in water. A typical substitution is the addition of water resulting in hydrolysis. The nucleophiles of OH^- and H_2O are the principal species responsible for the abiotic dehydrohalogenation of haloaliphatic compounds in water (Reinhard, 1988). However, a variety of other species can displace the halogen. Many of the nucleophiles under anaerobic conditions, i.e., hypoxic natural waters, are more nucleophilic than those found in oxygenated waters (Table 4.8). In hypoxic waters, the sulfur nucleophiles are generally the most powerful. Sulfides react with halogenated aliphatic compounds via substitution to produce mercaptans.

Dehydrohalogenation results in elimination of the halogen, forming an alkene. The number of substituted groups influences the dehydro-

TABLE 4.7   Abiotic Transformation of Halogenated Aliphatic Compounds

Hydrolysis reactions
Substitution reactions
Dehydrohalogenation reactions
Hydrogenolysis reactions
Dihalo-elimination reactions
Coupling reactions
Reduction reactions

SOURCE: Reinhard, 1988.

TABLE 4.8   Nucleophilic Ions Capable of Abiotic Displacement of
Halogen from Haloaliphatic Hydrocarbons

Oxygenated Waters	Anaerobic Waters
H_2O	SO_3^{2-}
OH^-	$S_2O_3^{2-}$
Cl^-	NH_3
Br^-	NO_2^-
SO_4^{2-}	S_n^{2-}
HCO_3^-	$R-C_6H_{13}S^-$
	$C_6H_5S^-$

SOURCE: Reinhard, 1988.

halogenation rate. The rate of dehydrohalogenation is increased as
more chlorine atoms are attached to the carbon. Polychlorinated species
hydrolyze less readily. When there are multiple halogens, hydrolysis by
substitution is more likely than the dehydrohalogenation. Dehydrohalogenation reactions do not appear to occur when chlorine atoms are on
adjacent carbon atoms, but they do occur when bromine is on adjacent
carbon atoms. Brominated compounds usually undergo dehydrohalogenation more rapidly than their chlorinated analogs.

In the absence of biological activity these reactions proceed slowly.
The half-lives for the monochloro- and monobromo-alkanes are about
1 month at 25°C. The abiotic half-life for 1,2-dichloroethane can be as
long as 37 years. The presence of stronger nucleophiles such as HS^-
can significantly reduce these half-lives (Table 4.9). Microbial
enzymes also catalyze these reactions, significantly reducing the
half-lives.

A variety of transition metals, including nickel, iron, chromium,
and cobalt, can reduce halogenated aliphatic compounds. As a result

TABLE 4.9   Abiotic Half-Lives for Dehalogenated Aliphatic
Hydrocarbons*

Compound	Nucleophile type	Half-life, years
1,2-Dibromoethane	OH^-, H_2O	3.4
1,2-Dichloroethane	OH^-, H_2O	37
1,2-Dibromoethane	HS^-	0.12
1,2-Dichloroethane	HS^-	6.5

SOURCE: Reinhard, 1988.
*Measured at 25°C, pH 7.0.

of this redox reaction, the metals are oxidized. The reduced products and metals mediating such reactions have been tabulated by Vogel (1987) (Table 4.10).

The transition metal reduces a halogenated compound, removing the halogen and creating an alkyl radical that readily picks up a hydrogen atom from water, resulting in the formation of an alkane.

$$-\underset{\underset{Cl}{|}}{C}-C- + M_R \rightarrow -\underset{\underset{H}{|}}{C}-C- + M_{ox} + Cl^-$$

For multiple halogenated aliphatic compounds, it is possible for the alkyl radical to lose a second halogen from an adjacent carbon atom, resulting in the formation of an alkene. The reduction of polyhalogenated alkanes can result in both alkanes and alkenes.

Microbial metabolism

Microbial degradation of halogenated aliphatic compounds can use one of several metabolism modes. These include oxidation of halogenated alkanes for an energy source, cometabolism under aerobic conditions, and reductive dehalogenation under anaerobic conditions. Unfortunately, all halogenated aliphatic compounds will not respond to all metabolism modes for degradation. Thus the choice must be based on the nature of the contamination and site conditions. The redox condition within the contaminated area, soil-water chemistry, and the available electron acceptors play an important part in this decision.

With molecular oxygen as the electron acceptor, the one- to three-atom substituted halogenated aliphatic compounds are transformed by three types of enzymes: oxygenase, dehalogenases, and hydrolytic dehalogenases (Semprini, 1992). With oxygenase the transformation products are alcohols, aldehydes, or epoxides. Dehalogenase transformation products are an aldehyde and glutathione. The glutathione is required as a cofactor for the nucleophilic substitution by dehalogenases enzyme. Hydrolytic dehalogenases will hydrolyze the aliphatic compound, yielding alcohols as a transformation product. The higher halogenated compounds, particularly when all available valences on carbon are substituted (tetrachloride or tetrachloroethylene), have not been transformed under aerobic systems. They must be transformed by reductive dehalogenation.

Halogenated aliphatic compounds may be either oxidized or reduced, depending on their structure and the existing environmental conditions. Reduction is possible because of their electron negative

TABLE 4.10 Metal Medicated Reduction of Halogenated Aliphatic Compounds

Compound	Products	Reductant
	Methanes	
Chloromethane	Alkylated co-complex	Co(I) chelates
Dichloromethane	Methane	Cr(II)SO$_4$
	C1-alkylated B$_{12}$	B$_{12}$Co(III)
		(methylcobalamin)
Trichloromethane	Methane	Cr(II)SO$_4$
	Dichloromethane	Fe(II)
	Dichloromethane	Fe(II)P
	C1-alkylated	B$_{12}$-Co(III)
Tetrachloromethane	Methane	Cr(II)SO$_4$
	Chloroform	Fe(II)P
	C1-alkylated B$_{12}$	B$_{12}$-CO(III)
Bromomethane	Methane	CR(II)SO$_4$
	Alkylated co-complex	Co(I)-complex
Dibromomethane	Methane, ethane	Fe(II)P
Tribromomethane	Br$_2$-alkylated B$_{12}$	B$_{12}$-Co(III)
	Ethanes	
Chloroethane	Alkylated co-complex	Co(I)-complex
1,1-Dichloroethane	Ethane, ethanol	Cr(II)SO$_4$
1,1,1-Trichloroethane	Ethane, ethanol,	Cr(II)SO$_4$
	ethane, chloroethene	
	1,1-Dichloroethane	Fe(II)
	1,1-Dichlorethane	Fe(II)P
Hexachloroethane	Tetrachloroethane	Fe(II)P
	Tetrachloroethane	Cr(II)SO$_4$
Bromoethane	Ethane	Ni(I)
1,1-Dibromoethane	Ethane, ethanol	Cr(II)SO$_4$
1,2-Dibromoethane	Ethane	Fe(II)
	Ethane	Fe(II)P
	Propanes	
1-Chloropropane	Alkylated co-complex	Co(I)-complex
1,1-Dichloropropane	Propane, propanol,	Cr(II)SO$_4$
	propene	
1-Bromopropane	Alkylated co-complex	Co(I)-complex
1,2-Dibromo-3-chloro-	Propene, allyl	Cr(II)SO$_4$
propane	chloride	

SOURCE: Vogel, T. M., Criddle, G. S., and McCarty, P. L., "Transformation of Halogenated Aliphatic Compounds," *Environ. Sci. Technol.,* Vol. 21, No. 8, pp.

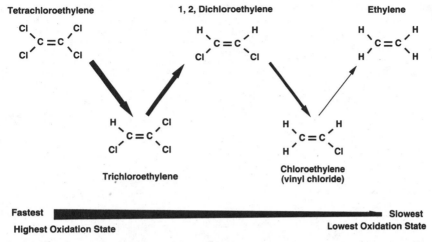

Figure 4.37 Rate of dehalogenation under anaerobic metabolic mode.

character. As a result, polyhalogenated aliphatic compounds often behave as an electron acceptor or the oxidant in the redox reaction. The more halogenated a compound is, the more oxidized the compound is, and the more susceptible it is to reduction. This principle explains why the rate of dehalogenation decreases under the anaerobic metabolism mode as tetrachloroethylene is dehalogenated to vinyl chloride (Fig. 4.37). Since vinyl chloride is more reduced, the thermodynamic equilibrium tends to stabilize vinyl chloride as the end product in aquifers that have negative redox potentials.

Aerobic metabolism.   Chlorinated two-carbon-atom compounds were once considered nonbiodegradable. However, in 1986, their mineralization by methane-utilizing bacteria was reported (Vira, 1991). Biotransformation of some halogenated aliphatic compounds has been demonstrated under aerobic conditions (Table 4.11). Aerobic bacteria have been isolated from contaminated soil that can transfer 1,2-dichloroethane to chloroethanol. Chloroethanol is readily mineralized to carbon dioxide. *Xanthobacter autotrophicus* will use 1,2-dichloroethane as a sole carbon source (vanderWijngaard, 1993b). *Ancylobacter aquaticus* also degrades 1,2-dichloroethane and with a higher affinity than *X. autotrophicus*. *A. aquaticus* has better properties for use in packed-bed reactors, since it adheres well to surfaces without excessive slime production (vanderWijngaard, 1993).

Pure cultures of *Pseudomonas* and *Hyphomicrobium* can use dichloromethane as sole carbon and energy sources (Semprini, 1992). *Methylosinus trichosporium* also will transform 1,2-dichloroethane as well as 1,1-dichloroethane and 1,1,1-trichloroethane. *Pseudomonas*

TABLE 4.11   Aerobic Biotransformation of Halogenated Aliphatic Compounds

Compound	Product	Culture	Matrix
Methanes			
Dibromomethane			
Bromomethane			
Dichloromethane	CO_2	M/P	
Trichloromethane	CO_2		S/A
Ethanes			
1,1-Dichloroethane			
1,2-Dichloroethane	CO_2	P	
1,2-Dichloroethane	Chloroethanol	P	
1,1,1-Trichloroethane			
1,1,2-Trichloroethane			
Bromoethane			
1,2-Dibromomethane			
Hexachloroethane	Tetrachloroethane	M	
Bromoethane		M	
1,2-Dibromoethane	Ethane		S/A
1,2-Dibromoethane	CO_2		S/A
Ethenes			
Monochloroethylene			
Chloroethene	CO_2	P	
Trichloroethylene	CO_2	P	S/A
cis-Dichloroethylene			
trans-Dichloroethylene			
Propanes			
1-Chloropropane		P	
1,2-Dichloropropane		P	
1,2-Dibromo-3-chloropropane	Propanol		S/A
1,2,3-Trichloropropane			

SOURCE: Modified from Vogel, 1987.
M = mixed culture, P = pure culture, S/A = soil/aquifer.

fluorescens can transform 1,1,2-trichloroethane as well as 1,2-dichloroethane. A *Mycobacterium* strain has been isolated from soil that can use vinyl chloride or ethylene as a sole energy source (Hartman, 1985; Ensign, 1992). Under aerobic conditions, many soil organisms can oxidize vinyl chloride. Most of the halogenated aliphatic compounds are eventually mineralized to carbon dioxide. The aerobic degradation capabilities of these microorganisms for chlorinated aliphatic compounds have provided for successful treatment processes with seeded bioreactors (see Chap. 9).

Under conditions supporting aerobic metabolism, some halogenated aliphatic compounds are degraded by cometabolism. These aerobic organisms generate oxygenase enzymes of broad-substrate specificity that oxidize halogenated aliphatic compounds. These include micro-organisms that belong to the genera *Alcaligenes, Mycobacterium, Pseudomonas, Nitrosomonas, Xanthobacter,* and *Ancylobacter. Nitrosomonas* derives energy from the oxidization of ammonia (see Chap. 3). *Nitrosomonas europaea* can cometabolize many halogenated aliphatic compounds. These include dichloromethane, dibro-momethane, trichloromethane (chloroform), bromoethane, 1,2-dibro-moethane (ethylene dibromide), 1,1,2-trichloroethane, 1,1,1-trichloro-ethane, and 1,2,3-trichloropropane (Vanelli, 1990). However, tetrachloromethane (carbon tetrachloride) is not degraded by *N. europaea. N. europaea* catalyzes the aerobic transformation of vinyl chloride (monochloroethylene), *cis-* and *trans-*dichloroethylene, *cis-*dibromoethylene, and trichloroethylene (Vanelli, 1990). Neither tetra-chloroethylene (perchloroethylene) nor *trans-*dibromoethylene is degraded by *N. europaea.*

During the cometabolism of chlorinated alkenes other genera derive their energy from other organic compounds such as propane, phenol, and toluene. *Pseudomonas cepacia G4 Phe1* is one of the toluene-utilizing microorganisms that can degrade trichloroethylene by cometabolism. Even under extreme variations in pH and tempera-ture, significant rates of degradation are measured (Shields, 1991). Optimum rate of degradation occurred at pH 7 with rates of 20 and 45 percent of the optimum at pHs of 4 and 9, respectively (Fig. 4.38*a*). Trichloroethylene degradation rate decreased by 30 percent at 4°C compared with 30°C (Fig. 4.38*b*). Oxygen concentration above 31 mg/L was detrimental to the degradation (Fig. 4.38*c*).

Phenol-oxidizing microorganisms have demonstrated effective transformation of *cis-* and *trans-*dichloroethylene and trichloroethyl-ene in laboratory and in situ field studies (Hopkins, 1993*a*). The phe-nol-oxidizing microorganisms appear to have a much higher capacity to degrade trichloroethylene than the methanotrophs. Trichloro-ethylene degradation of 90 percent has been achieved with 99.8 per-cent removal of injected phenol. Separate laboratory studies suggest that the addition of noncompetitive external reducing power may sig-nificantly increase the transformation potential. Trichloroethylene transformation capacities (Tc) were enhanced by the addition of aliphatic compounds, the greatest enhancement being with formate or lactate (Table 4.12) (Hopkins, 1993*b*).

In situ studies have demonstrated that phenol-utilizing micro-organisms can be readily stimulated. Phenol addition was a good pri-mary substrate, achieving degradation of cis- and trans-dichloroethyl-ene and trichloroethylene in situ (Hopkins, 1993a). The rate of chloroethylene transformation was increased with increased phenol

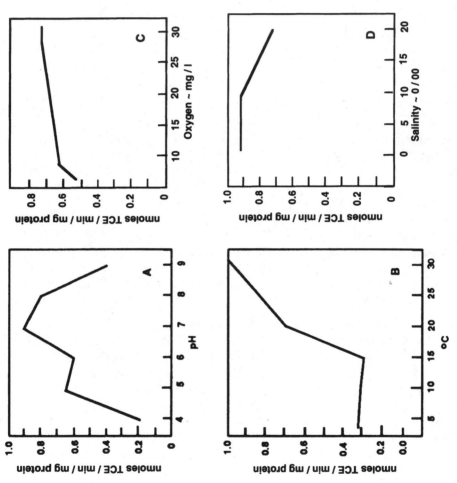

Figure 4.38 Effects of various parameters on the rate of trichloroethylene (TCE) degradation by *Pseudomonas cepacia G4 Phel.* (*Shields, 1991.*)

TABLE 4.12   Enhancement of Trichloroethylene Transformation
Capacities with an External Source of Reducing Power

Compound added	Amount added, mM	Tc, g of TCE per g of cells	Ty, g of TCE per g of phenol
Formate	10	0.38	0.17
Formate	20	0.31	0.14
Formate	40	0.25	0.11
Lactate	3.3	0.38	0.17
Acetate	5	0.27	0.12
Benzoate	1.3	0.20	0.09
Tryptophan	0.9	0.19	0.08
Catechol	1.5	0.14	0.06
Salicylate	1.4	0.04	0.02
Phenol	1.4	0.03	0.01
No added compound		0.24	0.11

concentrations. Phenol was injected as a pulse every 1 h out of 8 h. Approximately 0.003 g of trichloroethylene and 0.003 g of cis-dichloroethylene were removed per g of phenol utilized. These corresponded to 0.002 g of trichloroethylene and 0.002 g of cis-dichloroethylene removed per g of oxygen consumed (Hopkins, 1993a).

Other *Pseudomonas* species, *P. putida* and *P. mendocina,* also can oxidize trichloroethylene when using toluene as the primary substrate (Mahaffey, 1990). Another group capable of oxidizing halogenated aliphatic hydrocarbons are the propane-oxidizing bacteria (Ensign, 1992). Specific species that conduct these oxidations include *Mycobacterium vaccae, Rhodococcus erythyropolis, Alcaligenes denitrificans* subsp. *xylooxidans,* and a *Xanthobacter* strain (Ensign, 1992; Mahaffey, 1990).

Alcaligenes denitrificans and *Rhodococcus erythyropolis* can cometabolize trichloroethylene, dichloroethylene, and vinyl chloride. A *Xanthobacter* has been reported to degrade trichloroethylene, vinyl chloride, cis- and trans-1,2-dichloroethylene, 1,3-dichloropropylene, and 2,3-dichloropropylene. However, tetrachloroethylene was not degraded and 1,1-dichloroethylene was ineffectively degraded (Ensign, 1992).

Xanthobacter autotrophicus and *Ancylobacter aquaticus* are facultative methylotrophs. The methylotrophs are capable of growth on a variety of organic compounds; however, they cannot use methane as carbon and energy sources. All methanotrophs can grow on methane, many are also able to utilize methanol and formaldehyde, and a few can use a wider range of organic compounds. Most methanotrophs are obligate methophiles, which means that they are incapable of growth on compounds that contain carbon-carbon bonds.

The responsible enzyme of methanotrophic bacteria, monooxygenase, catalyzes the incorporation of one oxygen atom from molecular oxygen into methane to produce methanol. The lack of substrate specificity of the monooxygenase enzyme results in its ability to oxidize a broad range of compounds, including halogenated aliphatic compounds. It can hydroxylate many alkanes and aromatic compounds and form epoxides from alkenes (Fig. 4.39) (Semprini, 1992). The epoxides are unstable and yield an acid. Since the products of these reactions are not further metabolized by methanotrophs, a community of microorganisms is necessary for mineralization.

Biotransformation of trichloroethylene by bacteria using methane as a primary substrate has now been confirmed by many investigators. Trichloroethylene has been successfully degraded aerobically to carbon dioxide with 0.6 percent methane in air, although the extent and rate of transformation was less than for the dichloroethylenes. There is, however, a potential for toxicity problems. Trichloroethylene oxidation products are toxic to *Methylosinus trichorsporium* (Tsien, 1989), and perchloroethylene appears to inhibit trichloroethylene degradation (Palumbo, 1991). Trichloroethylene concentrations of 50 mg/L appear to inhibit methane utilization by methanotrophs (Semprini, 1992).

Another serious limitation is that methanotrophs have not been reported to transform perchloroethylene or higher halogenated

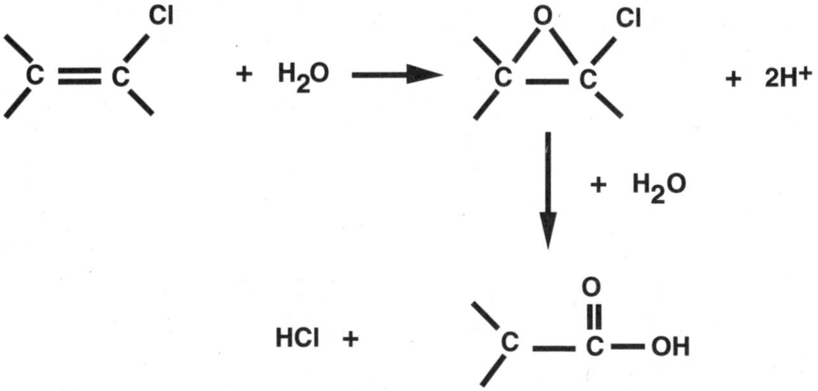

Figure 4.39 The epoxidation of halogenated aliphatic compounds by methanotrophic microorganisms.

aliphatic compounds. The less chlorinated the compound, the greater the rate and extent of transformation as expected from thermodynamic principles. The lower the oxidation state of the compound, the easier it is to oxidize. Since trichloroethylene is more oxidized than vinyl chloride, it is more difficult to further oxidize this compound. As noted in Fig. 4.37, the opposite is true when attempting to reduce an oxidized compound. The higher the degree of oxidization, the easier it is to perform reduction of that compound.

Although these oxygenase-generating organisms can oxidize halogenated aliphatic compounds, the engineering applications are not simple. First, the oxygenase enzymes require the presence of a reductant to catalyze the oxygenation of substrates. Sustained oxidations in obligate lithotrophs are dependent upon continued oxidization of substrate. Therefore, the cometabolite must always be present for sustained reactions. However, biotransformation activity is also inhibited by excessive methane and high oxygen concentrations (above 30 mg/L). High methane, the primary substrate, hinders the reaction rate since it will compete for the monooxygenase enzyme, making the enzyme unavailable for the target compounds. This principle should apply to all cometabolism systems. Elevated concentrations of primary substrate can be expected to reduce the rate of biotransformation of the target compounds.

A more difficult problem to control is that the epoxide transformation product is frequently an inactivator of the oxygenase and other cellular proteins (Ensign, 1992). In addition to the epoxide, other transformation products can include dichloroacetic acids, trichloroacetic acid, 1,1,1-trichloroethanol, chloral, and glyoxylic (Uchiyama, 1992). Some of these compounds, such as the chlorinated acetic acids and chloral, have mutagenic and hepatocarcinogenic properties. Heterotrophic bacteria may play an important role in the mineralization of these compounds.

The transformation of trichloroethylene by methanotrophs can be influenced by the water-soil chemistry. The nature of various solutes in groundwater has significantly affected the transformation (Henry, 1990). Several metals influence the production of soluble methane monooxygenases and therefore influence degradation of halogenated aliphatic compounds. Another consideration is the influence of hydrogen peroxide as an oxygen source. Hydrogen peroxide may be inhibitory to methanotrophs (Semprini, 1992). Treatability studies are advisable before conducting site applications. This and other details must be delineated before entering field operations.

TABLE 4.13   Aerobic Cometabolism in Situ Treatment
Response on Halogenated Aliphatic Compounds

Chemical	Extent of transformation, %
Vinyl chloride	90–95
trans-1,2-Dichloroethylene	85–90
cis-1,2-Dichloroethylene	33–45
Trichloroethylene	10–29

SOURCE: McCarty, 1990.

Engineering the biological process to overcome these limitations has been evaluated in several studies, including a pilot in situ bioremediation study. The field in situ study was conducted at the Moffett Naval Air Station, Mountain View, Calif. (Roberts, 1989). To maintain the cometabolism activity of methanotrophs, methane gas was injected in the subsurface as the primary substrate. During field pilot studies essentially complete transformation of vinyl chloride and 30 percent of trichloroethylene was achieved (Table 4.13).

Anaerobic metabolism.   Many halogenated aliphatic compounds are transformed under anaerobic conditions. In the presence of a consortium of microorganisms, these compounds will be mineralized. One of the predominant mechanisms for transformation of halogenated aliphatic compounds is reductive dehalogenation. Reductive reactions result in the replacement of a halogen by hydrogenolysis or dihalo-elimination. The reductive process is usually through cometabolism. There are exceptions to the need for cometabolism.

Chloromethane serves as carbon and energy source for a strictly anaerobic homoacetogenic bacterium, and two anaerobic mixed cultures use chloromethane as a substrate (Braus-Stromeyer, 1993). This degradation was under methanogenic conditions, suggesting that the chloromethane is first transformed to products that can be used by methanogenic bacteria. A proposed pathway for chloromethane degradation to acetate is based on the interaction of two microorganisms. The first microorganism (DMA) is dependent on some growth factor from a second microorganism (DMB). Formate is a key intermediate in the transformation (Fig. 4.40).

Halogenated aliphatic compounds are transformed by reductive dehalogenation even when present at low concentrations of 10 to 200 μg/L (Bouwer, 1981, 1988). In reductive dehalogenation the halogenated organic compound serves as the electron acceptor. The rate of dehalogenation under anaerobic conditions should be linked to the rates of primary substrate oxidation. The reductive dehalogenation is carried out by electrons from the oxidation of the primary substrate. The control of these reactions for bioremediation requires an under-

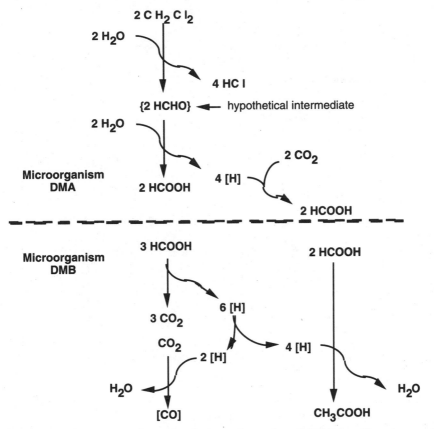

Figure 4.40 Proposed pathway for the transformation of dichloromethane as a sole energy source under anaerobic conditions. (*Braus-Stromeyer, 1993.*)

standing of the redox conditions and the influence and availability of specific electron acceptors and donors on the overall metabolism mode.

The availability of electron acceptors in anaerobic systems affects reductive dehalogenation by competing with the halogenated compounds for reducing potential. For example, nitrate and sulfate can inhibit the dehalogenation of some halogenated compounds. The influence of electron acceptors is explained by thermodynamic principles. Recalling that microorganisms will couple half reactions that yield the greatest free energy provides the basis for understanding these principles.

The potential free energy from coupling various reductive dehalogenation reactions with different electron acceptors is provided in Fig. 4.41. Based upon Fig. 4.41, one can predict the relative suitability of the compounds to transformation over the range of electron acceptor conditions. Each arrow in the diagram points in the favor-

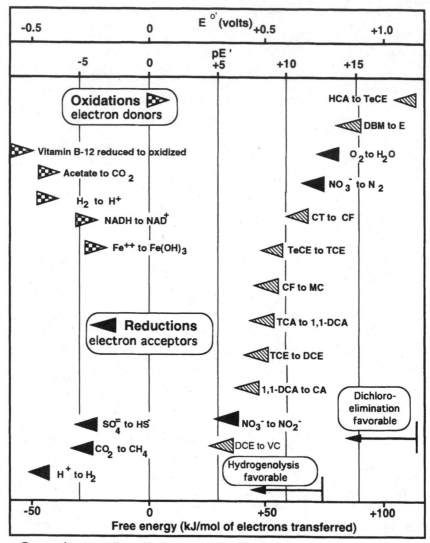

Bases of arrows align with the potentials of the half-reaction shownin volts.

COMPOUND	ABBREVIATION
METHANE	CH_4
CHLOROMETHANE	CM
DIBROMOMETHANE	DBM
DICHLOROMETHANE (METHYLENE CHLORIDE)	MC
TRICHLOROMETHANE (CHLOROFORM)	CF
TETRACHLOROMETHANE (CARBON TETRACHLORIDE)	CT

DICHLOROETHYLENE	DCE
TRICHLOROETHYLENE	TCE
TETRACHLOROETHENE (PERCHLOROETHYLENE)	TeCE
ETHANE	E
CHLOROETHANE	CA
1,2-DIBROMOETHANE	EDB
PENTACHLOROETHANE	PCA
HEXACHLOROETHANE	HCA
DICHLOROETHANE	DCA

Abbreviations used for chemical species

Figure 4.41 Half-reaction potentials for reduction and oxidation of halogenated aliphatic compounds: potential electron acceptors and electron donors. Bases of arrows align with the potentials of the half reaction shown in volts. (*Modified from Vogel, 1987.*)

able reaction direction, and the arrow's base shows the potential of the half reaction. Electron acceptors which are located farther to the right are stronger than those to the left. Thus reactions located to the right are more favorable. As redox conditions become more reducing, lower pE, one would expect more of the compounds shown to undergo transformation. More compounds will be transformed under methanogenic conditions than under denitrification. It also should be noted that several biologically active donors, and even ferrous ion, have lower reduction potentials than do most of the halogenated aliphatic compounds. As a result, they can be involved in halogen removal by reduction.

Hexachloroethane (HCA) and 1,2-dibromoethane (EDB) are stronger electron acceptors than oxygen. Therefore, the reduction of hexachloroethane (HCA) to tetrachloroethane (TeCE) and 1,2-dibromoethane (EDB) to ethane (E) is favorable under all electron acceptor conditions, including oxygen. They have higher reduction potential than oxygen and can be dehalogenated in aerobic systems. On the other hand, the sequential reduction of tetrachloroethane (TeCE) to trichloroethylene (TCE) to dichloroethylene (C12CE) is not favorable under aerobic conditions. They have reduction potentials below that of oxygen and nitrate. Thus one would not expect these reactions to proceed extensively in the presence of nitrate. However, in the presence of sulfate or carbon dioxide, one would expect many of the reactions to proceed. The reduction potential of methane is significantly lower.

The work of Bouwer (1981, 1988) and that of EPA's Research Laboratory, Ada, Okla. (Sewell, 1990) illustrate the influence of electron acceptors on the dehalogenation of aliphatic compounds. In situ organisms responsible for dehalogenation reactions at the Traverse City site, Traverse, Mich., were significantly influenced in their ability to dehalogenate. The addition of nitrate to natural soil microcosms completely blocked the dehalogenation of tetrachloroethylene (Fig. 4.42).

Sulfate also influences these reactions. With sulfate as the electron acceptor, the dehalogenation of tetrachloroethylene proceeded at a slower rate than the carbon dioxide acceptor system. More important, sulfate appears to block the dehalogenation of trichloroethylene as illustrated by its buildup from the tetrachloroethane transformation (Fig. 4.43). In the absence of sulfate the trichloroethylene is removed within 40 days after its peak concentration level (Fig. 4.43).

The specific influence of sulfate and other electron acceptors cannot be generalized. Indigenous microorganisms at other sites were blocked by sulfate in their ability to dehalogenate tetrachloroethane. Different microbial systems and different soil-water chemical conditions undoubtedly cause shifts in the thermodynamic equilibrium. These findings illustrate the need for site-specific treatability studies.

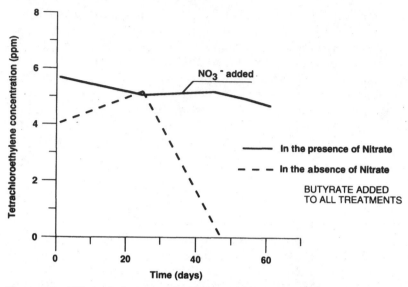

Figure 4.42 Effect of nitrate on dehalogenation of tetrachloroethylene (landfill microcosms). (*U.S. EPA Research Laboratory, Ada, Okla.*)

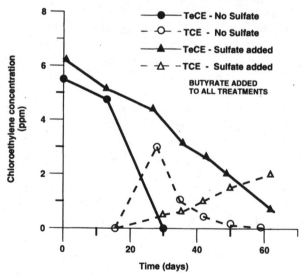

Figure 4.43 Sulfate effect on tetrachloroethylene transformation. Butyrate added to all treatments.

Bouwer's studies on electron acceptor influence are even more informative since a greater number of compounds were evaluated. The study used a contact system consisting of a fixed biofilm reactor that was operated with a 2.5-h detention time (Bouwer, 1981, 1988).

**TABLE 4.14   Halogenated Aliphatic Compounds
That Are Anaerobically Transformed**

Compound	Percent removal
Electron Acceptor—Nitrate	
Chloroform	0
Tetrachloroethylene	0
Dibromochloropropane	14
1,2-Dibromomethane	23
1,1,1-Trichloroethane	30
Hexachloroethane	80
Bromodichloromethane	90
Bromoform	>99
Carbon tetrachloride	>99
Electron Acceptor—Sulfate	
Chloroform	0
Tetrachloroethylene	13
1,2-Dibromomethane	63
1,1,1-Trichloroethane	72
Bromodichloromethane	96
Dibromochloropropane	98
Carbon tetrachloride	>99
Bromoform	>99
Hexachloroethane	>99
Electron Acceptor—Carbon Dioxide	
Tetrachloroethylene	86
Chloroform	95
1,1,1-Trichloroethane	>99
Carbon tetrachloride	>99
Dibromochloropropane	>99
Bromodichloromethane	89
Bromoform	>99
Ethylene dibromide	>99
Hexachloroethane	>99

SOURCE: Bouwer, 1981, 1988.

The percentage of compound transformation was measured under identical contact conditions, but with different electron acceptors. The transformation was evaluated under conditions of denitrification, sulfate reducing, and methanogenesis. According to the thermodynamic considerations presented in Fig. 4.41, more compounds are transferred under methanogenic conditions than under denitrification (Table 4.14). When nitrate was the electron acceptor, no transformation of chloroform or tetrachloroethylene occurred. Under sulfate-reducing conditions, some tetrachloroethylene was removed with increased removals of other halogenated aliphatic compounds. Under methanogenic conditions, effective removal occurred for most of the halogenated organic compounds, including chloroform.

Much discussion has been devoted to the electron acceptor. However, an inadequate type and amount of electron donor also will hinder anaerobic dehalogenation. Chlorinated aromatic compounds can serve as both the electron acceptor and the donor. This is not true with chloroethenes. The electron donor must be other organic matter, such as hydrocarbons associated with a fuel spill, sewage, sewage sludge, or selected organic compounds.

Many compounds will serve as electron donors for dehalogenation of aliphatic hydrocarbons (Table 4.15). The volatile fatty acids, pro

TABLE 4.15   Some Electron Donors That Support Anaerobic Dehalogenation

Methanol	Crotonate
Ethanol	Lactate
Glucose	Formate
Sucrose	Acetate
Benzoate	Butyrate
Toluene	

duced under methanogenic conditions, appear to be best at enhancing dehalogenation. For the U.S. Coast Guard Air Station site, in Traverse City, Mich., butyrate provided the best response (Sewell, 1990). Butyrate at 0.1 mmol concentration was the difference between effective dehalogenation of chloroethenes and insignificant dehalogenation (Fig. 4.44). Toluene, associated with gasoline spills, has been reported to possibly enhance dehalogenation by serving as an electron donor. However, other studies have found toluene and benzene to moderately inhibit methanogenesis (Grbić-Galić, 1987).

Enhanced dehalogenation in organic-carbon-rich environments has been attributed to nonmethanogenic fermentation (Baek, 1988). These fermenters are capable of reductive dehalogenation. As some methanogens consume hydrogen as an electron donor, fermenters need to reoxidize a reduced electron carrier, such as $NADH_2$, in their metabolism. Thus $NADH_2$ is oxidized to NAD and H_2 in the presence of methanogens. This phenomenon, called "interspecies hydrogen transfer," is found in many microbial communities. Through this symbiotic process, fermentation is enhanced and the development of CO_2

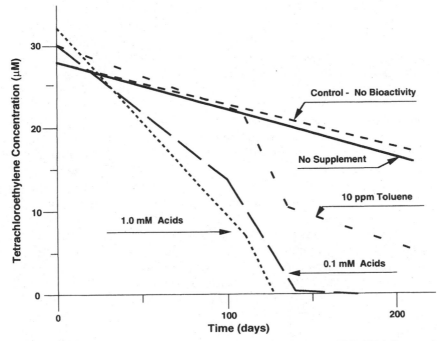

Figure 4.44 Effect of butyrate on anaerobic dehalogenation. (*U.S. EPA Research Laboratory, Ada, Okla.*)

reducers is encouraged. Nonmethanogenic fermenters play two important roles: (1) supplying hydrogens for methanogenesis and (2) saturating the carbon-carbon bond of vinyl chloride.

The reductive dehalogenation of chloroethylene compounds can result in the accumulation of vinyl chloride (Fig. 4.37). The final dechlorination step from vinyl chloride to ethane is usually rate-limiting. However, field applications of methanogenic dehalogenation don't always result in vinyl chloride buildup. One example is the in situ dechlorination of chloroethylenes at the Traverse City site, Traverse City, Mich. In situ studies conducted by U.S. EPA Research Laboratory, Ada, Okla., demonstrated that anaerobic dechlorination is an effective in situ process (Sewell, 1990). No vinyl chloride was detected in monitoring wells during the dechlorination program. Ethane was detected as the final dechlorination product.

Laboratory studies have confirmed that tetrachloroethylene can be completely reduced to ethane with little vinyl chloride residual (DiStefano, 1991). This transformation occurred in methanol-enriched cultures that began as predominantly methanogenic. Methanogenic activity ceased as tetrachloroethylene levels increased, but dechlorination of tetrachloroethylene continued. Acetogenic microorganisms were suggested as the possible active dechlorinators. Pure cultures are known to transform carbon tetrachloride to chloroform and dichloromethane under both sulfate-reducing and methanogenic conditions (Semprini, 1991). An acetogen was found to dehalogenate chloroform. *Clostridium* sp. is reported to transform trichloroethane, chloroform, and carbon tetrachloride (Semprini, 1991). Thus a methanogenic condition is not the only anaerobic metabolism mode possible for dehalogenation of chlorinated aliphatic hydrocarbons.

In situ bioremediation using anoxic conditions has been evaluated at Moffett Field Naval Air Station, Mountain View, Calif. (Semprini, 1991). An indigenous microbial community was rapidly stimulated by introducing acetate as a growth substrate into an aquifer that contained nitrate and sulfate as electron acceptors. The biostimulation resulted in complete nitrate utilization, yielding a transitory buildup of nitrite. After nitrate removal, sulfate reducers or other anaerobic microorganisms such as *Clostridium* brought about dehalogenation of trichloroethane, Freon-11, Freon-113, and chloroform. The required acetate was close to the stoichiometric ratios of nitrate to acetate of 1 mg nitrate per mg of acetate.

Although no vinyl chloride has been found in field projects with anaerobic systems, vinyl chloride buildup may occur at other sites. Treatability studies must consider optimization procedures to minimize buildup of these more reduced compounds under anaerobic conditions.

One procedure is to use anaerobic followed by aerobic treatment. Natural ecological succession could result in this multiple mode of biological response at some sites. The anaerobic activity brings about dehalogenation and methane production. The methane can support methanotrophic activity at the aerobic boundary of the anaerobic plume. Aerobic oxidation of vinyl chloride also will occur without methane as a primary substrate.

Since halogenated aliphatic compounds can undergo a variety of abiotic and biotic transformations, several potential degradation pathways are possible. In addition, the biotic transformation depends on the presence and absence of various electron acceptors. The different pathways for the possible transformation of 1,1,1-trichloroethane are illustrated in Fig. 4.45. This compound can undergo two abiotic transformations as well as a biologically mediated reductive dehalogenation under anaerobic conditions.

One pathway illustrates the abiotic process of dehydrohalogenation of 1,1,1-trichloroethane to acetic acid. Acetic acid can be rapidly mineralized by numerous organisms to carbon dioxide. Another pathway

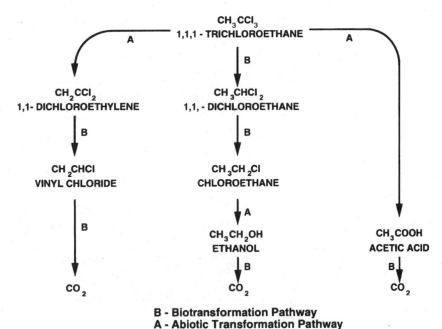

B - Biotransformation Pathway
A - Abiotic Transformation Pathway

Figure 4.45 Pathways for the transformation of 1,1,1-trichloroethane under methanogenic conditions. (*Vogel, 1987.*)

is the abiotic dehydrohalogenation to 1,1-dichloroethylene, which can then be transformed microbially under methanogenic conditions to vinyl chloride. Vinyl chloride can then be mineralized to carbon dioxide. Another pathway is the microbial reduction of 1,1,1-trichloroethane to chloroethane, which can then be abiotically hydrolyzed to ethanol. Ethanol is then rapidly mineralized by organisms to carbon dioxide. The validity of these pathways has been confirmed by field observations on the presence and occurrence sequence of these contaminants in groundwater.

Summation

1. Aerobic and anaerobic microbial systems have been applied to the degradation of halogenated aliphatic hydrocarbons.

2. Halogenated aliphatic compounds will undergo abiotic transformations in the environment.

3. Transition metals can reduce halogenated aliphatic compounds.

4. Under microbial activity halogenated aliphatic compounds can be either oxidized or reduced, depending on the degree of halogenation and environmental conditions.

5. All halogenated aliphatic compounds will not be transformed by all metabolism modes.

6. Some halogenated aliphatic compounds can be used as the sole source of carbon and energy under aerobic conditions.

7. Some halogenated aliphatic compounds can be transformed by cometabolism under aerobic conditions.

8. Methanotrophs have been demonstrated to transform some halogenated aliphatic compounds, in situ, under cometabolism.

9. Many halogenated aliphatic compounds can be transformed under anaerobic conditions by reductive dehalogenation.

10. Under anaerobic conditions, the electron acceptors (nitrate, sulfate, or carbon dioxide) significantly affect the degree of reductive dehalogenation.

11. Nitrate has blocked the anaerobic dehalogenation of some halogenated aliphatic compounds.

12. Sulfate also influences the degree of dehalogenation; however, the specific influence cannot be generalized.

13. The presence of an adequate type and concentration of electron donor is necessary for successful reductive dehalogenation.

5

Microbial Detoxification of Specialty Chemicals

Introduction

The microbial degradation of many organic compounds is discussed in Chap. 4. They respond similarly to microbial degradation for each chemical category. They also are found present as a mixture on many contaminated sites. There are, however, a few chemicals that for one or more reasons don't fall into a generalized discussion. This may result from the uniqueness of their chemical structure or from their use patterns by society and interaction with the environment. This category includes the specialty chemicals that are designed for very specific uses. They are the pesticides and their subcategories of insecticides, herbicides and fungicides, polychlorinated biphenyls, and azo dyes. Several chemicals of each are examined with regard to their treatability by bioremediation.

Pesticides

The demand for specialty chemicals will continue to generate numerous new chemicals every year. Probably none have had any more impact on society and the environment than agricultural chemicals. Pesticides have, on one hand, significantly increased agricultural yields while impacting detrimentally on the food chain of wildlife and humans. By design these chemicals are toxic to some form of life. As a result, some are not suitable for bioremediation; however, many can be biologically degraded.

Chemical nature

Most pesticides are organic and can be classified according to common structural groups. For discussing microbial transformation, it is best to group pesticides according to these structural similarities. These generalized categories are presented in Table 5.1, which includes examples of pesticides in each category. Typical structures for each category are provided in Fig. 5.1.

Degradation potential. The microbial degradation of pesticides is not a recent finding but one that has been long recognized. Early research on pesticide degradation was directed at the failure of repeated soil applications to control agricultural pests. This phenomenon is considered a problem by the agricultural industry, but an ecological benefit by environmental scientists and engineers. Repeated applications result in the development of microbial consortiums that effectively transform the applied pesticides. The exact mechanism for microbial adaptation to pesticides is not understood. It could be simple population growth or genetic changes. Microorganisms can acquire the genetic material that encodes the biochemical mechanisms necessary to deal with a new potential substrate. Alternatively, microorganisms may transform a compound to remove toxicity to their well-being rather than use it as an energy source.

The degradation principles for pesticides, insecticides, and herbicides are no different from those for the organic compounds discussed in Chap. 4. They are presented in a separate section because these compounds frequently occur as contaminants separate from other organic compounds and many are extremely difficult to bioremediate. Thus, from an engineering standpoint they are usually treated individually and will be considered separately.

Over 125 herbicides, over 300 insecticides, and over 325 fungicides are registered in the United States (Kearney, 1972). These chemicals represent a variety of dissimilar chemical structures, resulting in some that are extremely resistant and some that degrade to compounds of higher toxicity to those that are mineralized. The persistence of these compounds in soil is a function of many variables, and there has been great interest in measuring their persistence in soil for evaluating useful life. Alexander (1977) has reported some persistence data with emphasis that actual periods of persistence for chlordane, DDT, dieldrin, and heptachlor are probably much longer than that illustrated in Table 5.2.

As with many chemicals, pesticides are subject to both abiotic and biotic transformation processes. Many of the abiotic transformations result in partial degradation to products that can be further degraded

TABLE 5.1   Pesticide Categories and
Examples as Based on Chemical Structure

Halogenated Aliphatic Pesticides

  DBCP (1,2-dibromo-3-chloropropane)
  1,2-D (1,2-dichloropropane)
  *cis-, trans-*1,3-Dichloropropane
  2-3-Dibromobutane
  Dalapon (2,2-dichloropropionic acid)

Chlorinated Cyclicaliphatic Pesticides

  Lindane (γ-hexachlorocyclohexane)

Halogenated Aromatic Pesticides

  DDT (dichlorodiphenyltrichloroethane)

Chlorinated Phenoyalkanoates

  2,4-D (2,4-dichlorophenoxy acetic acid)
  2,4,5-T (2,4,5-trichlorophenoxy acetic acid)
  2,4-DB
  2,4-DEP
  2,4-DEB
  2,4,5-TES
  MCPA
  MCPB
  MCPES
  Silvex
  Sesone
  Erbon
  Dichloroprop
  Mecoprop

Halogenated Aniline-Based Pesticides

  Acylanilides:
    Propanil
    Karsil
    Drcryl
    Metalaxyl
    Metolachlor
    Alachlor
    Butachlor
    Propachlor
    Ramrod
    Trifluralin

Cyclodiene Pesticides

  Heptachlor
  Chlordane
  Dieldrin
  Aldrin
  Amitraz
  Mirex

TABLE 5.1   Pesticide Categories and
Examples as Based on Chemical Structure
(Continued)

Carbamate Pesticides
  Aldicarb
  Carbofuran
  Carbaryl

Pyrethroid Pesticides
  Permethrin
  Deltamethrin
  Fastac
  Fenvalerate
  Fluvalinate

Organophosphonate Pesticides
  Glyphosate
  Parathion

by microorganisms. Hydrolytic reactions are the most significant abiotic transformations. Hydrolytic reactions can be base- or acid-catalyzed and often occur through interactions with reactive chemical groups on mineral surfaces, reactive organic compounds, and inorganic metals (e.g., Cu^{2+}). The hydrolysis reactions may be necessary for microbial degradation. However, microorganisms can catalyze the hydrolysis reactions through cometabolism. Degradation may result in accumulation of metabolites, incorporation of the pesticide compound into the soil organic fraction, or mineralization.

An important variable for the bioremediation of pesticide-contaminated soil is the availability of the pesticide to soil microorganisms. This availability is a function of the adsorption affinity of the organic compound to soil. The available pesticide for biodegradation is equivalent to the amount of pesticide dissolved in the aqueous phase (Anderson, 1984). By increasing the ratio of water to soil solids, the total amount of available pesticide for degradation is increased. At moisture levels below 50 percent field capacity the rate of pesticide degradation continues to decrease. For field capacities at 25 percent and below, degradation rates are reduced 80 to 90 percent (Skipper, 1990). Usually insignificant rate changes occur in the range of 50 to 100 percent field capacity. The second most important variable is total biomass or number of microorganisms. Increasing the number of microorganisms by nutrient addition and other environmental controls results in a corresponding increase in the rate of pesticide degradation.

Pesticide concentration also affects the degradation rate. For soil concentrations below 5 µg/L the rate is first-order. At higher concentrations a biphasic breakdown occurs (Smith, 1989).

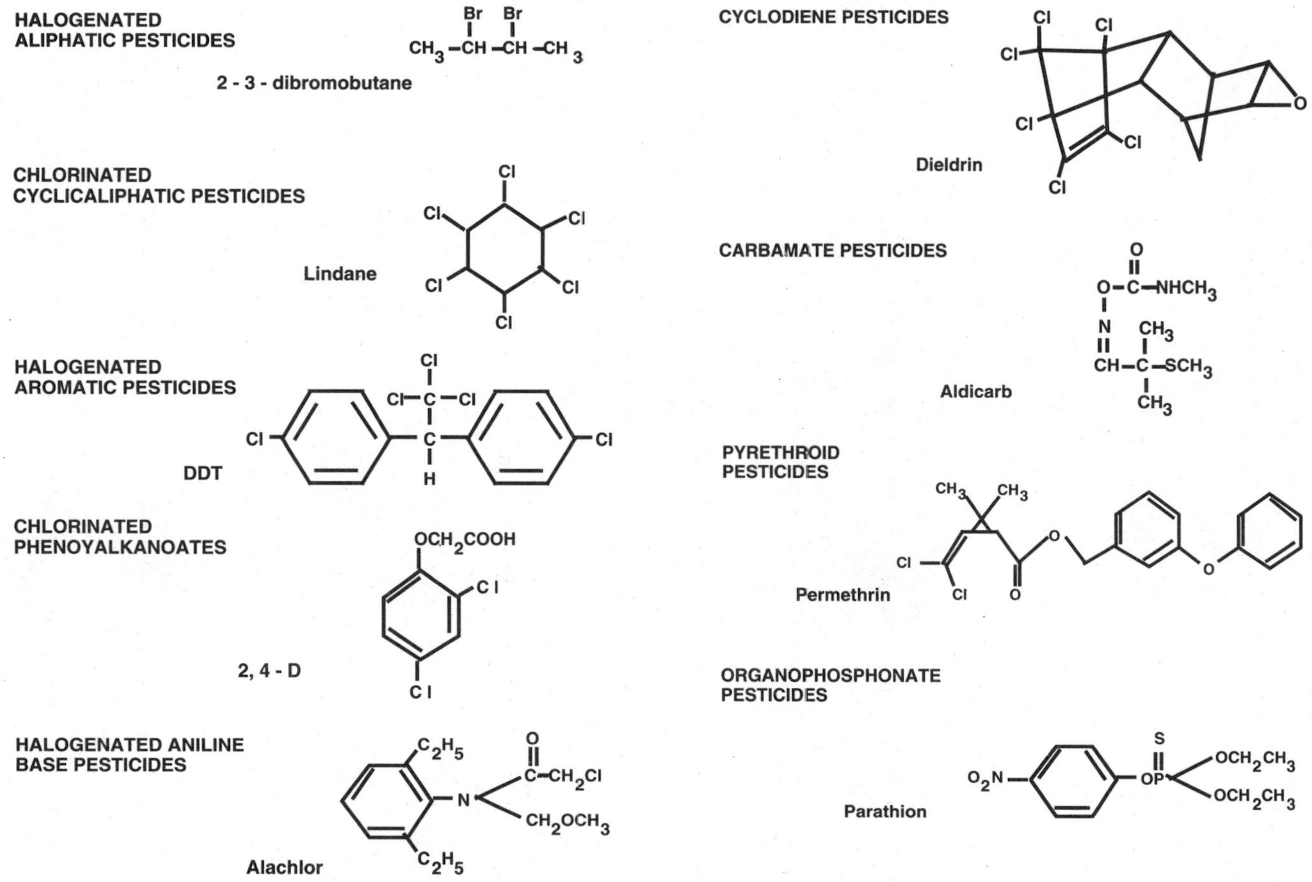

Figure 5.1 Chemical structures of several pesticides.

169

TABLE 5.2   Persistence of Select Pesticides

Chemical	Time still detectable	Approximate half-life
Chlordane	21 years	2–4 years
DDT	24 years	3–10 years
Dieldrin	21 years	1–7 years
Heptachlor	16 years	7–12 years
Toxaphene	16 years	10 years
Dalapon	10 weeks	
DDVP		17 days
Methyldemeton S		26 days
Thimet		2 days

SOURCE: Alexander, 1977.

For many sites contaminated with specialty compounds such as pesticides, PCBs, and dyes, soil may not contain significant populations of degrading organisms. No isofenphos-degrading microorganisms are found in soil that did not have exposure to this pesticide, whereas soils contain between 6000 and 12,000 isofenphos-degrading organisms that had previous isofenphos applications (Racke, 1990). The degradation of these compounds is mediated by a small fraction of the total soil population. For example, the ratio of 2,4-D-degrading microorganisms to total microorganisms is between $1:6 \times 10^2$ and $1:3 \times 10^5$ (Racke, 1990). For these difficult-to-degrade compounds, development of degrading organisms may require longer periods of months or years. Bioremediation of these compounds is usually aided by developing specific seed cultures and inoculation of the soil or the bioreactor.

Attempts to bioaugment soil for enhanced pesticide degradation appear to improve the overall rate of degradation. Parathion degradation in soil has been enhanced by the inoculation of an adapted two-member culture (Maloney, 1988). Similar results have been indicated for a phenyl carbamate herbicide and the chlorinated phenyl alkanoate 2,4,5-T. Analog enrichment with aniline has a very positive effect on the removal of 3,4-dichloroaniline from contaminated soils (Brunner, 1985). Some organisms that are reported to degrade specialty compounds are listed in Table 5.3.

For many pesticides, such as lindane, heptachlor, DDT, aldrin, and endrin, degradation is more rapid under anaerobic than under aerobic conditions. The greater success of anaerobic over aerobic conditions for degradation of these chemicals is partially related to the advantage of reductive dehalogenation by anaerobes versus aerobes.

Although there are many pesticides, insecticides, and herbicides with very dissimilar chemical structures, there are only several kinds of reactions that initial their degradation (Alexander, 1977). The degradation reactions can be characterized as follows:

TABLE 5.3   Microorganisms from Soil That Demonstrate Pesticide Degradation

Bacteria	
Achromobacter:	
2,4-D	Racke, 1990
Carbofuran	Racke, 1990
Arthrobacter:	
EPTC	Racke, 1990
Esofenphos	Racke, 1990
2,4-D	Racke, 1990
Alcaligenes:	
Isofenphos	Reed, 1990
2,4-D	Racke, 1990
Flavobacterium:	
PCP	Kaufman, 1977
EPTC	Racke, 1990
2,4-D	Racke, 1990
Methylomonas:	
EPTC	Racke, 1990
Pseudomonas:	
Alachlor	Reed, 1990
Isofenphos	Reed, 1990
Carbofuran	Racke, 1990
2,4-D	Racke, 1990
Rhodococcus:	
EPTC	Racke, 1990

Fungi	
Trichoderma spp.:	
PCP	Lamar, 1990
Trametes versicolor:	
PCP	Kaufman, 1977
Fusarium:	
EPTC	Reed, 1990
Isofenphos	Reed, 1990
Aspergillus sp.:	
EPTC	Reed, 1990
Phanerochaete sordida:	
PCP	Lamar, 1990
Phanerochaete chrysosporium:	
PCP	Glaser, 1988

Actinomycetes	
Streptomyces:	
Terbufos	Reed, 1990
Alachlor	Reed, 1990

1. Addition of a hydroxyl group

2. Oxidation of an amino group

3. Oxidation of a sulfur molecule

4. Addition of an oxygen to a double bond

5. Addition of a methyl group

6. Removal of a methyl group

7. Removal of a chlorine

8. Chlorine migration

9. Reduction of a nitro group

10. Replacement of a sulfur with an oxygen

11. Cleavage of an ether linkage

12. Metabolism of side chains

13. Hydrolysis

Typical examples of the microbial initiated degradation of pesticides are illustrated in Fig. 5.2. The hydroxyl group addition results in the replacement of a hydrogen atom carbon bond with the hydroxyl group, OH, as illustrated by reaction I. For benzene ring pesticidal compounds, the hydroxyl addition results in cleavage of the aromatic ring as typical with benzene degradation. The oxidation of amino groups on pesticidal compounds results in conversion of the $-NH_2$ group to $-NO_2$, reaction II. Oxidation of sulfur molecules in pesticidal compounds yields $-SO_2$, reaction III.

A common microbial reaction is the addition of oxygen to a double bond resulting in the formation of an epoxide group, reaction IV. The epoxide group can be fairly resistant to microbial attack and toxic to microbial cell material. Methyl groups can be added or removed as typical of the methylation of arsenic pesticides, reaction V, and the cleavage of methyl from the nitrogen atoms of herbicides, reaction VI. For alkyl groups of several carbon chains, multi groups may be cleaved off. A common reaction of halogenated compounds is the removal of chlorine with replacement by hydrogen or hydroxide, reactions VII and VIII. Microorganisms also can remove a chloride and hydrogen atom as in reaction IX. The action of microorganisms on chlorinated benzene rings can result in the migration of chlorine to another ring position (reaction X). Reduction of nitro groups is also common (reaction XI).

One major class of insecticides contain sulfur-phosphorus double bonds. With these compounds, the sulfur can be replaced with an oxygen atom, reaction XII. Another class of pesticides contains the ether

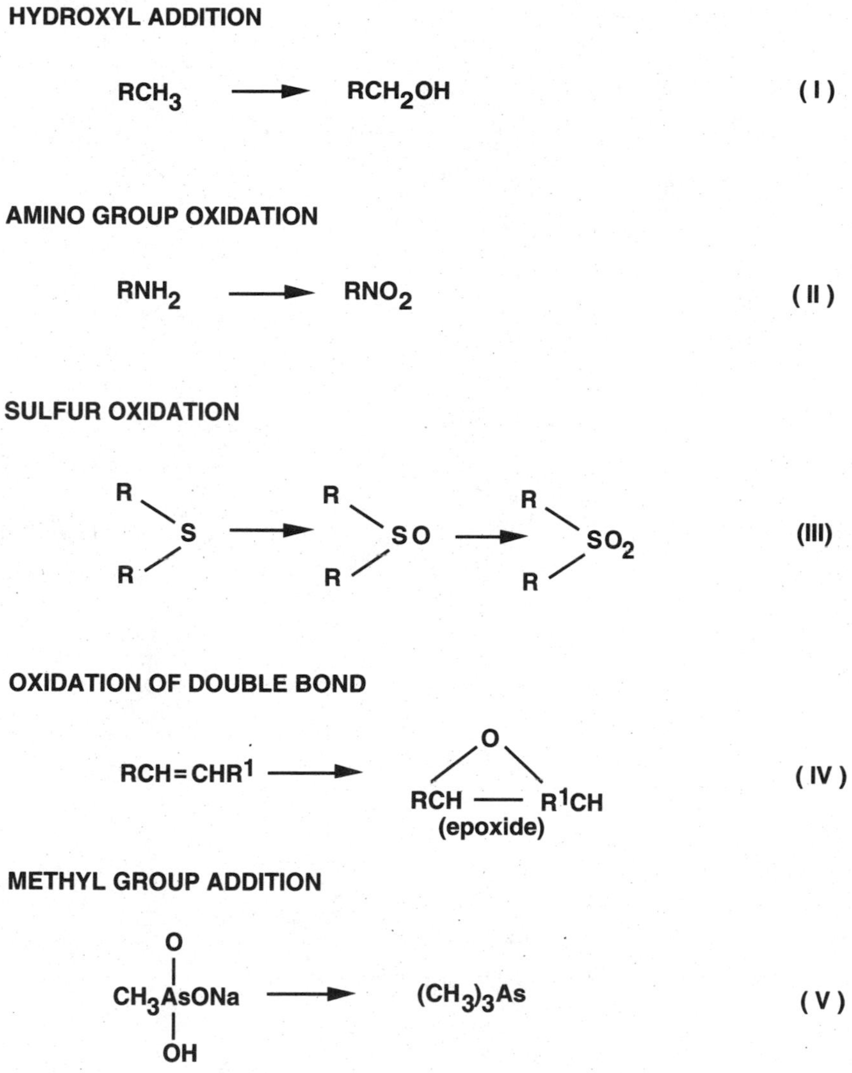

HYDROXYL ADDITION

$$RCH_3 \longrightarrow RCH_2OH \qquad\qquad (\,I\,)$$

AMINO GROUP OXIDATION

$$RNH_2 \longrightarrow RNO_2 \qquad\qquad (\,II\,)$$

SULFUR OXIDATION

$$\begin{array}{c} R \\ \diagdown \\ \diagup \\ R \end{array}\!\!S \longrightarrow \begin{array}{c} R \\ \diagdown \\ \diagup \\ R \end{array}\!\!SO \longrightarrow \begin{array}{c} R \\ \diagdown \\ \diagup \\ R \end{array}\!\!SO_2 \qquad\qquad (III)$$

OXIDATION OF DOUBLE BOND

$$RCH\!=\!CHR^1 \longrightarrow \overset{\displaystyle O}{\overset{\displaystyle \diagup\ \diagdown}{RCH \!-\! R^1CH}} \qquad (\,IV\,)$$
(epoxide)

METHYL GROUP ADDITION

$$\overset{\displaystyle O}{\underset{\displaystyle OH}{CH_3\overset{|}{\underset{|}{As}}ONa}} \longrightarrow (CH_3)_3As \qquad\qquad (\,V\,)$$

Figure 5.2 Initial reactions in pesticide degradation. (*Alexander, 1977.*)

linkage which can be broken as illustrated by equation XIII. As with aliphatic hydrocarbons, two carbon atoms can be cleaved by β oxidation, reaction XIV. Hydrolysis reactions, as discussed in Chap. 3, are also common with pesticides, reaction XV.

Complex formation.   Not all microbial reactions result in mineralization or even transformation to simpler compounds. Some pesticides,

METHYL GROUP REMOVAL

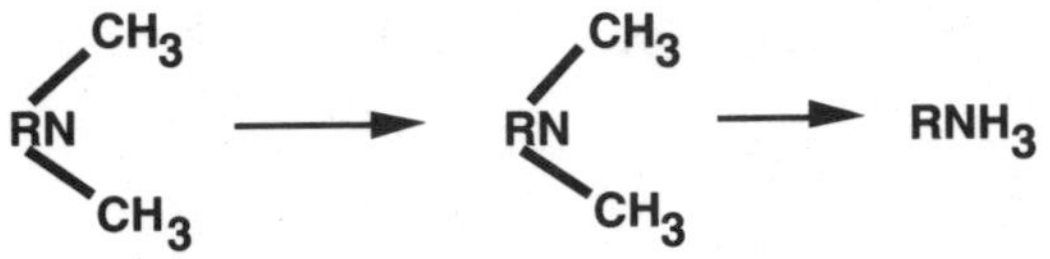

$$RN\begin{smallmatrix} CH_3 \\ CH_3 \end{smallmatrix} \longrightarrow RN\begin{smallmatrix} CH_3 \\ CH_3 \end{smallmatrix} \longrightarrow RNH_3 \qquad (VI)$$

CHLORINE REMOVAL

$$RCH_2Cl \longrightarrow RCH_2OH \qquad (VII)$$

$$RCCl_3 \longrightarrow RCHCl_2 \qquad (VIII)$$

CHLORINE AND HYDROGEN REMOVAL

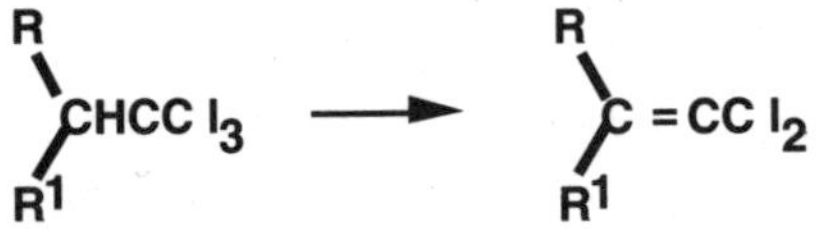

$$\begin{smallmatrix} R \\ R^1 \end{smallmatrix}CHCCl_3 \longrightarrow \begin{smallmatrix} R \\ R^1 \end{smallmatrix}C = CCl_2 \qquad (IX)$$

CHLORINE MIGRATION

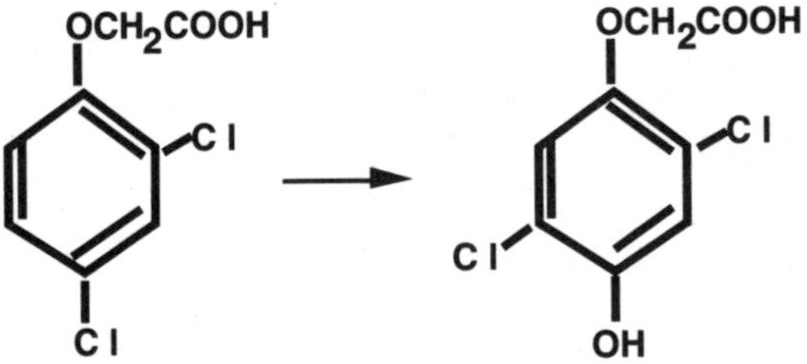

$$(X)$$

Figure 5.2 (*Continued*).

those with halogenated phenol and aniline structures, become complexed with soil humic material. In this form they become very stable. These coupling complexes are resistant to most analytical extraction procedures. Acid and base extractions, and thermal treatment do not recover the coupled hazardous compound (Bollag, 1992). This process removes the compound from an exposure transport pathway, immobilizing it in a soil fixation process. Risk potential is reduced. However, the health significance of these complexes is not understood.

NITRO GROUP REDUCTION

$$RNO_2 \longrightarrow RNH_3 \qquad\qquad (XI)$$

SULFUR REPLACEMENT WITH OXYGEN

$$(XII)$$

ETHER CLEAVAGE

$$ROR' \longrightarrow ROH \; + \; R'H \qquad (XIII)$$

BETA OXIDATION

$$RCH_2CH_2CH_2COOH \longrightarrow RCH_2\overset{\displaystyle O}{\overset{\|}{C}}CH_2COOH \qquad (XIV)$$
$$\downarrow$$
$$RCH_2COOH$$

HYDROLYSIS

$$R\overset{\displaystyle O}{\overset{\|}{C}}OR' \; + \; H_2O \longrightarrow R\overset{\displaystyle O}{\overset{\|}{C}}OH \; + \; R'OH \qquad (XV)$$

Figure 5.2 *(Continued).*

The formation of natural humus results from the incorporation of natural organic acid precursors and aromatic structures into an irreversible complex. These base compounds are biological degradation by-products of natural organic matter. This process is called humification. Humic compounds account for the bulk of organic chemicals in soil and water. They are extremely complex and of a varied structure that has not been well defined. They have been classified into two categories: humic acids and fulvic acid. One suggested structure for fulvic acid is illustrated in Fig. 5.3. The phenolic structure is a major component of humic material, and hazardous chemicals having this structure are potential candidates for binding with natural soil material. Many of the specialty chemicals discussed in this chapter will complex with humic material.

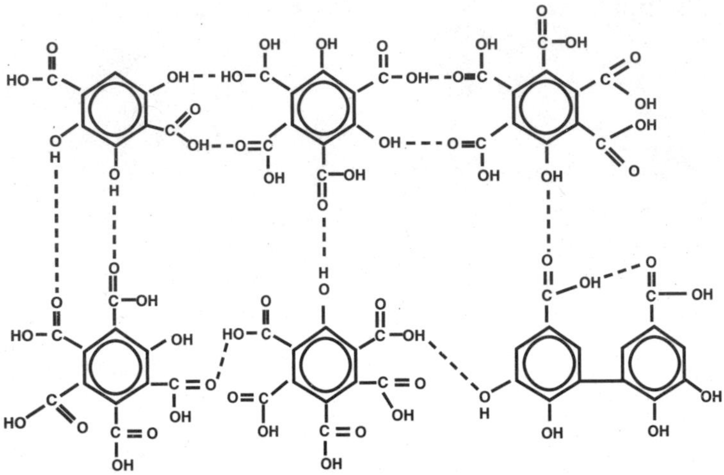

Figure 5.3 Suggested chemical structure of soil humic material: fulvic acid. (*Schnitzer, 1972.*)

The first step in humification is adsorption. Adsorption mechanisms include van der Waals forces, hydrogen bonding, charge transfer, electrostatic attractions, and hydrophobic interactions. The longer the pesticides remain in soil, the stronger this attachment becomes. The weaker reversible adsorption forces are changed to strong covalent bonding, called oxidative coupling. Oxidative coupling is one of the major reactions in the humification process.

Oxidative coupling for a phenolic pesticide is a free-radical reaction where an electron and a proton are lost from the phenolic structure. This resonance-stable free radical can react with another free radical, resulting in coupling as illustrated for 2,4-dichlorophenol and syringic acid (Fig. 5.4).

Oxidative coupling is mediated by many catalysts, including metal oxides, clay minerals, and microbial enzymes. For example, peroxidase enzymes as produced by many organisms can catalyze this coupling. The enzymatic coupling of phenolic compounds to soil humic material is a general phenomenon. The herbicide 2,4-D yields the metabolite of 2,4-dichlorophenol that is coupled with humic derived compounds of syringic acid, vanillic acid, and vanillin (Bollag, 1992). Dimer, trimer, tetramer, and pentamer polymerization products are yielded. The fungus *Rhizoctonia praticola* couples 4-chlorophenol, 2,4- and 2,6-dichlorophenol, 2,3,6- and 2,4,5-trichlorophenol, 2,3,5,6-tetrachlorophenol, and pentachlorophenol to syringic acid. This organism also can couple substituted anilines to humic material. Fig. 5.5 illustrates the coupling of 4-chloroaniline with guaiacol. The herbicide 2,4-DCP also is coupled with fulvic acids.

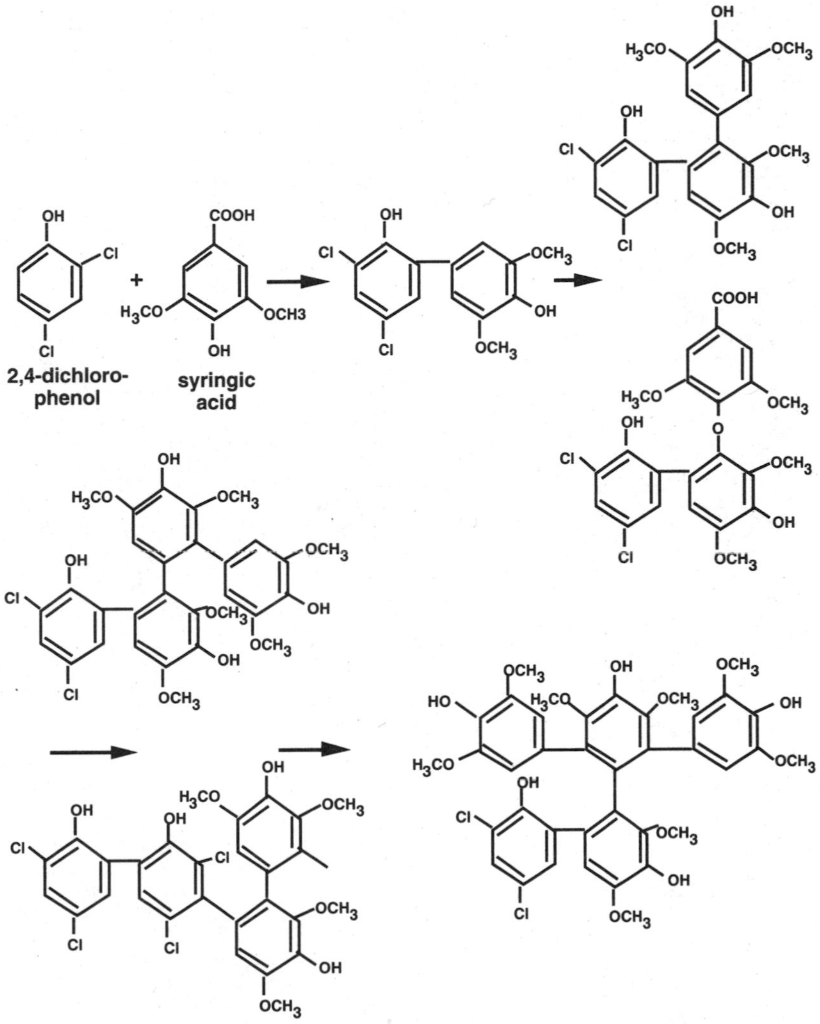

Figure 5.4 Enzyme oxidative coupling of 2,4-dichlorophenol with syringic acid, a humic acid monomer. (*Bollag, 1980.*)

Many factors influence this coupling. They include properties of the pesticide such as nature and position of substituted groups and molecular weight. Other important considerations include pH, soil chemistry, microbial activity, and presence of chemical enhancements such as fulvic acid and catechol. These coupling reactions are very stable. Atrazine remained associated with the soil after 9 years (Bollag, 1992). Coupled pesticides are released only gradually and at a very low rate. When released, they can be degraded or reincorporated into humus. They are essentially immobilized and detoxified, since binding decreases the availability of interaction with biota.

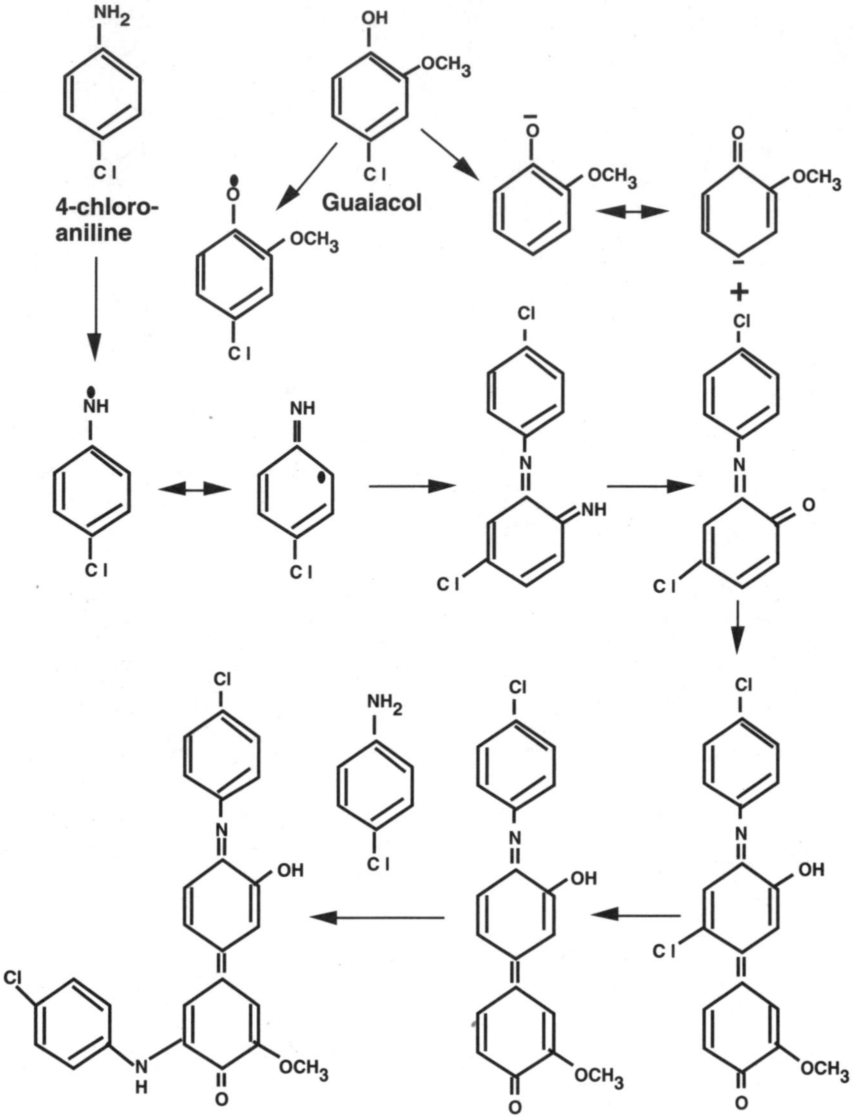

Figure 5.5 Microbial coupling of 4-chloroaniline with humic-type compounds. (*Simmons, 1989.*)

Halogenated aliphatic pesticides

The halogenated aliphatic pesticides are broad-range fumigants that have been used in agriculture. Some such as 1,2-dibromo-3-chloropropane (DBCP) are now banned in the United States. Other aliphatic pesticides include 1,2-dichloropropane (1,2-D), 1,3-dichloropropene, and 2,3-dibromobutane. Some of these aliphatic pesticides are readily hydrologized, such as *cis-* and *trans-*1,3-dichloropropane, and others such as DBCP are not (Castro, 1968). Several are known to temporarily block nitrification.

As with many chlorinated aliphatic compounds, aerobic metabolism can be used to degrade methyl bromide, 1,2-D, and DBCP with the autotrophic nitrifying bacteria *Nitrosomonas europaea* and *Nitrosolobus multi-formis* (Racke, 1990). The ammonia-oxidizing enzyme ammonia monooxygenase is capable of cometabolizing these aliphatic pesticides. As discussed in Chaps. 3 and 4, ammonia is required as a reductant for this cometabolism. Depletion of ammonia stops the degradation. These aerobic dehalogenations result in the production of ethylene, ethylene oxide, and formaldehyde.

Microbial consortiums may be necessary for the degradation of many of these complex compounds. For one degradation pathway, two microorganisms are responsible for the degradation of Dalapon (2,2-dichloropropionic acid) (Slater, 1984). The two-member consortium consisted of *Pseudomonas putida* and *Flavobacterium* sp.

The anaerobic dehalogenation of aliphatic pesticides has been established for some time. These degradation reactions are typical of those discussed in Chap. 4. Reductive dehalogenation studies at several pH ranges illustrate that pH levels of 7.5 and 8.0 resulted in the most rapid dehalogenation (Castro, 1968).

Chlorinated cyclic aliphatic pesticides

As discussed in Chap. 4, the microbial degradation of cycloalkanes varies dramatically with the type of substitution. The saturated, highly chlorinated compounds are extremely persistent under aerobic conditions. Those containing chlorine represent a major class of pesticides. Lindane (γ-hexachlorocyclohexane, γ-HCH) is one example of these chlorinated cyclic-aliphatic compounds. Lindane has been used extensively since 1940 as an insecticide for livestock. It is known to persist in soils up to 11 years (MacRae, 1969).

Lindane has been reported to be degraded by fungi. The white rot fungus *Phanerochaete chrysosporium* has been demonstrated to mineralize lindane (Kennedy, 1990). The chlorinated cyclic-aliphatic pesticides are subject to anaerobic dehalogenation by a variety of anaerobic microorganisms (Marks, 1989). These include α, β, and δ isomers of hexachlorocyclohexane (MacRae, 1969). The anaerobic metabolism of lindane has been demonstrated with *Clostridium rectum, Clostridium butyricum, Clostridium pasteurianum, Pseudomonas putida, Clostridium sporogenies, Bacillus coli,* and *Citrobacter freundii* (Jagnow, 1977; Ohisa, 1980; Kearney, 1972). The first steps of degradation are dehydrogenation and dehydrochlorination.

Lindane has been demonstrated to degrade under anaerobic conditions by reductive dehydrodechlorination (Kennedy, 1990). However, chlorobenzene and α-hexachlorocyclohexane can result from lindane degradation. The α-, β-, and δ-HCH isomers are dechlorinated more slowly than γ-HCH (Jagnow, 1977). Although the best dechlorination was under strict anaerobic conditions, dechlorination was also achieved by aerobically grown *C. freundii* and facultative bacteria when subjected to anaerobic conditions. The facultative anaerobic bacteria included *Bacillaceae* species that are abundant in soil.

Cometabolism may be necessary for degradation (Ohisa, 1980). As frequently found with anaerobic dehalogenation, the addition of alanine, leucine, pyruvate, leucine-proline mixtures, formate, and glucose was necessary for lindane degradation. These substrates function as electron donors for lindane reduction. The proposed pathway for lindane degradation yields a chlorinated aromatic hydrocarbon that is then mineralized, as discussed in Chap. 4 (Fig. 5.6).

Pentachlorophenol and DDT

The degradation of the halogenated aromatic pesticides under aerobic conditions is a function of the degree of chlorination. The higher the degree of chlorination the more resistant they become to aerobic degradation. However, many of these compounds are degraded under anaerobic conditions. As discussed in Chap. 4, anaerobic dehalogenation of iodo-, bromo-, fluoro-, and chloroaromatic compounds is well documented. Two widely utilized chlorinated aromatics are pentachlorophenol and 1,1,1-trichloro-2,2-bis(*p*-chlorophenyl) ethene (DDT). DDT is also referenced by its abbreviated name dichlorodiphenyl-trichloroethane. Both compounds can be found in animal tissue and waters worldwide.

Of the chlorinated aromatic pesticides, pentachlorophenol (PCP) has been the most widely distributed in the United States and has had the most varied uses. The U.S. Environmental Protection Agency had registered over 500 products containing PCP as the active agent. Of the PCP production, 80 percent was used for the preserving of wood (Saber, 1985). PCP was frequently carried in a solvent such as mineral spirits or diesel fuel. Significant PCP accumulation in the food chain has resulted from its years of use. Tissue of human populations from industrialized societies is reported to have averaged 10 to 20 ppb (Saber, 1985).

Pentachlorophenol readily degrades in the environment by both chemical and biological processes. Bacteria and fungi are active degraders of PCP. Microbial species include the bacteria of *Pseudomonas* sp. and *Flavobacterium* sp. and fungi of *Phanerochaete chrysosporium, P. sordida, Trichoderma* spp., and *Trametes versicolor* (Kaufman, 1977; Lamar, 1990; Glaser, 1988).

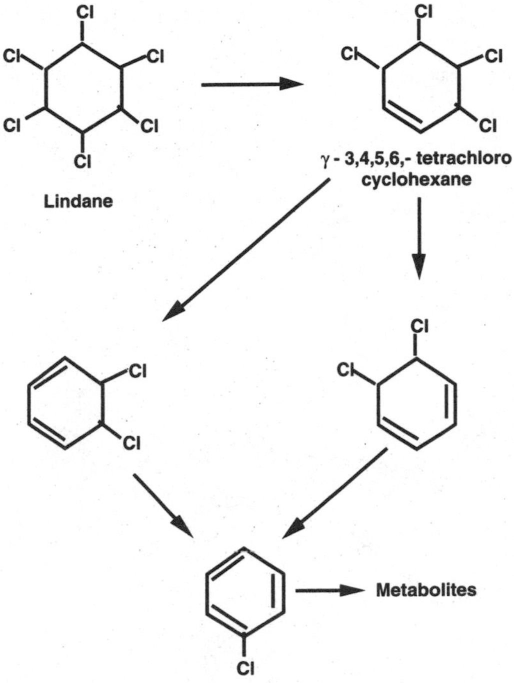

Lindane

γ - 3,4,5,6,- tetrachloro cyclohexane

Metabolites

Figure 5.6 Anaerobic degradation of lindane. (*Adopted from Mac Rae, 1969.*)

Bacterial degradation of PCP is more rapid in soils containing higher organic content and moisture. Anaerobic metabolism affects a higher rate of degradation than aerobic modes. Reductive dehalogenation is the most significant pathway. Numerous mono-, di-, tri-, and tetrachlorinated phenol transformation products occur. Ultimately PCP degradation leads to ring cleavage (Fig. 5.7).

Pentachlorophenol is rapidly mineralized to carbon dioxide by *P. chrysosporium,* although the transformation product of 1,4-tetrachlorobenzoquinone is also produced. Pentachloroanisole (PCA) is a major by-product of PCP transformation in soil by both *P. chrysosporium* and *P. sordida* (Lamar, 1990). The transformation of PCP in soil is a two-step reaction as based on time lag. The first step yields rapid transformation of PCP and accumulation of PCA. This step represents a 60 to 70 percent conversion of PCP to PCA. In the second step, both PCP and PCA are reduced. The first transformation step for *P. chrysosporium* was significantly faster than that of *P. sordida.* The second step was much slower.

Fungi appear to have significant differences in their sensitivity and growth in the presence of PCP. *Phanerochaete* spp. was very sensitive at PCP levels of 5 μg/g. *P. chrysosporium* and *P. sordida* could tolerate 25 μg/g of PCP. A ranking of species growth rate is as follows: *P. chrysosporium* > *P. sordida* > *P. laevis*>*P. chrysorhiza* = *P. san-*

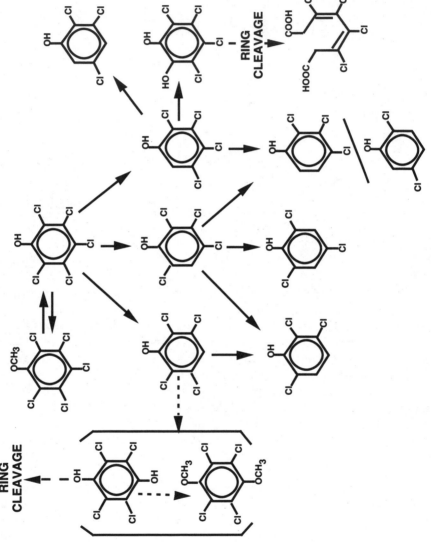

Figure 5.7 Proposed pathway for PCP degradation in soil. (*Adapted fro Kaufman, 1977.*)

guinea>*Inonotus circinatus* = *P. filamentosa* (Lamar, 1990). All strains are capable of growth at temperatures of 14 to 28°C. *P. filamentosa* and *I. circinatus* are inhibited at temperatures 30°C. Successful field bioremediation of PCP has been conducted with both bacteria and fungi in solid-phase and slurry-phase reactors.

Lignin-degrading fungi have been documented to mineralize pentachlorophenol (PCP) through carbon-14 studies. However, not all PCP may be mineralized. Fungi and other organisms can form polymers of pentachlorophenol with humic acid. In addition to polymerization, transformation products such as pentachloroanisole (PCA) have been noted to accumulate. PCA results from the methylation of PCP. Degradation by microbial consortiums containing specific PCP degraders mixed with significant populations of indigenous organisms did not result in the accumulation of PCA. This suggests that a consortium of organisms aids in the complete mineralization.

DDT is a persistent pesticide in the environment. It readily accumulates in fish and animal tissue, entering the human food chain. The degradation of DDT requires an initial anaerobic condition for reductive dehalogenation. Over 300 strains of microorganisms have been shown to convert DDT to DDD [1,1-dichloro-2,2-*bis*(*p*-chlorophenyl)-ethane] (Rochkind, 1986). These include *Pseudomonas* spp., *P. aeruginosa, P.* sp., *Clostridium perfringens, Bacteroides* sp., *Escherichia coli, Enterobacter aerogenes,* and *Proteus vulgaris.*

Under anaerobic methanogenic and sulfur-reducing conditions, some persistent pesticides can be detoxified. By combining metabolism under both anaerobic and aerobic conditions, it is possible to completely mineralize DDT. DDT is metabolized under anaerobic conditions to yield *p*-dichlorodiphenylmethane (Fig. 5.8) (Slater, 1984). The chlorine atom is removed by a series of reductive dehalogenations. *Aerobacter aerogenes* can mediate this series of dehalogenations (Focht, 1987). Ring fission of the metabolites then occurs by the separate action of other microbes. A *Pseudomonas* bacterium and dehalogenating fungus has been confirmed to bring about ring fission. The same *Pseudomonas* bacterium does not metabolize DDT aerobically but reduces it anaerobically to 4,4-dichlorodiphenylmethane which it then can oxidize aerobically to 4-chlorophenylacetic acid. The latter compound is mineralized by another organism, *Arthrobacter.* Although a sequence of anaerobic and aerobic processes is difficult to control in situ, it can be done in surface reactors.

Chlorinated phenoxyalkanoates

Herbicides consisting of phenoxyalkanoate structures are widely used for broad-leaved weed control by homeowners and in agriculture and highway maintenance. They include compounds such as 2,4-D (2,4-

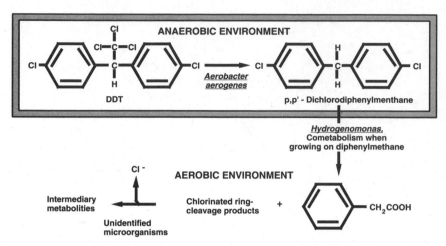

Figure 5.8 Proposed scheme for DDT degradation involving anaerobic and aerobic metabolism modes. (*Slater, 1984.*)

dichlorophenoxyacetic acid) and silvex [2-(2,4,5-trichlorophenoxy) propionic acid].

Most of the chlorinated phenoyalkanoates are degraded by soil microorganisms. However, little degradation is experienced at temperatures below 10°C and at low moisture (Smith, 1990). The major microbial transformations are ring hydroxylation, beta oxidation of any long-chain aliphatic acid unit, and cleavage of the ether linkage and ring structure (Kearney, 1972). The common soil bacteria *Arthrobacter* sp. cleaves the ether linkage of 2,4-D (Fig. 5.9). This oxidative mechanism yields 2,4-dichlorophenol and alanine (Tiedje, 1969). As with the aerobic oxidation of the benzene ring, metabolism of 2,4-D yields a catechol structure as an intermediate. Both 4-chlorocatechol and 3,5-dichorocatechol are reported as oxidation products. These chlorocatechols are degraded by several pathways that result in one or more forms of dichloromuconic acid and chloromaleylacetic acid that degrade to products such as succinic acid.

Almost all chlorinated phenoyalkanoate pesticides tested under anaerobic conditions are subjected to reductive dehalogenation as discussed in Chap. 4 for halogenated aromatic structures. These compounds are attacked by anaerobic organisms, resulting in the removal of chlorine atoms and cleavage of the ether linkage yielding chlorophenols (Figs. 5.10 and 5.11). The chlorophenols are mineralized, as discussed in Chap. 4.

One environmental health concern with phenoxyalkanoate pesticides, particularly 2,4,5-T, is the presence of TCDD (2,3,7,8-tetrachlorodibenzo-*p*-dioxin). TCDD results as an impurity in the chemical production of several chlorinated aromatic formulations.

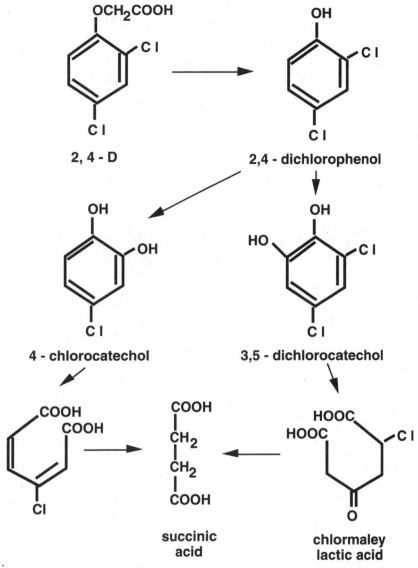

Figure 5.9 Aerobic degradation of 2,4-D. (*Tiedje, 1969.*)

Halogenated aniline-base pesticides

The chlorinated aniline-base pesticides contain a chlorinated aniline moiety C_6H_nNH—R. Many of these herbicides are used for controlling weeds. Although half-life in soil is reported in days, this half-life may involve only minor transformations. The metabolism of this group of pesticides varies from complete mineralization to little transformation.

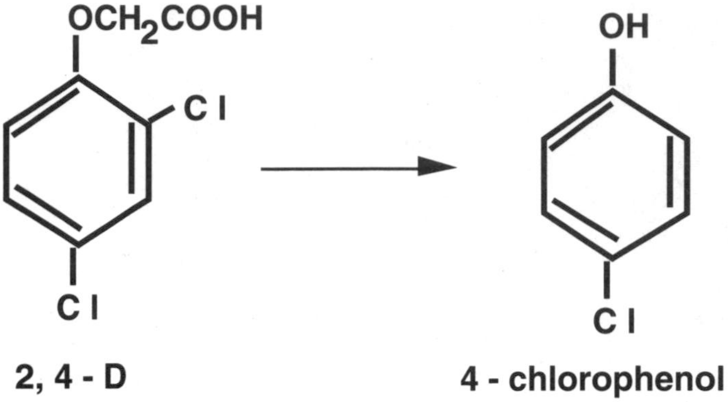

OCH₂COOH

OH

C l

C l

C l

2, 4 - D

4 - chlorophenol

Figure 5.10 Anaerobic dehalogenation of 2,4-D. (*Mikesell, 1985.*)

The two most widely used aniline-base herbicides are alachlor, 2-chloro-2′, 6′-diethyl-*N*-(methoxymethyl) acetanilide, and metolachlor, 2-chloro-2′-ethyl-6′-methyl-*N*-(1-methyl-2-methoxyethyl)-acetanilide. No microbial strains have been reported to mineralize these two herbicides.

Many phenylcarbamates, phenylureas, and acylanilides are transformed by soil microorganisms (Kearney, 1972). Those that degrade significantly yield aniline molecules. Acetanilide herbicides like propanol, Karsil, or the urea and carbamate herbicides where the N atom is a distributed secondary amine can be degraded. For many of these compounds the aniline moiety is destroyed. However, for some acylanilide herbicides the microbial transformation yields aniline condensation products of potential concern (Kearney, 1972). Microorganisms cleave these compounds to form an organic acid and an aniline derivative. The aniline derivative is not mineralized but becomes covalently bound to other soil organic matter. The metabolite 3,3′,4,4′-tetrachloroazobenzene (TCAB) has now been identified as a degradation product for the herbicides propanol, Karsil, and dicryl (Kearney, 1972). The environmental significance of this compound is not clear. Propanol (3′,4′-dichloropropionanilide) is readily metabolized in soil, but an aromatic metabolite, 3,4-dichloroaniline, persists in the soil (Villarreal, 1991). The metabolism of these compounds also is hindered by their high soil-adsorption property.

The acylanilide fungicide metalaxyl can be transformed by an actinomycete and two fungal strains (Zheng, 1989). However, none of these microorganisms could use metalaxyl as a sole carbon source. The metalaxyl transformation by fungal species *Syncephalastrum racemosum* and *Cunninghamella elegans* appear to be a result of cometabolism. The fungal transformations, however, did not destroy the aniline and aromatic moiety.

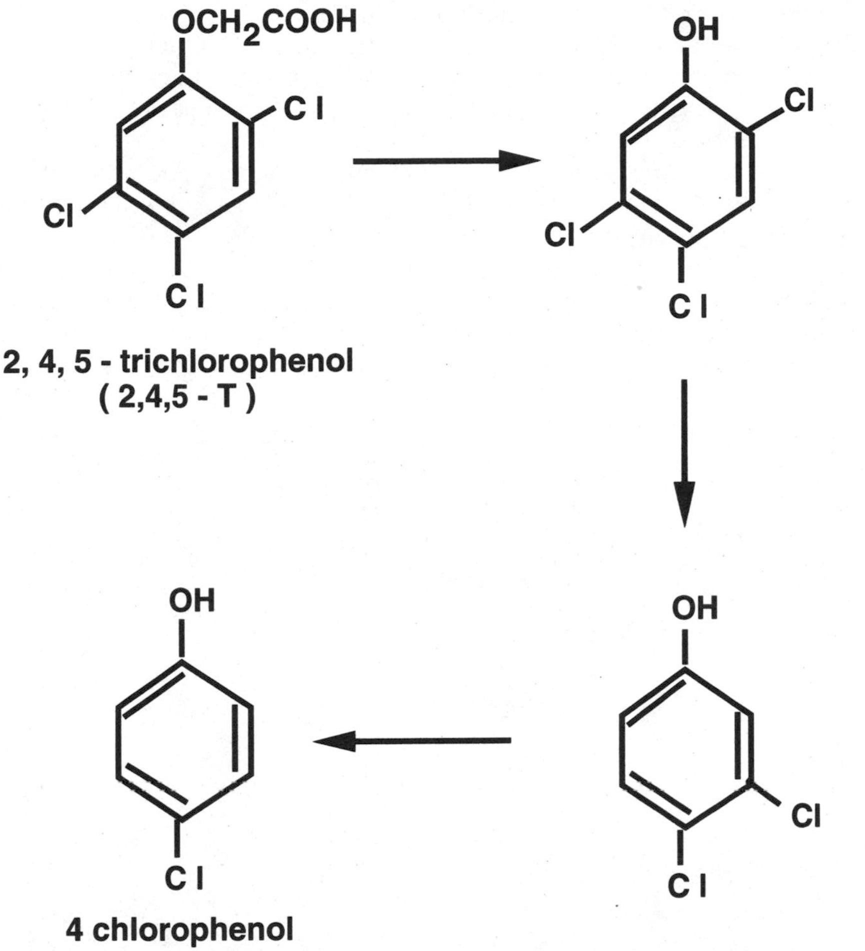

Figure 5.11 Anaerobic dehalogenation of 2,4,5-T. (*Mikesell, 1985.*)

The herbicide metolachlor [2-chloro-N-(2-ethyl-6-methylphenyl)-N-(2-methoxy-1-methylethyl)-acetamide] appears to be even more stable than metalaxyl, and the related herbicides of alachlor [2-chloro-2′,6′-diethyl-N-(methoxymethyl)-acetanilide], butachlor [2-chloro-2′,6′-diethyl-N-(butoxymethyl)-acetanilide], propachlor (2-chloro-N-isopropylacetanilide), and Ramrod (2-chloro-isopropylacetanilide). Although some microbial transformations occur, no mineralization of any of these herbicides has been observed by various investigators (Saxena, 1987).

The resistance of these pesticides to microbial mineralization has been attributed to multiple factors. They include the presence of the $2^1,6^1$-dialkyl substituents on the aniline ring, the presence of N-alkyl groups, stearic hindrance resulting from the alkyl substituents at the 2,6 position, and presence of trisubstituted nitrogen atom such as tri-fluralin (α, α, α-trifluoro-2,6-dinitro-N,N-dipropyl-p-toluidine) (Saxena, 1987). At present, it is clear that these pesticides are not candidates for bioremediation.

Microbial consortiums will mineralize propachlor. The mineralization of propachlor has recently been reported by a microorganism resembling species of either the *Moraxella* or *Xanthobacter* genera (Villarreal, 1991). It appears that this organism can use propachlor as a sole source of carbon and energy. Catechol and a chloramide are intermediates (Fig. 5.12). This organism was obtained from a site contaminated with several herbicides including propachlor.

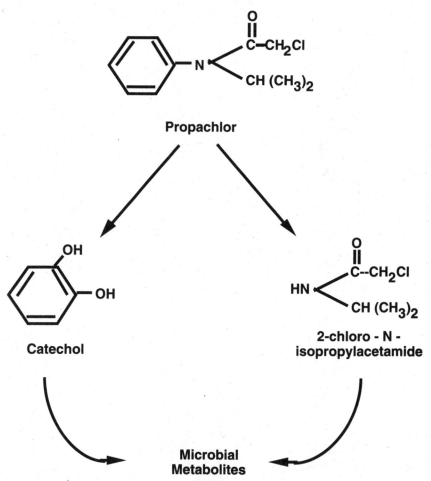

Figure 5.12 Proposed degradation of propachlor. (*Villarreal, 1991.*)

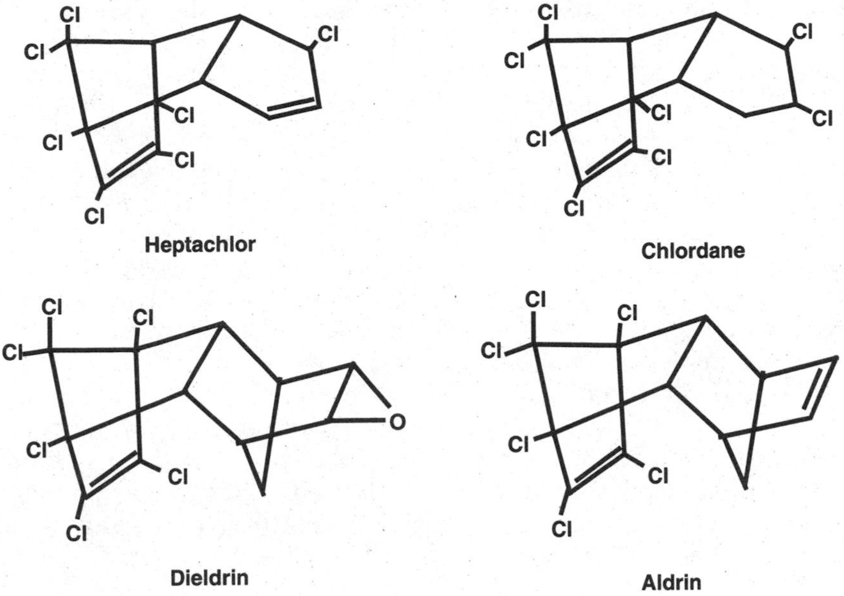

Figure 5.13 Cyclodiene pesticides.

Cyclodiene pesticides

The cyclodiene insecticides are persistent pesticides and have been extensively used. Although their use has been minimized in recent years, they are still used for termite control and similar applications. These compounds are structurally similar. Heptachlor and chlordane are identical except that the double bond between carbons in heptachlor has been replaced with a single bond and the addition of another chlorine atom (Fig. 5.13). Dieldrin is similar to aldrin except that dieldrin has an oxygen atom to form an epoxide.

Aldrin is oxidized to dieldrin (Kearney, 1972). Cultures of *Pseudomonas* sp. metabolize dieldrin, yielding many metabolites, including aldrin and a ketone derived by rearrangement of the epoxide ring, an aldehyde, and an acid. For all these transformations, the chlorinated structure persists. Endrin is similarly metabolized, yielding ketones and aldehydes with five to six chlorine atoms.

Heptachlor is oxidized to heptachlor epoxide and chemically hydrolyzed to hydroxychlordane. Microbial dehalogenation of heptachlor produces chlordane, which can undergo epoxidation to yield chlordane epoxide. For many of these cyclodiene pesticides, degradation does not result in detoxification of the compound. The metabolite is similar in its toxic action to its precursor.

Some of these compounds, aldrin, dieldrin, and heptachlor, may be degraded by cometabolism (Alexander, 1977). Cometabolism has been demonstrated to be important for the transformation of the pesticide Amitraz [1,5-di-(2,4-dimethylphenyl)-3-methyl-1,3,5-triazapenta-1,4-diene] (Slater, 1984). Degradation of some cyclodiene pesticides is more likely under anaerobic than under aerobic conditions.

Mirex has been used for the control of fire ants in the southern United States. Although mirex has been banned, massive spraying of farmland and pasture fields occurred for two decades. Studies have been performed on soil samples that contained many years of mirex contamination in search of an aerobic and anaerobic consortium that would degrade the compound. Significant degradation has not been found with any microbial system (Aslanzadeh, 1985; Kennedy, 1990).

Recent studies utilizing fungus have demonstrated some progress on the degradation of the cyclodiene insecticides. The white rot fungus *Phanerochaete chrysosporium* was capable of extensive degradation of chlordane. Mineralization appears as the potential end result as measured by carbon dioxide yield (Kennedy, 1990). Unfortunately, heptachlor, dieldrin, aldrin, and mirex were not degraded to any measurable degree by the fungus.

Nitrogen pesticides: carbamate, carbamothioate, and dicarboximide

Carbamate pesticides have been developed as an alternate to the more stable halogenated pesticides. One example is aldicarb, which has been utilized as an insecticide and nematicide for agriculture crops. Aldicarb degrades relatively fast (Lightfoot, 1987). The degradation of aldicarb involves both biotic and abiotic reactions. Heterogeneous catalysis and chemical hydrolysis are influenced by soil composition, pH, and temperature (Lightfoot, 1987).

The first step in aerobic microbial degradation is the oxidation of aldicarb to aldicarb sulfoxide and then to aldicarb sulfone. These products are hydrolyzed to form noncarbamate oximes and nitriles (Lightfoot, 1987). Under anaerobic conditions the degradation rate of aldicarb is significantly faster than that for aerobic conditions (Smelt, 1983; Bromilow, 1986).

Carbofuran (2,3-dihydro-2,2-dimethyl-7-benzofuranyl-N-methyl carbamate) has been widely used in agriculture. It is chemically hydrolyzed under alkaline conditions and can be biologically degraded. *Achromobacter* sp. has been reported to hydrolyze carbofuran within 42 h under aerobic conditions as a sole source of nitrogen when glucose was added as a carbon source (Ramanand, 1988). *Achromobacter* sp. uses carbofuran as its sole nitrogen source (Racke, 1990). Degradation proceeds by primary hydrolysis at the carbamate linkage, leading to the intermediates of carbofuran phenol and methyl isocyanate. Soil-containing mixed cultures were capable of mineralizing the hydrolyzed carbofuran. Under anaerobic conditions, no appreciable degradation of carbofuran was achieved.

Carbendazin (MBC) degradation has been attributed to fungi and bacterial populations. Identification of fungi found 80 percent of the fungal isolates to be *Alternaria alternata, Bipolaris, Ulocladium* sp., or *Acremonium faliforme* (Aharonson, 1990). Although fungi are capable of carbendazim degradation, bacteria were considered the major responsible organism. The pesticide has been degraded within 12 days in laboratory cultures.

The carbamothioate herbicides are readily degraded by microorganisms. Common soil organisms of *Flavobacterium* sp. and *Methylomonas* sp. are active in the degradation (Skipper, 1990). These compounds include EPTC (S-ethyl dipropylcarbamothioate), vernolate, pebulate, and cycloate.

The dicarboximide fungicides undergo varied microbial transformations. Iprodione and vinclozolin degrade more readily than myclozolin and procymidone. Transformation products include 3,5-dichloroaniline. Transformation was very pH-dependent with no degradation below pH 5.5. Optimum degradation was at pH 6.5. Above pH 7.5, abiotic chemical hydrolysis became significant in the degradation.

Fungi have been identified in the degradation of diphenamid (dimethyl-2,2-diphenyl acetamide). The more efficient fungal degraders were identified as *Fusarium, Aspergillus,* and *Penicillium* sp. (Aharonson, 1990). However, bacteria are considered as the primary degraders in soil. The rate of degradation by fungi is much slower, requiring months versus days for bacteria.

The acylanilide herbicides vary in their degree of biodegradability. Alachlor and metolachlor are reported as recalcitrant to mineralization. The alkyl substitutions at the ortho positions of the aromatic ring are believed to prevent enzyme attach (Villarreal, 1991). Other acylanilide herbicides are cleaved to form organic acids and an aniline derivative. The aromatic portion can be incorporated into the soil humic structure. The herbicide propachlor (2-chloro-N-isopropyl-acetanilide) is degraded by a microorganism pair belonging to the genera *Moraxella* and *Xanthobacter* (Villarreal, 1991).

Pyrethroid pesticides

The pyrethroid pesticides are replacing the more toxic chemicals as a result of their low mammalian toxicity. They represent over 25 percent of the world's insecticide applications. These pesticides include permethrin, deltamethrin, Fastac, fenvalerate, and fluvalinate.

Pyrethroid pesticides are reported to degrade under aerobic and anaerobic conditions (Maloney, 1988). Permethrin is the most rapidly degraded pyrethroid and deltamethrin the least. Transformation

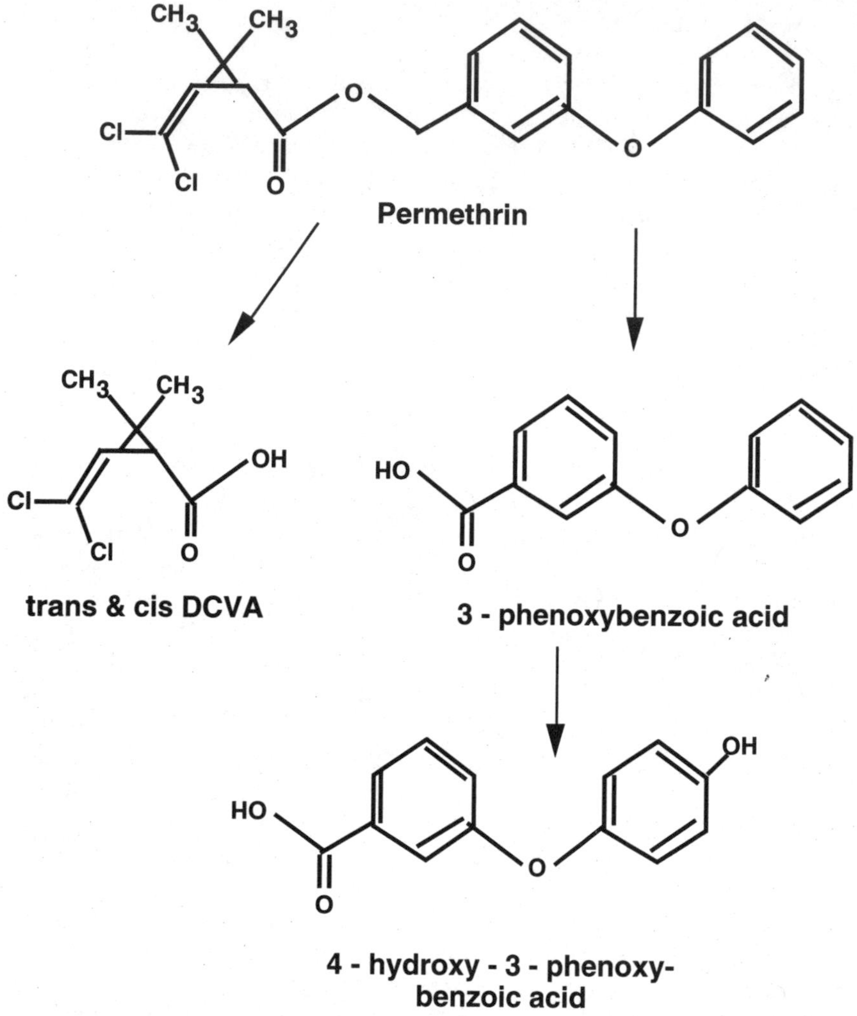

Figure 5.14 Aerobic degradation of permethrin. (*Maloney, 1988.*)

products included 3-phenoxybenzoic acid, and 4-hydroxy-3-phenoxy-benzoic acid (Fig. 5.14) (Maloney, 1988). Organisms responsible for these transformations were identified as *Pseudomonas fluorescens, Achromobacter* sp., and *Bacillus cereus.*

Organophosphonate pesticides

Many stable carbon-to-phosphorus bond compounds are used as pesticides, flame retardants, and antibiotics. Although the organophosphate pesticides have a high acute toxicity, they are rapidly metabo-

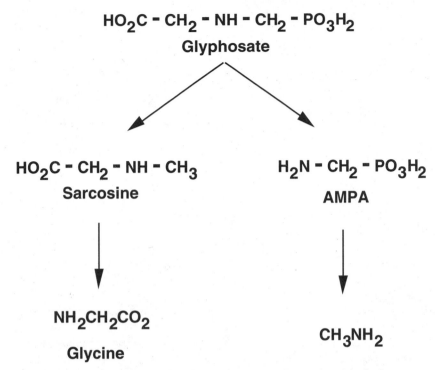

Figure 5.15 Glyphosate degradation by bacteria: *Pseudomonas* sp., *Arthrobacter* sp., *Flavobacterium* sp., and *A. atrocyaneus*. (*Pipke, 1988 and Jacob, 1988.*)

lized. A number of bacteria have been found that degrade the phosphonates (Jacob, 1988). Mixed soil cultures and pure bacterial cultures have been demonstrated to degrade the herbicide glyphosate [N-(phosphonomethyl) glycine] (Pipke, 1988).

The initial step in organophosphate metabolism is the cleavage of the organophosphate bond. Bacteria degrade glyphosate by two pathways, leading to the intermediate production of either glycine or aminomethylphosphonate (AMPA) (Fig. 5.15) (Jacob, 1988). In glycine production, the C—P bond of glyphosate is cleaved. For the AMPA pathway, glyphosate is degraded by cleaving its carboxymethyl carbon-nitrogen bond.

Parathion (*o,o*-diethyl-*o*-*p*-nitrophenyl phosphorothioate) has been extensively used as an insecticide. This organophosphate insecticide is extremely toxic. Degradation of parathion appears to require cometabolism and a mixed microbial consortium (Slater, 1984). At minimum, four microbes, none of which could grow individually on parathion, are necessary for its degradation. One of the organisms in this consortium was capable of hydrolyzing parathion to a diethyl thiophosphate and *p*-nitrophenol but could not use either of these compounds to support

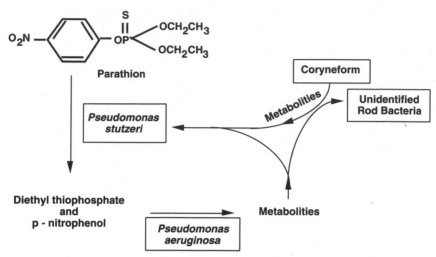

Figure 5.16 Microbial consortium necessary for the degradation of parathion. (*Maloney, 1988.*)

growth. A second organism that could not perform this hydrolyzing mechanism on parathion could grow on the *p*-nitrophenol. The joint activities of *Pseudomonas stutzeri* and *Pseudomonas aeruginosa* degraded a portion of the parathion (Fig. 5.16). Enhancement in biodegradation by bioaugmentation with specific organisms has been documented for the degradation of parathion (Brunner, 1985).

The insecticide isofenphos is mineralized by several organisms, including *Arthrobacter* sp. and *Pseudomonas* sp. Arthrobacter can use isofenphos or its hydrolysis products, isopropyl salicylate and salicylate, as sole carbon sources (Racke, 1990a).

Polychlorinated Biphenyls

Polychlorinated biphenyls (PCBs) were manufactured between 1929 and 1978 for a variety of commercial purposes. Uses included glues, plasticizers, hydraulic fluids, heat transfer fluids, and carbonless copy paper. Although PCB use in the United States is now banned, industrial production was extensive during the last six decades. It is estimated that of the 600,000,000 kg manufactured, 15 percent entered the ecological system. PCBs are known to bioaccumulate with measurable levels in almost all mammalian systems (Brunner, 1985).

Chemical nature

Polychlorinated biphenyls are also known under their trade name Aroclor. A variety of Aroclors exist and differ in the percentage of chlorine by weight. Most were designated by numbers that represent the number of carbon atoms and a percentage of chlorine by weight. For example, Aroclor 1260 contains 12 carbon atoms and is 60 percent by weight chlorine.

The chlorine can be substituted at a number of positions on the biphenyl structure. For example, Aroclor 1232 is a mixture of the following congeners (percent by weight) (Hogan, 1988):

	Percent
Biphenyls	0.1
mono-cl	31
di-cl	24
tri-cl	28
tetra-cl	12
penta-cl	4
hepta-cl	0.1

The mono-, di-, and tri-chloro compounds make up 83 percent of the mixture. Thus each Aroclor is a complex mixture of chlorinated biphenyl molecules. Theoretically, 209 different chlorinated biphenyls (congeners) are possible (Quensen, 1990). Actual congeners of PCBs are probably represented by more than 100 forms. PCBs are very insoluble in water and are generally found in soil and sediments at sites of their release. For example, the solubility of Aroclor 1260 is only 2.7 ppb. This insolubility contributes to its slow biodegradation in nature. As with many of the chlorinated aromatic compounds, degradation is also hindered by the need for specific organisms and/or metabolic modes.

Degradation potential

Both aerobic and anaerobic metabolism modes affect some biotransformation of PCBs. However, PCBs chlorinated with four or more chlorine atoms are resistant under aerobic conditions. No single organism is responsible for the degradation of multiple chlorinate PCBs. Degradation is achieved through a consortium of select organisms. The cometabolic transformation steps of 4,4'-dichlorobiphenyl (4,4-DCBP) to 4-chlorobenzoate by two species of *Acinetobacter* sp. has been well established (Adriaens, 1989). The proposed degradation pathway is illustrated in Fig. 5.17. The first step of the pathway is a hydrolytic ring dehalogenation.

The aerobic biodegradation of PCBs appears to be limited to congeners with five or fewer chlorine atoms and two adjacent unsubstituted carbon atoms (Quensen, 1990). Aerobically, many of the higher chlorinated PCBs are only hydroxylated and not dehalogenated. The biode-

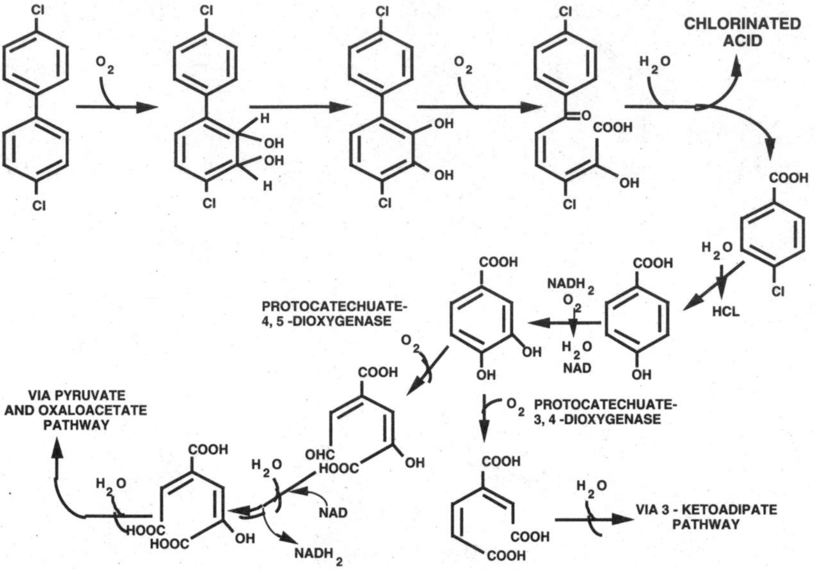

Figure 5.17 Cometabolism degradation of 4,4'-dichlorobiphenyl (4,4' DCBP). (*Adriaen, 1989.*)

gradation potential decreases with the degree of chlorine substitution. No confirmation on the aerobic degradation of Aroclor 1260 exists.

Analog substrate enrichments have produced varied results for PCB degradation. The addition of biphenyl as an analog substrate had significant effect on the degradation of Aroclor 1242 (Fig. 5.18) (Brunner, 1985). Although analog enrichment was the largest single factor for increasing biodegradation of PCBs, this was not the only factor of significance. Natural soil microflora may not provide the necessary microflora to degrade PCBs. The addition of *Acinetobacter* with the biphenyl substrate reduced the adaptation period compared with noninoculated soil (Fig. 5.18).

Analog enrichment, however, has not yielded positive results in studies with Aroclor 1254 (Rhee, 1993). The degradation of PCBs by Hudson River microorganisms illustrated that the sediments without biphenyl enrichment dechlorinated faster and more extensively than the biphenyl-enriched sediments. However, this difference was not significant during the first 12 months of degradation. The dechlorination followed first-order kinetics with rate constants of -0.016 and -0.011 Cl/month for the nonenriched and biphenyl enriched systems, respectively.

Of greater significance was that the decrease in higher chlorinated congeners yielded an increase in lower chlorinated congeners. Although the congener composition changed, no significant decrease

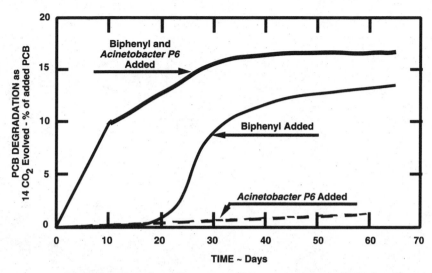

Figure 5.18   The effect of analog substrates and bioaugmentation on PCB Arochlor 1242 degradation (100 mg/kg of PCB). (*Brunner, 1985.*)

in the total molar concentration of chlorobiphenyls occurred (Rhee, 1993). A threshold concentration may exist below which no dechlorination occurs.

Since *Acinetobacter* cannot dehalogenate any of the products that it forms from cometabolism of PCBs, further degradation of these compounds is carried out by other soil microflora. A diverse group of organisms has been identified as participating in PCB's degradation (Table 5.4). PCBs are cometabolized to chlorinated benzoates and

TABLE 5.4   PCB-Degradation Organisms

Pseudomonas putida
Pseudomonas cepacia
Pseudomonas testosteroni
Pseudomonas sp.
Pseudomonas aeruginosa
Acinetobacter sp.
Alcaligenes odorans
Alcaligenes faecalis
Alcaligenes eutrophus
Alcaligenes denitrificans
Arthrobacter sp.
Corynebacterium sp.

SOURCE: Unterman, 1988; Hernandez, 1991; Kohler, 1988; Kohler, 1988a; Clark, 1979.

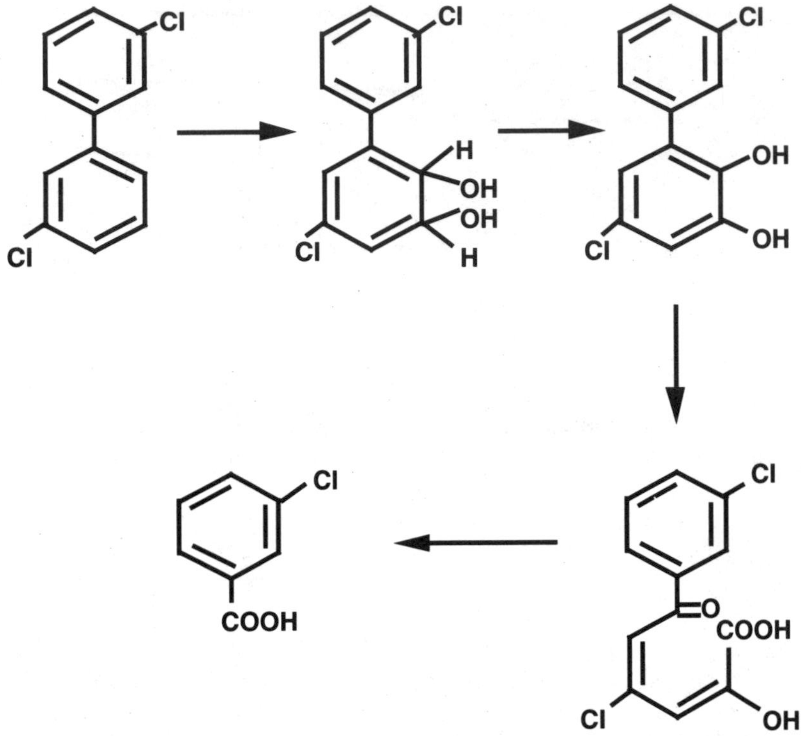

Figure 5.19 Aerobic degradation of polychlorinated biphenyl by *Alcaligenes* sp. (*Remeke, 1984.*)

dihydrodiols, resulting in a mixture of halogenated benzoates and dihydrodiols.

The complete degradation of PCB is not simple. Various microbial strains have distinct congener preferences. For example, *A. eutrophus* and *P. putida* strains prefer congeners with substitutions at the 2,5 position, whereas *Corynebacterium* is capable of handling congeners substituted in both para positions (Unterman, 1988). Two proposed oxidations for PCBs by *Alcaligenes* sp., *Corynebacterium* sp., *A. eutrophus,* and *P. putida* are illustrated in Figs. 5.19 and 5.20. In soils, *Pseudomonas putida* has degraded PCBs at 500 ppm of Aroclor 1242 and at 50 ppm of Aroclor 1254 (Unterman, 1988).

Anaerobic reductive dehalogenation is the only known biodegradation process for the more highly chlorinated PCB mixtures (Morris, 1992). The dehalogenation is usually associated with syntropic communities. Only one obligate anaerobic that is capable of aryl reductive dehalogenation has been isolated (*Desulfomonile tiedjei*) (Morris, 1992).

Anaerobic metabolic modes have a significant advantage over aerobic modes for PCBs. As discussed in Chap. 3, thermodynamics would

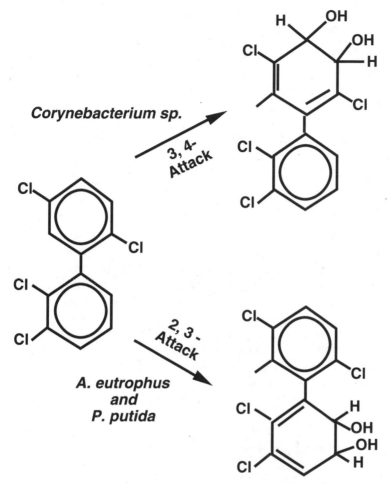

Figure 5.20 The microorganism-specific nature of PCB degradation attack.
(*Unterman, 1988.*)

favor anaerobic dehalogenation of PCBs over aerobic systems. This is
indeed the case. Even dechlorination of the more highly chlorinated
Aroclor 1260 occurs to a significant extent under anaerobic conditions
(Quensen, 1990).

Under anaerobic metabolism the degree of dechlorination decreases
with the number of substituted chlorine atoms. The dechlorination is
influenced by the position of the chlorine atom. For Aroclors 1242 and
1248, the meta and para positions are dechlorinated, but the ortho
position appears stable (Quensen, 1990). At least one microorganism,
DCB-1, appears to obtain energy from the dechlorination of meta-
chlorobenzoate (see Fig. 3.8).

There are different microbial species capable of anaerobic dechlorination of PCBs. The microbial dechlorination potential appears to vary with sites, indicating the potential need for select microbial seeding.

Availability of electron acceptors in anaerobic communities affects reductive dehalogenation by competing with the halogenated compound for reducing potential. For example, sulfate inhibited the dehalogenation of PCBs (Morris, 1992). Sulfate reduced dechlorination activity by 50 percent compared with microbial activity with CO_2 as the electron acceptor. With addition of CO_2 or nitrate, dechlorination rates were the highest (Morris, 1992). Pyruvate or H_2 also will stimulate dechlorination of PCB. Hydrogen has been chosen for use since it eliminated the growth of many pyruvate fermenters. These findings are supported at other sites, but exceptions do exist.

The presence of sediment appears necessary for significant PCB dechlorination (Morris, 1992). The role of sediment in the enhancement of PCB dehalogenation is unknown. It may provide some carbon or growth factor for the microbial community. It also may support immobilized cells. PCB dechlorination by microbial community found in Hudson River sediments included sulfate reducers and methanogens (Morris, 1992).

The most effective biodegradation of PCBs would occur under a sequential anaerobic-aerobic treatment system. The anaerobic step would remove chlorine atoms that limit aerobic degradation, yielding dehalogenated products with significantly less chlorine substitution. These products would then respond more favorably to aerobic degradation. Such a system would affect degradation on more PCB congeners and achieve degradation of those PCBs, such as Aroclor 1260, that cannot be degraded under aerobic systems.

Azo Dyes

Chemical nature

Azo dyes, the largest chemical class of dyes, are used extensively for textile dyeing and paper printing. Their chemical structures are based on azobenzene and the azo naphthol derivatives (Fig. 5.21). Several of the azo dyes are mutagenic and carcinogenic, including 4-phenylazoaniline and N-methyl and N,N-dimethyl-4-phenylazoanilines (Spadaro, 1992).

Degradation potential

The degradation of azo dyes is varied and they seldom degrade under aerobic conditions. The substituted groups of azo, nitro, and sulfo are frequently recalcitrant to bacteria. The only aerobic degradation has been attributed to the fungus *Phanerochaete chrysosporium* and the

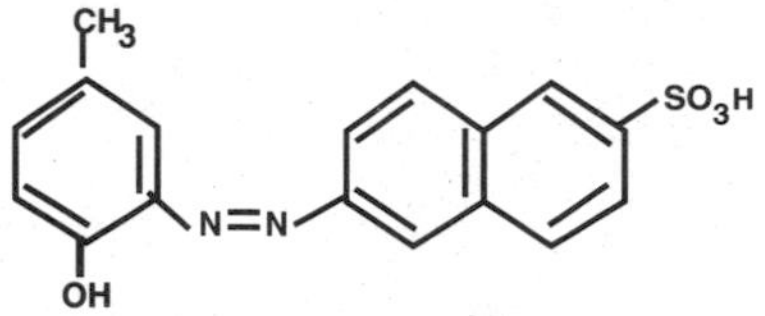

Acid Yellow 21

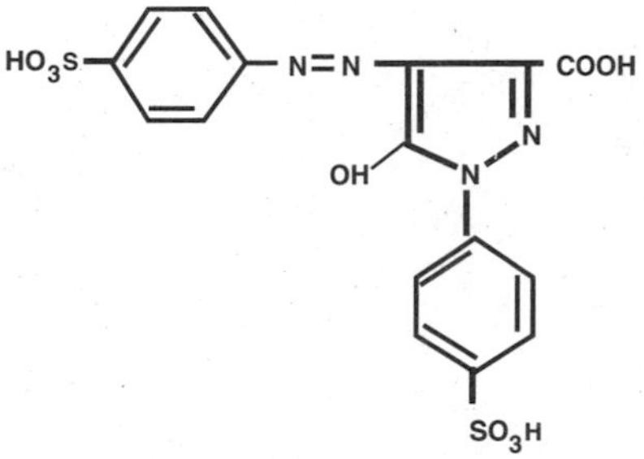

Tartrazine = Acid Yellow 23 = FD&C Yellow 5

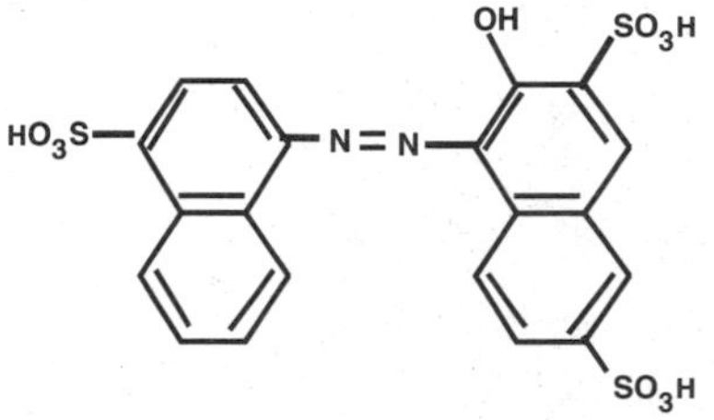

Amaranth = Acid Red 27 = FD&C Red 2

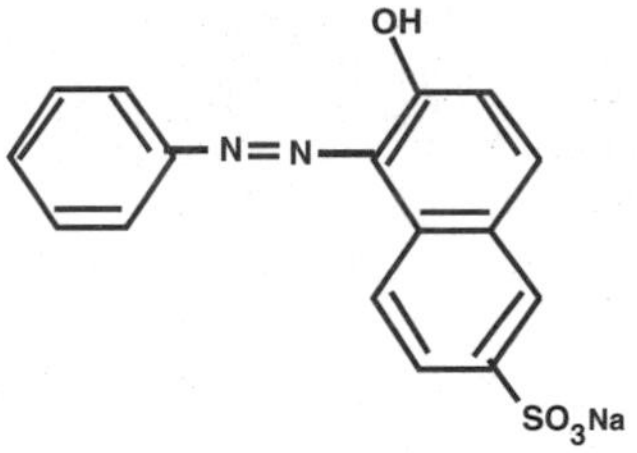

Acid Orange 21

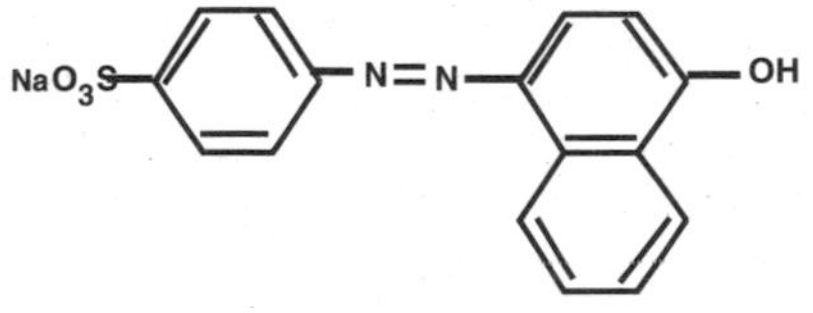

Orange I

Figure 5.21 Typical chemical structure of azo dyes.

action of *Pseudomonas* strains adapted to growth on carboxylated azo dyes (Pasti-Grigsby, 1992). *Phanerochaete chrysosporium* is reported to degrade three azo dyes, congo red, orange II, and tropaeolin O (Cripps, 1990).

Under anaerobic conditions, many bacteria can reduce azo compounds to amines through cometabolism (Haug, 1991). These microorganisms include *Clostridium* spp., *Bacillus subtilis, Streptococcus* spp., *Enterococcus* spp., and *Proteus* spp. (Pasti-Grigsby, 1992). Anaerobic transformation begins with deductive fission of the 920 linkage, resulting in aromatic amines. These aromatic amines, however, can be highly toxic and carcinogenic. They accumulate with little or no further degradation.

One of the first reported mineralizations of azo dye is through a combined metabolism mode of anaerobic-aerobic sequence (Haug, 1991). The sulfonated azo dye mordant yellow 3 (MY 3) has been mineralized by this dual mode. The bacterial degradation is initiated by

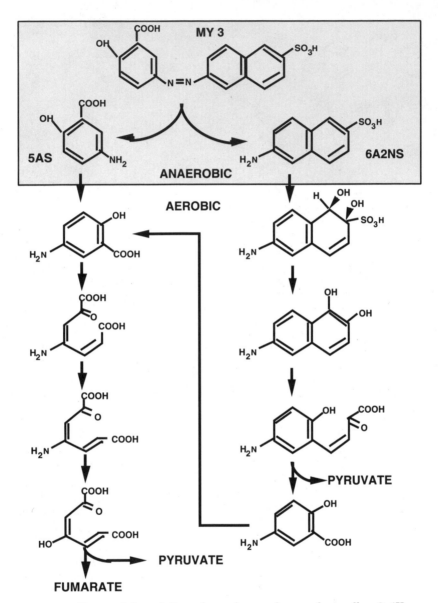

Figure 5.22 Proposed degradation scheme for azo dye mordant yellow 3. (*Haug, 1991.*)

the reduction of the azo linkage yielding aromatic amines. This anaerobic-aerobic microbial system consists of facultative organisms since the alternating presence and absence of oxygen causes no toxic effects. Two organisms act synergistically for mordant yellow dye degradation (Fig. 5.22). After reduction of the azo linkage under anaerobic conditions, forming 6-aminonaphthalene-2-sulfonic acid (6A2NS), two organisms under aerobic metabolism use 6A2NS as the sole source of carbon and energy (Haug, 1991). At present this is one of the few degradation pathways that has been delineated for the complete mineralization of an azo dye. This anaerobic-aerobic microbial consortium was also capable of mineralizing the azo dyes of amaranth, 4-hydroxyazobenzene-4'-sulfonic acid, and acid yellow 21. Tartrazine, however, could not be significantly transformed.

Reduction of the azo bond, yielding aromatic amines, is enhanced in the presence of glucose. This organic compound is either acting as an electron donor or causing the rapid reduction of oxygen that reduces the redox potential for improved reduction. The addition of 1-butanol caused a threefold enhancement in the degradation of MY3. However, other potential primary substrates (methanol, ethanol, acetone, isopropanol) had no enhancing effect (Haug, 1991).

P. chrysosporium under nitrogen-limiting conditions and high oxygen concentrations can mineralize several azo dyes. Relative rates of mineralization are illustrated in Table 5.5. Aromatic rings substituted with a phenolic amino or acetamide group are degraded to a greater extent (Spadaro, 1992). Enzymes other than lignin peroxidase are involved in the breakdown of these dyes. Initial dye loss may be due to adsorption by the mycelia (Cripps, 1990).

TABLE 5.5   Mineralization of Azo Dyes by P. chrysosporium

Dye	Percent mineralized in 12 days
4-Phenylazophenol	28–38
4 Phenylazo-2-methoxyphenol	21–48
Disperse yellow 3	43
4-Phenylazoaniline	26
N,N-Dimethyl-4-phenylazoaniline	30–46
Disperse orange	40–43
Solvent yellow 14	5–23
Tropaeolin O	No rate data
Orange II	No rate data
Congo red	No rate data

SOURCE: Spadaro, 1992; Cripps, 1990.

Site Characterization
for Bioremediation

Introduction

Any treatment process must be well managed for success. A well-managed process requires process control, and this is achieved only when site evaluations have been properly performed. Site characterization requires an intense investigation, particularly for in situ treatment. Bioremediation is a scientific intense technology that relies on the application of many disciplines (Table 6.1). The integration of each discipline is an important part of a project's initial planning phase.

The project manager orchestrates the necessary inputs from each of these disciplines. This requires the knowledge to place limits on the extensiveness of each while assuring adequate data for successful design and operation. Site characterizations must define the potential biological systems, the site characteristics that impact on biological

TABLE 6.1 Bioremediation
Requires the Integration of
Many Scientific and
Engineering Disciplines

- Biochemistry
- Microbiology
- Geochemistry
- Hydrogeology
- Water chemistry
- Soil science
- Unit operations
- Unit processes
- Materials handling

reactions, and provide information for process control. Site characterization must provide adequate data for designing a facility that provides control of the degradation process. This includes the delivery of necessary chemicals and nutrients at appropriate rates throughout the treatment zone.

First, one must determine the appropriateness of bioremediation. The applicability of bioremediation is established after considering the following:

1. The nature of the contaminants

2. The concentration and distribution of contaminants

3. The site's microbial activity

4. The site's hydraulic characteristics

5. The site's redox potential

6. The soil and water characteristics

7. Availability of nutrients

Site assessments must address these questions and provide data to optimize the desired biochemical transformations. Sufficient data are collected so engineering controls can be designed to maintain optimum conditions. These controls are directed at the principles of containment and mass transfer. Containment is achieved in surface reactors or by hydraulic intervention in the subsurface. Physical barriers may be combined with hydraulic intervention. The principles of materials handling and mass transfer are applied to manipulate and control the environment for enhanced biodegradation.

The importance of proper evaluation of those parameters that impact on process operation cannot be overemphasized. Fully understanding the biological system is of little value if you cannot optimize and maintain that system's performance in the field. Proper engineering is a must, and information for achieving this resides on the data obtained during site characterization.

The effort required to characterize a site will vary with each project and will vary on the same site. The data-collection effort is dependent on a number of variables, including the health and ecological significance of the contamination, the potential for litigation exposure, and even political sensitivity. Most site characterization efforts are determined by the contaminant's location and nature. The amount of data required is also a function of the alternate treatment systems to be evaluated. Certain data are important for all bioremediation processes. However, significant differences exist in data needs for on-site versus in situ treatment. Surface treatment offers wider latitude in process variables that can be manipulated to optimize treatment efficiency. For surface biological treatment, the design and operational controls are easier to engineer and require less site information than design for the heterogeneous geological-chemical domain of the subsurface. The following discussion has divided site characterization for bioremediation into two levels of intensity. One level is directed at the needs for surface biological treatment, and the second level at the needs for in situ bioremediation.

Site Surveys

The nature of the contamination determines the microbial metabolism and the need for primary and secondary energy sources. Contaminants that can be readily used as a primary energy source with or without the presence of oxygen provide for the simpler treatment systems. The application of cometabolism or an analog substrate complicates the engineering design and operation. For most contaminated sites, the initial site assessment has established the nature of the organic or metal contaminants and the extent of contamination and has obtained general information on hydrogeology. These data are usually limited to the information provided in Table 6.2 and can be classified as survey information. Such data provide little information for designing or even evaluating the appropriateness of bioremediation. Ideally, the contamination and its distribution have been characterized by horizontal isoconcentration contours (Fig. 6.1). In some situations, these survey data may not adequately delineate the quantity and location of contaminants. One should always consider information on the nature of the spill, its age and duration, and the physical and chemical properties that influence transport and distribution. The nature of the chemical spill will aid in establishing possible locations for the various contaminants in the subsurface.

TABLE 6.2   Initial Site-Assessment Data

Nature of the contaminants:
  Volatile organic compounds
  Semivolatile organic compounds
  Pesticides and herbicides
  PCBs
  Metals
  Cyanide
  Hydrocarbons
Contaminant concentrations and distribution
General groundwater hydrology:
  Flow direction
  Gradient
  Confined or unconfined
General geological system

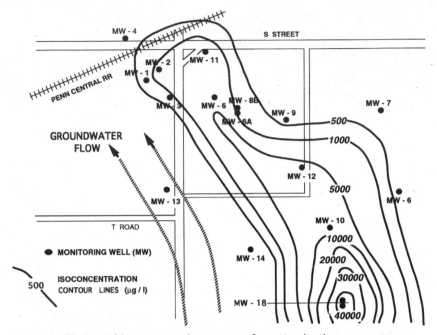

Figure 6.1 Horizontal isoconcentration contours for contamination.

Field characterization mainly relies on soil samples and well-water samples. The selection of samples should be based on existing knowledge of the site and the contaminant distribution properties. Samples should be collected not only from the points of highest contamination but also from the outer fringe of the plume as well as from transition zones between high and moderate concentration levels. The highest concentration point may yield low microbial counts owing to toxicity. However, significant microbial counts may exist within the moderately contaminated area.

Many sampling methods are available, and the best collection technique is usually site- and project-specific. Typical sampling devices are provided in Table 6.3. During sampling, changes in the sample's redox potential, pH, temperature, moisture, etc., should be prevented. Store samples in appropriate sealed containers and hold at 5 to 10°C. Do not freeze samples for microbial counts.

Survey information is obtained from piezometric measurements and existing well data. These data typically include flow direction, gradient, and the general nature of the aquifer, confined or unconfined. Groundwater flow rates may have been adequately established during a site's initial survey. In some cases, however, flow rates are significantly different from those originally estimated. It is typical for a year or more to elapse before the engineering team and regulatory agencies proceed

TABLE 6.3   Field Sampling Devices

Hand-driven	Power-driven
Auger/dry tube cover	Bucket auger
Barrel auger	Cable-tool drill rig
Dutch auger	Continuous flight power
Post-hole auger	auger (hollow-stemmed)
Screw-type auger	Core sampler
Split-spoon sampler	Rotary drill rig
Tube-type sampler	Split-spoon sampler

SOURCE: Guide for Conducting Treatability Studies under CERCLA: Biodegradation Remedy Selection, EPA/540/0-00/000, 1993.

after a site's initial assessment, and in some cases the pollution has moved well beyond the expected contaminant boundaries.

The geological data assembled during a site's initial assessment are usually very limited. Information is usually collected from sources such as the U.S. Geological Survey, state geological surveys, and local governmental agencies. Although these sources can provide good basic information concerning the underlying geology of a site, they cannot provide the data necessary to design and operate a successful bioremediation project.

The first aspect of any bioremediation program is to evaluate the contaminant's potential for biodegradation and delineate the metabolic systems capable of providing degradation. Once the microbial processes for degradation of the target compounds are known, the site is evaluated to determine if these microbial processes are feasible or even operating at a low intensity within the contaminated area. Finally, the site-specific characteristics that are important to bioremediation are defined, and the necessary environmental modifications are delineated to optimize the degradation process.

Site characterization for designing a bioremediation project can be divided according to two purposes. The first gathers information that is necessary for optimizing the biological process. The second gathers information necessary for controlling the environment to maintain the optimum conditions. Thus one phase of the evaluation is directed at data relative to the biological principles of bioremediation and the second phase is relevant to engineering design, process controls, and mass transfer of agents to the contaminated soil or groundwater.

Assessments for Biological Processes

For evaluating the biological process, the site characterization data must yield answers to the following:

- Potential sources of carbon and energy for the microbial organisms
- Electron acceptor availability and the redox condition
- Existing microbial activity and potential toxicity
- Availability of nutrients
- Status of the site's environmental parameters significant to microbial activity

Carbon and energy

Of prime importance is the availability of carbon and energy in the contaminated soil or groundwater. Some contaminants can be used as a primary energy source by bacteria with or without the presence of oxygen. Other chemical constituents require applications of a primary substrate for cometabolism to stimulate the necessary enzyme production for degradation of the target compounds.

A literature review on the biodegradation of compounds present at a site will usually provide information to determine the suitability of bioremediation. This review must include information on the environmental conditions under which these transformations take place. Chapters 4 and 5 provide this type of information for many compounds frequently found in contaminated sites.

The organic contaminants may be available at high enough concentrations to supply the bacteria with all the carbon and energy they need for growth. Some compounds may not supply sufficient energy for microbial growth, and others provide no energy. These conditions require the supplementation with a carbon and energy source.

Various parameters have been used to determine the amount of carbon and energy available to bacterial populations. Some methods include:

- Summation of the available carbon in the individually measured contaminants
- Total organic carbon (TOC) measurement
- Dissolved organic carbon (DOC) measurement
- Biochemical oxygen demand (BOD) measurement
- Modified oxygen uptake test
- In situ respiration test

Data collected from soil and well samples on contaminant concentrations can be used to establish the quantity of organic pollutants in various sections of the site and groundwater plume. However, the estimate on total contaminant load is only as good as the site data.

This is a function of the heterogeneity of a site and the number of observation points for characterization. Other measurements such as total organic carbon (TOC) and biochemical oxygen demand (BOD) can provide a comparative check on the estimated contaminant load.

Measurements of BOD and TOC and respiration tests have the added value of establishing the amount of organic material that may be present other than the contaminants of concern. Microorganisms will degrade both target organic compounds and other natural organics existing in either groundwater or soil. For this reason one must establish the available organic content through either measurement of biochemical oxygen demand, total organic carbon, or respiration studies. The total amount of organic matter subjected to bioremediation is considered when determining the amount of nutrients and electron acceptor that must be delivered to the site. Respiration studies or a modified biochemical oxygen demand provide a better representation of available organic matter than total organic carbon. Approximately 2 to 3 percent of the organic carbon in most soil can be associated with biomass (Turco, 1990). This cellular mass is a complex structure that is mainly available to the higher microbial forms such as protozoa. Thus it is usually of minor consequence to the oxygen needs for bacterial respiration. Another significant organic source is stable humic material that is not readily available to bacteria.

A frequently useful test is a modification of the standard BOD test for wastewater. Known weights of contaminated soil and/or groundwater are placed in BOD dilution water and oxygen uptake is monitored over weeks. The BOD bottle is reaerated as needed to maintain a rear saturated dissolved oxygen level. The total oxygen demand and utilization rate are determined after several weeks of monitoring has established little additional uptake. Commercial respiration equipment is available that provides improved control and monitoring of both oxygen utilization and carbon dioxide production. Computer-interfaced respiration units can collect data over weeks or months for cumulative oxygen uptake.

A test that has found successful field application in the unsaturated zone is the in situ respiration test (Hinchee and Ong, 1992). This test is gaining acceptance in the design of bioventing systems (bioremediation through air extraction or air injection). The respiration test consists of ventilating the contaminated soil with air and periodically monitoring the depletion of oxygen and production of carbon dioxide over time after the air is turned off (Fig. 6.2).

Typically one to four gas monitoring points are placed 5 to 7 m apart in the contaminated soil. A monitor control, noncontaminated soil, is used to establish the natural background respiration and to account for any inorganic sources of carbon dioxide. The monitoring

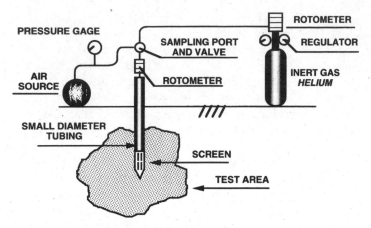

Figure 6.2 Schematic setup of equipment for the in situ respiration test. (*Hinchee and Ong, 1992.*)

tubes can be slotted stainless steel or nylon tubing with attached screens similar to gravel-packed aquarium tank aerators.

Air and an inert gas (helium) is injected for 24 h to provide oxygen to the soil. Injection rates range from 1.7 to 2.8 m³/h (60 to 100 ft³/h) of air with helium as 1 to 2 percent of the mixture. The air injection is turned off and the soil gas monitored for oxygen, carbon dioxide, total hydrocarbon, and helium over time. Before reading, the monitoring point is purged for several minutes until the readings remain constant. The test is usually completed when oxygen in the soil decreases to 5 percent (4 to 5 days). A relatively constant helium concentration is interpreted as evidence that the gas sampled was the gas injected. A significant drop in helium indicates that the data are probably useless.

Typical oxygen-utilization rates are illustrated for several sites in Fig. 6.3. In situ biodegradation rates as based on oxygen utilization range from 0.4 to 19 mg of hydrocarbon per kg of soil per day (Hinchee and Ong, 1992).

Electron acceptor

Site characterization should establish the existing electron acceptors within the site. The electron acceptor establishes the most favorable reaction and therefore the microbial metabolism mode for degradation. Obtaining the desired reactions is achieved by controlling the electron acceptor. For many bioremediation projects, oxygen is the electron acceptor of choice. However, some compounds degrade more favorably under one of the anaerobic modes. Controlling one or more of these specific electron acceptors is particularly important for the

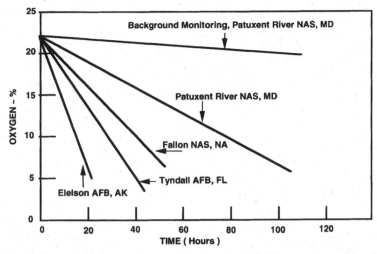

Figure 6.3 In situ oxygen utilization at several field sites contaminated with hydrocarbon fuels. (*Hinchee and Ong, 1992.*)

dehalogenation of chlorinated organic compounds. The primary inorganic electron acceptors are:

- Oxygen (O_2)
- Nitrate (NO_3^-)
- Sulfate $(SO_4^=)$
- Carbon dioxide (CO_2)

Common soil minerals such as manganese (IV) and iron (III) oxides also may serve as alternative electron acceptors. The importance of such alternative electron acceptors depends primarily on their availability.

An indication of the available electron acceptors within a site can be estimated from redox measurements. The redox potential expresses the electron availability as it affects oxidation states. Probes are available to measure redox potentials. Commercially available monitoring units will typically record dissolved oxygen (DO), redox potential, pH, temperature, and conductivity. These units can be placed in wells or have groundwater pumped to surface-installed sensors. Sample exposure to the atmosphere must be prevented prior to the redox measurement. Indirect measurements for the redox level, such as dissolved oxygen, nitrogen, and sulfate species, can be used to supplement redox measurements.

Redox potentials usually decrease with depths below the surface because of the limited diffusion of oxygen. For contaminated sites, redox potentials are usually a result of the site's biological activity.

TABLE 6.4   Classification of Aquifers According to
Redox Potential

Soil classification	Redox potential, millivolts
Oxidized soil	>400
Moderately reduced soil	100–400
Reduced soil	−100–100
Highly reduced soil	−300– −100

TABLE 6.5   Typical Redox
Potential Ranges for Various
Modes of Metabolism, Millivolts

Aerobic respiration	+800
Denitrification	+750
Sulfate reduction	−220
Methane fermentation	−240

SOURCE: Bitton, 1984.

Thus redox potential can vary within a contamination zone. A classification of groundwater aquifers according to redox potential has been proposed by Bitton, (1984) (Table 6.4). Typical ranges of redox potential for various modes of microbial metabolism are presented in Table 6.5.

One may have the option of using either an aerobic or an anaerobic mode of metabolism in designing a bioremediation system. Some organic compounds can be degraded under either aerobic or anaerobic conditions. For these contaminants, selection of the microbial metabolism may well depend upon the existing redox potential in the aquifer. It may be undesirable to change an existing redox potential because of cost or inducing undesired chemical precipitation in the aquifer.

Microbial activity

For most contaminated sites, the existing microbial population is probably the most suitable for the bioremediation project. Given adequate time, the exposure of soil microorganisms to the contaminants results in a preselection of the most efficient microorganism. This preselection develops because those organisms possess the necessary metabolic pathways and enzymes to gain energy from the contaminants. The period for this preselection of indigenous bacteria to develop has not been adequately established, but many consider 1 to 2 years appropriate. This time factor is also a function of the specific microorganism. For example, the methanotrophs are present in most all soil samples and can be readily stimulated when their carbon and energy source, methane, is provided.

Not all organisms are as common or ubiquitous as the methanotrophs, and the distribution of microorganisms isn't homogeneous. Sites contaminated with low-energy-yielding compounds and those requiring cometabolism may acquire inoculation. Bioaugmentation, introduction of bacteria from other sources (for example, contamination areas with good biological response, activated sludge, enriched cultures), may increase the rate of bioremediation.

A site's microbial activity is directly related to the available energy and electron acceptors as long as excessive toxicity isn't present. A measurement of the natural microbial activity is one procedure for evaluating potential toxic conditions at a site. Low bacteria counts can indicate a potential toxicity problem. Groundwater has typical bacterial counts ranging from 10^3 to 10^8 counts per liter. Typical microbial counts for soil vary between 10^3 and 10^7 counts per gram of soil. Counts below 10^3 organisms per gram of soil in contaminated zones may indicate toxic conditions. Microbial activity is normally greater near the surface and decreases to low levels at depths of over 200 ft.

The potential for microbial activity can be measured by counting or enumeration techniques. Special instances may occur, however, when it is necessary to know if specific microbial transformations are occurring. For these conditions it is more appropriate to look for predominant microbial species or their metabolites. An example would be the measurement of methane for determining the activity of methanogens. Evaluation of field data for microbial degradation products and the utilization of specific substrates (such as methane) should always be used as a tool for characterizing the type of microbial activity. This technique is far more successful than identification of microbial species.

Several techniques for measuring bacterial populations are available. Standard culturing techniques are frequently used to determine the bacterial populations from subsurface samples. In all cases, aseptic techniques should be used to minimize false results that may occur during sample handling and preparation for bacterial enumeration. The American Society of Agronomy (No. 337, 2d edition) spread plate method has been used to enumerate aerobic soil bacteria. The basic procedure is as follows:

- Peptone water, neutralized with NaOH to pH 7, is used as dilution water.

- Bacteria are extracted from the soil by vortexing, and the soil is allowed to settle. The supernatant is withdrawn and diluted in a peptone solution.

- Serial dilutions are made of the supernatant to provide a range of concentrations for the spread plates.

- Duplicate plates are inoculated with aliquots from each dilution.
- Counts of cells per gram of soil are calculated from the colonies counted on the plates. Corrections for serial dilutions and for soil moisture content are made.

A colony results from a clump of several bacteria. Thus the counts are understated and a function of the degree of mixing during extraction. This extraction step is significantly less than 100 percent and can yield variable results if a consistent technique is not employed. As little as 10 percent of the soil microorganisms may be capable of growth on the selected growth media. When evaluating groundwater, the soil extraction step is eliminated. However, one must recognize that most of the bacteria can be expected to be attached to the aquifer soils and not dispersed in the groundwater.

Enumeration of bacterial densities under anoxic and anaerobic conditions is significantly more difficult owing to the need to exclude oxygen from the cultures and maintain other appropriate gases. A gas mixture of N_2:H_2:CO_2 of (85:10:5) has been employed. Anaerobic glove boxes and transfer techniques have been used for the culturing and examination of these microorganisms. The U.S. EPA has utilized modified coring devices and anaerobic glove boxes to ensure the maintenance of totally anaerobic and aseptic conditions during sample handling (Figs. 6.4 and 6.5). A soil sample is collected by first drilling a borehole to the desired depth of the sample location with an auger. Then the sample is collected with a core barrel. The soil core is extruded through a sterile paring device that removes the outer layer of soil. This can be performed in an anaerobic glove box for controlled atmospheric exposure. This central core of soil should be uncontaminated and have minimum exposure to oxygen.

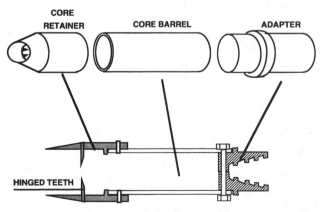

Figure 6.4 Modified coring device for collection of soil samples. (*Wilson, 1985.*)

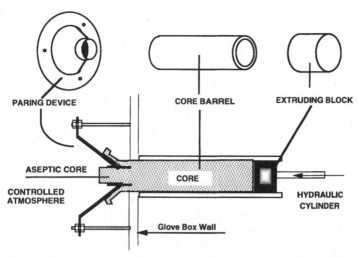

Figure 6.5 Coring device to recover soil cores aseptically. (*Wilson, 1985.*)

Standard protocols are very limiting to the needs of some site investigations. For example, evaluating the use of nitrate as an electron acceptor over oxygen requires specialized laboratory protocols. It may be desirable to count nitrate-reducing organisms that are the active degraders of target compounds during bioremediation rather than performing a total plate count. These procedures require enrichment cultures with controlled gases in culture head space. An example of such a protocol is one used for monitoring nitrate-reducing organisms during the degradation of BTEX (Mikesell, 1991). The generalized procedure is as follows:

- Bacteria are extracted from soil by vortexing in a liquid medium and allowed to settle. The supernatant is withdrawn and serial dilutions performed in liquid mediums (composition below).

- Plates are inoculated by plating a minimal medium containing 10 mM nitrate.

- Plates are incubated in glass desiccators containing a small piece (2 by 2 cm) of BTEX-saturated filter paper.

- Anoxic conditions are established in an anaerobic jar containing a H_2-CO_2 atmosphere. Oxygen is removed with a palladium catalyst.

- The plates are incubated at 30°C.

- BTEX is replenished after 1 day and colonies counted after the second day of incubation.

- Counts of colonies per gram of original soil are calculated after correction for serial dilutions.

Liquid medium	Quantities per liter
0.02 g	$MgSO_4 \cdot 7H_2O$
0.01 g	$CaCl_2$
5 mg	$FeSO_4 \cdot 7H_2O$
25 mg	$NaMoO_4 \cdot 2H_2O$
140 mg	Nitrate - N
5 mM	Potassium phosphate buffer (pH 7.1)

Direct cell counts are also used to enumerate bacterial populations. One technique is epifluoresence microscopy. The preparation of soil samples for direct counting techniques is similar to the technique used for preparing plate counts. The bacterial cells are extracted from the aquifer material and suspended in solution. Serial dilutions are prepared to give an adequate number of cells per unit volume. The generalized procedure for epifluoresence microscopy is as follows:

■ The cell suspension is stained for 2 min with 0.01 percent acridine orange, which binds to DNA and fluoresces under ultraviolet light.

■ The suspension is rinsed with distilled water and filtered on a 0.22-μm-pore-size filter that has been stained black to provide increased contrast during visual examination.

■ The number of cells collected on the filters are counted. Several microscopic fields are counted to provide statistically significant counts. A conversion factor based on the size of the field, the area of the filter, the dilution rate, and the amount filtered is then used to translate the direct microscopic counts to cells per milliliter. This concentration can then be converted to cells per gram of soil.

These direct count procedures usually yield values of 10 to 100 times greater than the plate counting technique.

Nutrient conditions

Site characterization should delineate the potential source of nutrients with specific emphasis on nitrogen (inorganic forms and total Kjeldahl nitrogen) and phosphorus (Table 6.6). Monitoring can utilize procedures that provide instantaneous direct readouts, or samples can be collected and delivered to a laboratory. Measurements of nitrate, nitrite, ammonia, sulfite, and sulfate are not only important for nutrient evaluation but they also aid in establishing redox conditions and potential electron acceptors. Monitoring of these ions throughout the bioremediation project helps establish possible shifts in the oxidation-reduction potential and biological response.

**TABLE 6.6 Site
Characterization for
Nutrient Conditions**

Nitrogen:
  Ammonia
  Nitrite
  Nitrate
  Total Kjeldahl nitrogen
Phosphate
Sulfate

The nutrient most frequently added for bioremediation is nitrogen. It is usually added as a nitrogen source for cellular growth, but it can also serve as an alternate electron acceptor. As a nutrient source, nitrogen is commonly added as urea or as an ammonia salt. These forms are inexpensive and easy to handle, and the ammonium ion is readily assimilated in bacterial metabolism. However, several chemicals have been used for nutrient supplementation (Table 6.7). In aerobic systems, ammonium ion will cause an additional oxygen demand, since it will be oxidized to nitrate. Slow-release nutrient preparations and garden and lawn fertilizers also have been used.

Phosphorus is the second most likely nutrient to be required followed by potassium. Phosphorus can be supplied in one of several forms for augmentation to groundwater (Table 6.7). Common forms are orthophosphoric and polyphosphate salts.

Other environmental factors

Finally, there are several basic environmental parameters that influence bioremediation. These include but are not necessarily limited to:

- Moisture
- Temperature
- pH
- Osmotic pressure and salinity
- Alkalinity

TABLE 6.7 Nutrient Sources

Ammonium sulfate	Potassium monophosphate
Ammonium nitrate	Polyphosphate salts
Disodium phosphate	Orthophosphoric salts
Monosodium phosphate	Phosphoric acids
Urea	Lawn and agricultural fertilizers

- Metal concentrations
- Radiation

The most important of these are moisture, pH, and temperature. Temperature significantly influences the rate of biodegradation, but it is costly to control. Moisture and pH should be optimized for both the desired microorganisms and the desired chemical reactions. Measuring alkalinity helps establish the buffer capacity of groundwaters and allows one to calculate changes in carbonate and bicarbonate ions. Acid titration of soil suspended in water can be performed to evaluate the soil's buffering potential. This provides information on potential pH drop as a result of biological activity, and the release of CO_2 from the natural buffer versus that of biological origin. The control of pH in weakly buffered groundwater is important enough that automatic chemical feed systems should be considered for surface bioreactors.

Assessments for in Situ Bioremediation

Project design for in situ bioremediation not only requires all of the above assessments, but the scope of these evaluations is significantly expanded. Since in situ bioremediation requires the control of microbial processes in the subsurface, a detailed understanding of a site's physical, chemical, and hydrogeological characteristics is necessary. Providing environmental control in the heterogeneity subsurface is a difficult engineering feat even with the best hydrogeological data. Site characterization should accomplish the additional goals of providing data for:

- An understanding of the type of contamination and its location
- An understanding of the subsurface geology and hydraulics
- Hydrological isolation of the contamination zone
- Likely locations of minimum mass transport
- Designing a system for subsurface injection and extraction
- Anticipating the response of soil chemistry to injected chemicals
- Establishing a subsurface monitoring system for process control

Distribution of contamination

Both the concentration and distribution of contaminants are important in assessing the applicability of in situ bioremediation. Groundwater contamination is usually at concentrations below the level for supporting effective microbial populations. On the other hand, contaminated soils acting as the source of groundwater contamination may contain high concentrations. Although the adsorption of contaminants on soil reduces their availability for degradation, it also reduces the potential toxicity.

Significant concentration variations are usually experienced at a site. Understanding the distribution of a contaminant is important for proper site assessment as well as design of in situ treatment programs. These variations result from the quantity spilled, the time of the spill, as well as the properties of the chemical. Chemical properties cause concentration variations because of their effect on subsurface transport. Important properties are listed in Table 6.8. Their influence on chemical distribution has been discussed in many books and will not be presented here.

Organic contaminants in the subsurface will be distributed between the following phases:

1. A vapor phase

2. A free product phase

3. An adsorbed phase

4. A dissolved phase

TABLE 6.8 Chemical Properties Affecting Subsurface Distribution

Vapor pressure	Vapor pressure measures the tendency of a substance to evaporate. It applies to the vapor pressure of a chemical in equilibrium with its 100% solid or liquid form
Henry's law constant	Henry's law constant measures the distribution tendency of a chemical between its concentration in water and a vapor pressure. It applies to the volatilization tendency of chemicals from groundwater and is more important for describing distribution outside of the free product zone
Soil sorption coefficient	Soil sorption coefficient measures the affinity of a chemical for attachment on the surface of soil particles
Water solubility	Water solubility measures the distribution tendency of a chemical between its pure form and water phase. It controls the degree that a chemical moves into and with the groundwater
Density	Density determines the location of the major portion of any free project. For gasoline (sg 0.7) with specific gravities less than 1.0, the free product will be at the surface of the groundwater table. For chlorinated aliphatic hydrocarbons (sg 1.2–1.5) with specific gravities greater than 1.0, the free product will be at the bottom of the aquifer
Viscosity	Viscosity determines the amount of immiscible contaminants held in aquifer capillaries or soil pores

TABLE 6.9   A Typical Distribution for Gasoline

Vapor phase	25 gal
Free product phase	Varies
Adsorbed soil phase	962 gal
Dissolved groundwater phase	13 gal
Total	1000 gal
	(plus the
	free product)

It is also important to establish the smear zone, that area of the subsurface that experiences seasonal changes of the groundwater table. This zone frequently contains a significant mass of contaminants that have densities below that of water.

A typical mass distribution of contaminants is illustrated in Table 6.9. The success of any remediation action depends on recognizing this distribution and designing a system to address the contaminants in each. Any free product phase should be removed by pumping. Contaminants in the saturated groundwater can be treated through hydraulic intervention. Contaminants in the unsaturated zone can be treated by bioventing or other technologies.

Ideally the site's contamination load should be characterized by both horizontal and vertical isoconcentration contours (Figs. 6.1 and 6.6). Vertical isoconcentration contours are usually established by collection of samples clustered at a point but taken from different depths. For groundwater, three or more wells are clustered and screened at various depths. The vertical lines in Fig. 6.6 illustrate the location of well screens. Well clusters are usually placed along the center axis of a plume.

An understanding of the distribution and transport mechanisms for organic contaminants in groundwater is imperative in decisions on well placement and the development of contours. Many organic compounds are not soluble in water. These nonaqueous-phase liquids (NAPLs) may form pockets of undiluted compound. These pockets may either float on top of the groundwater (lighter nonaqueous-phase liquids, LNAPLs) or sink to the bottom of an aquifer (denser nonaqueous-phase liquids, DNAPLs). Thus a significant percentage of an organic contaminant can be immobilized at a site. If these pockets of contamination can be located, extraction can remove much of the source.

Both dissolved-phase and adsorbed-phase contamination will exit in the saturated groundwater zone and in the vadose (unsaturated zone). The vadose zone has been characterized as three zones. These zones consist of the soil belt, the intermediate belt, and the capillary fringe belt (Fig. 6.7). The soil belt is the uppermost zone where signif-

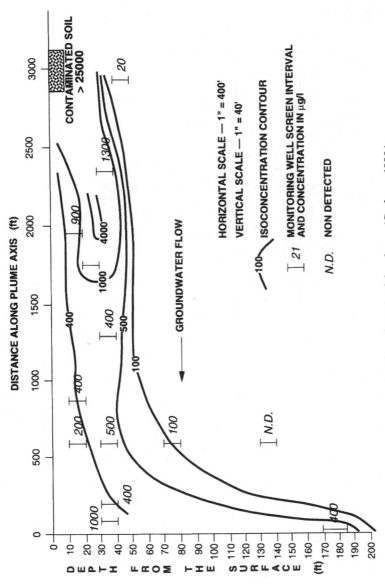

Figure 6.6 Vertical isocentration contours along the axis of the plume. (*Cookson, 1990.*)

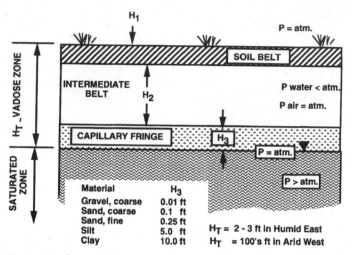

Figure 6.7 Zones characterizing the vadose zone. (*Adapted from EPA/530-SW-86-040.*)

icant biological and chemical activity exists. This includes the root zone for most plants. It is characterized by a high natural organic content and highly variable moisture content.

The intermediate belt extends from the soil belt to the capillary fringe belt. Its location will change as the water-table elevation changes. It has a relatively stable moisture content. The capillary fringe is at the base of the vadose zone. It has a very high moisture content ($\sim$100 percent) as a result of capillary rise of water from the saturated zone. It also contains a significant mass of contaminants with specific gravities less than that of water.

Microbial activity

In the subsurface microorganisms frequently exit in microenvironments. These microenvironments yield organisms of different metabolic capabilities and they are frequently spatially separated, thereby minimizing community interactions. One example illustrating these microenvironments is from the Moffett Field in situ pilot studies. Several methanotroph species were isolated within separate sections of the site that were active in the aerobic dehalogenation of chlorinated aliphatic compounds. Because of genic differences, one species could handle higher concentrations of the primary substrate, methane, than the other. These types of microbial variations can unfortunately hinder cleanup response if adequate treatability studies are not performed. Both microbial differences and site chemistry cause site-specific responses. One cannot be guaranteed that the performance achieved at one location can be duplicated at a second without treatability studies.

In most cases, it is desirable to minimize the spatial separation and promote interaction of a wide range of microorganisms. However, there will be situations where separation is desired, particularly where changes in electron acceptor are necessary to provide for complete mineralization.

In selecting the type of microbial activity to be promoted at a site for cleanup, several questions must be considered (Table 6.10). For some contaminated sites, either aerobic or anaerobic metabolism may be feasible. It is difficult and costly to change environmental parameters in the subsurface. For sites in a highly reduced state, anaerobic metabolism may be more cost-effective than aerobic respiration. A cost comparison for various electron acceptors is provided in Table 6.11.

TABLE 6.10 Considerations When Selecting Aerobic versus Anaerobic Systems

What are the feasible metabolic pathways for contaminant degradation?
What is the degree of mineralization by each pathway?
Will indigenous bacteria be appropriate?
What is the existing redox potential?
What is the nitrate concentration?
What is the sulfate concentration?
What is the potential of aquifer clogging?

TABLE 6.11 Cost Comparison of Various Electron Acceptors for in Situ Bioremediation

Electron acceptor	Oxygen from hydrogen peroxide	Oxygen by bioventing	Nitrate
Construction			
Construction	$ 35.0	$20.0	$ 90.0
Labor and monitoring	55.0	31.0	74.0
Chemicals	385.0	0.3	23.0
Electricity	19.0	5.0	9.0
Total costs per cubic yard	$494.0	$56.3	$196.0
Operation			
Labor and monitoring	$ 3.0	$7.7	$ 7.4
Chemicals	21.0	0.1	2.3
Electricity	1.0	1.3	0.9
Monthly total per cubic yard	$25.0	$9.1	$10.6

SOURCE: U.S. EPA Bioremediation in the Field, EPA 540/2-91-018.

For sites that are not aerobic, nitrate can be introduced as an alternate electron acceptor. If redox potential becomes sufficiently low, sulfate reduction and methanogenesis can occur. Methanogenic conditions are more conducive to the biodegradation of a variety of halogenated aliphatic and alicyclic and aromatic compounds than sulfate-reducing conditions. In fact, nitrate and high sulfate concentrations can inhibit the dehalogenation of chlorinated aromatics (Hickman, 1989).

Another consideration is the potential of aquifer clogging. Aerobic systems have a greater clogging potential. Increasing the redox potential can result in undesirable chemical precipitation. Potential chemical precipitation reactions are discussed in Chap. 7. As discussed in Chap. 3, the microbial growth yield is significantly higher for redox reactions with molecular oxygen as the electron acceptor than the other metabolism modes. This greater growth yield in bacterial populations usually results in a decrease of porosity and transmissivity. For tight aquifers, one may want to select anaerobic metabolism.

Soil and water characteristics

The geologic characteristics and nature of soil within and surrounding the contamination zone must be appropriately evaluated for in situ bioremediation. Soil samples should be characterized for physical properties as well as chemical composition. It is important to establish a site's heterogeneity relative to these characteristics (Table 6.12). Porosity is one of the most important soil characteristics with respect to interactions with environmental contaminants and moisture. Porosity is the volume of voids divided by the total volume of a unit volume of soil expressed as a percentage (Fig. 6.8). Porosity is primarily dependent upon particle size distribution and

TABLE 6.12   Important Soil Characteristics

Grain size distribution
Classification:
  Sand content
  Silt content
  Clay content
Porosity
Bulk density
Moisture content
Soil characteristic curve:
  Field capacity
  Saturation capacity
Soil organic matter
Redox status

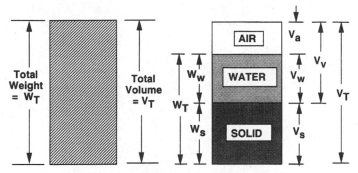

(a) Soil Element in Natural State,
(b) Three Phases of the Soil

Void Ratio	$e = V_V / V_S$
Porosity	$p = V_V / V_T$
Degree of Saturation	$s = V_W / V_V$
Volumetric Water or Moisture Content	$m_V = V_W / V_T$
Gravimetric Water or Moisture Content	$m = W_W / W_S$
Soil Specific Weight	$y = W_T / V_T$
Soil Bulk Density	$d = y / g$

g = acceleration of gravity

Figure 6.8 Soil volume and weight relationships.

TABLE 6.13 Porosity Values of Several Soil Types

Soil	Porosity, %
Coarse gravel	24–36
Fine gravel	25–38
Coarse sand	31–46
Fine sand	26–53
Silt	34–61
Clay	34–60

soil structure. Typical porosity values for several soil types are illustrated in Table 6.13.

Soil classifications are based on particle size and shape. The basis for soil classification is grain size as established by particle size distribution curves. A commonly used classification scheme is that of the U.S. Department of Agriculture. Sands are from 2 to 0.05 mm in

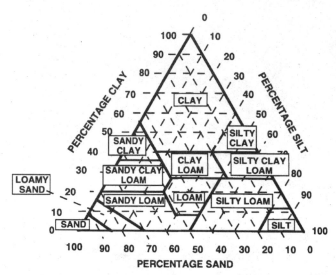

Figure 6.9 U.S. Department of Agriculture trilinear diagram for naming relatively fine-textured soils.

grain size, silts are from 0.05 to 0.002 mm, and clays are at a grain size below 0.002 mm. Soil is a blend of these classifications. The U.S. Department of Agriculture trilinear diagram for naming fine-textured soils is illustrated in Fig. 6.9.

In the intermediate soil belt of the vadose zone, the soil moisture will vary according to the soil classification, or more exactly its particle size distribution. Moisture content can be measured as a percent moisture based on volume, percent moisture based on weight (Fig. 6.8), or percent of field capacity. Field capacity is the moisture content at which water will no longer flow from soil under the force of gravity (Figs. 6.10 and 6.11). Saturation is the moisture content at which the soil pore space is completely occupied by water. At this condition the soil has a zero soil suction value.

As the moisture content of a soil decreases under the influence of gravity, the pressure within the soil water decreases. This decrease in soil water pressure is termed the soil suction or the matric potential. Field capacity occurs at a soil suction of 0.1 to 0.2 atm (bars). The field capacity and saturation are different for each type of soil and are established from a soil characteristic curve as illustration in Fig. 6.11.

The capillary rise of water in a soil pore is associated with the matric suction component of total water suction. The height of water rise and the surface curvature of the capillary have direct implications on the water content versus matric suction. Since the curved surface of the capillary water column is in hydrostatic equilibrium with the water table surface, the water pressure in the column is neg-

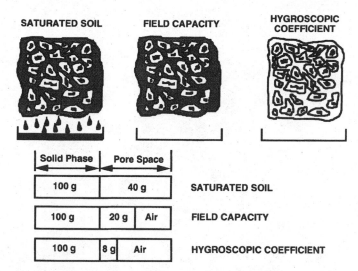

Figure 6.10 Depiction of soil moisture terminology. (*Brady, 1974.*)

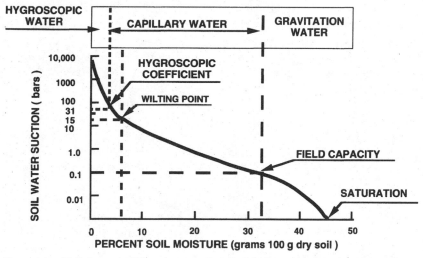

Figure 6.11 Soil characteristic curve: suction and moisture content curve for a loam. (*Brady, 1974.*)

ative. This negative pressure is expressed as the matric suction (see Chap. 3). The smaller the pore radius of a soil, the higher the capillary column and the higher the soil matric suction. Thus matric suction varies for soils according to grain size. This property results in differences in water distribution.

Applying these principles with information of the soil classification allows one to recognize likely areas of higher and lower moisture content in the subsurface. For example, three separated soils of identical

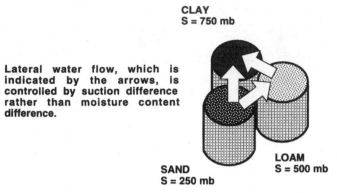

CLAY
S = 750 mb

Lateral water flow, which is indicated by the arrows, is controlled by suction difference rather than moisture content difference.

SAND
S = 250 mb

LOAM
S = 500 mb

Figure 6.12   Effect of suction values on water flow. (*U.S. Department of Agriculture, 1957.*)

moisture content will change if they are brought into good contact because of the suction difference established by their properties. Water will flow from sand to a loam and from a loam to a clay (Fig. 6.12). Site data that include soil classifications allow one to predict the location of areas of higher moisture content or zones of reduced conductivity for gas transport.

Several procedures are used to measure water suction. Thermocouple psychrometers have been used for total water suction. Matric water suction can be measured with tensiometers or axis-translation devices. Indirect measurement techniques are frequently used. One system uses a porous block that is brought into equilibrium with the matric suction in soil. The water content of the porous block is measured by changes in its electrical or thermal properties. Since matric suction is the major component of total water suction, these measurement techniques are adequate.

Samples of soil and associated groundwater from the contaminated area should be characterized for inorganic ions, dissolved gases, and organic content (Table 6.14). The major cations of importance are iron, manganese, magnesium, calcium, copper, nickel, and potassium. The major anions of importance are ammonium, nitrate, nitrite, sulfide, sulfate, phosphate, carbonate-bicarbonate distribution, and the dissolved gases of oxygen and methane. These gases provide information on the subsurface redox potential.

Direct monitoring of redox potential by electrode is difficult and requires skill to prevent changes due to atmospheric exposure. Redox status can be determined by measurements for the presence of oxidized or reduced ions. Lyngkilde (1991) formulated a guide for assigning redox status for probable metabolism modes to groundwater parameters (Table 6.15). This table should be used only as a guide, and

TABLE 6.14 Soil-Water Chemical Characteristics Important to in Situ Bioremediation

Inorganic Ions
Cations:
Ions
Manganese
Calcium
Copper
Nickel
pH
Anions:
Ammonia
Nitrate
Nitrite
Sulfite
Sulfate
Phosphate
Carbonate/bicarbonate

Dissolved Gases
Oxygen
Methane

Organic Compound
Total organic carbon, TOC
Biochemical oxygen demand, BOD

Biological
Microbial plate count

TABLE 6.15 Groundwater Criteria and Probable Redox Status*

Parameter	Aerobic	Nitrate-reducing	Mn-reducing	Fe-reducing	Sulfate-reducing	Methan-ogenic
Oxygen	>1.0	<1.0	<1.0	<1.0	<1.0	<1.0
Nitrate	---	---	<0.2	<0.2	<0.2	<0.2
Nitrite	<0.1	---	<0.1	<0.1	<0.1	<0.1
Ammonium	<1.0	<1.0	---	---	---	---
Mn(II)	<0.2	<0.2	>0.2	---	---	---
Fe(II)	<1.5	<1.5	<1.5	>1.5	---	---
Sulfate	---	---	---	---	---	<40
Sulfite	<0.1	<0.1	<0.1	<0.1	>0.2	---
Methane	<1.0	<1.0	<1.0	<1.0	<1.0	>1.0

SOURCE: Lynkilde, 1991.
*All values in mg/L.
---indicates that the value is higher than any given minimum.

samples should not be contacted with atmospheric gases before these ions are measured. For measuring ammonium, samples can be preserved by adding sulfuric acid. Samples for nitrate and nitrite determination can be preserved with mercury chloride. Samples for sulfate must be kept cool. Samples for sulfite analysis can be preserved with a strong basic antioxidating solution.

The organic content of soils can influence treatment rate. High organic soils may hinder the degradation of target compounds. Organisms will use high-energy-yielding compounds, leaving the more stable organic compounds to persist over long periods. High organic soils also can lead to buildup of microbial metabolites that hinder the degradation of target compounds. This is demonstrated by the stability of complex organic compounds in the high organic soils found in the humid, cold climates, compared with low organic soils found in dry, hot climates. In the forests of Maine, typical of the humid, cold climates, microbial breakdown of humic materials is slow. It is also found that in such soils, DDT persists for more than 30 years. In contrast to this is the low organic soils of Colorado, where 38 percent of DDT in agricultural soils degrades within 4 weeks (Focht, 1985).

Metabolic buildup is usually controlled as a result of other organisms that utilize these metabolites as an energy source. The pumping rate of injection and extraction wells also can control the buildup of metabolites as well as the availability of electron acceptors. Soils that are characterized by a high flux of water have demonstrated enhanced degradation rates when in the presence of nitrate or sulfate. The improved degradation rates are believed to be related to toxic accumulation of nitrite as a result of low water flux.

Sulfate, in some instances, has been reported to inhibit methanogenesis. This inhibition does not appear to occur in aquifers with high fluxes of water (Hickman, 1989). If natural hydraulic characteristics provide inadequate flux of water, recirculation with injection-recovery wells can provide improved flushing action.

The geochemistry of an aquifer is also important because of chemical interactions between minerals and contaminants, and minerals and the microorganisms. A number of adsorption and chemical reactions occur. Nutrient transport to the contaminated area can be retarded by adsorption on the subsurface matric. Ammonia ions, for example, can be lost by ion exchange with other cations. Phosphates have a high adsorption affinity for most soils, whereas nitrate has a lesser affinity. Nutrient adsorption on soil does not make the nutrient unavailable for biological activity, but it does limit the effective distance over which the nutrient can be transported in the subsurface.

Phosphates will react with metal ions, particles, and precipitate from solution. The precipitation of calcium phosphate has been attributed to decreased permeability at in situ bioremediation sites

(Rainwater, 1991). A second consideration is that the addition of nutrients and their associate ions may lead to dispersion or swelling of clay particles. Finally, the interaction of phosphate with calcium and magnesium can reduce the hydraulic conductivity of the subsurface.

There are also various metals of importance when considering bioremediation. Metals such as iron and magnesium can form precipitates that hinder the mass transfer of oxygen and nutrients to the subsurface. Certain metals such as copper and nickel are important when evaluating oxygen sources because of their ability to catalyze the degradation of hydrogen peroxide.

It is difficult if not impossible to predict potential subsurface reactions. Present information on chemical reactions relative to soil composition is very limited. Adequate measurements of many parameters are lacking and one cannot make useful comparison with previous data to judge a site's performance. At present the only means to evaluate these potential reactions is by laboratory treatability studies.

Assessments for Process Control

In situ bioremediation requires a detailed characterization of a site's hydrogeology for adequate process control. For in situ bioremediation, the goal is to provide a system that controls mass transfer into and out of a contaminated area. It is through proper engineering and operation that the microbial processes are manipulated in the subsurface. Although the principles and concepts for this condition are simple, achieving such controls is difficult in field applications when site characteristics are not well understood.

The subsurface is controlled by hydraulic injection and/or extraction of fluid, either water or air. Air injection or extraction is used to deliver oxygen to the unsaturated zone. This may be coupled with water extraction to lower the groundwater table and expose the smear zone to the gaseous oxygen. Within the saturated zone, water is injected and/or extracted to deliver the electron acceptor and other nutrients. The water table may be raised to include the smear zone. The goal of the injection program is to deliver these agents to the contaminated zone in the most efficient manner and to prevent the pushing of contaminants out of the treatment zone. This requires minimizing losses, minimizing clogging, and promoting a uniform distribution within the microbial growth area. Failure to provide a uniform distribution of agents to the contaminated zone results in zones of high microbial densities and zones of low densities, resulting in imbalanced microbial activities and rates of cleanup. Poor distribution expands the bioremediation cleanup time or has the potential of leaving unrecognized pockets of contamination.

Poor distribution is usually a result of the heterogeneity of the subsurface and its effect on fluid flow. The ability to move fluid through soil is again related to that important property of particle size distribution. The flow of viscous liquids through porous materials is described by Darcy's law. Flow Q varies directly with the loss of head or hydraulic gradient H. Thus

$$Q = \frac{k}{\mu(A)(H)}$$

where H is the hydraulic gradient (length/length), A is the cross-sectional area (length^2), μ is the fluid viscosity, and k is the intrinsic or absolute permeability (length/time). In the petroleum industry this permeability unit is the darcy:

$$\text{Darcy} = \frac{1 \text{ centipoise} \times \text{cm}^3/\text{s}/\text{cm}^2}{\text{atmosphere/cm}}$$

where 1 centipoise = 0.01 dyne $-$ s/cm^2
1 atmosphere = 1.0132×10^6 dynes/cm^2
1 darcy = 9.87×10^{-9} cm^2

Darcy's law is frequently used with the term hydraulic conductivity as

$$Q = K(A)(H)$$

where K is the hydraulic conductivity (length/time). Thus the hydraulic conductivity and the darcy are related by

$$K = \frac{(k)\ (\text{density of fluid})}{\text{viscosity}}$$

The laboratory or standard value of hydraulic conductivity is defined for pure water at 15.6°C. Pure water in a soil with an intrinsic permeability of 1 darcy would have a hydraulic conductivity of 8.61×10^{-4} cm/s at 15.6°C and 7.4×10^{-4} cm/s at 10°C. Units are usually cm/s or feet/day. However, other units are frequently used such as gallons per day per square foot. Some conversion values are

$$1 \text{ gal/day/ft}^2 = 0.0134 \text{ ft/day}$$

$$1 \text{ gal/day/ft}^2 = 4.72 \times 10^{-5} \text{ cm/s}$$

$$1 \text{ ft/day} = 3.53 \times 10^{-4} \text{ cm/s}$$

$$1 \text{ ft/day} = 7.48 \text{ gal/day/ft}^2$$

TABLE 6.16   Typical Values for Hydraulic Conductivity and Intrinsic Permeability

Soil type	Hydraulic conductivity, cm/s	Intrinsic permeability, darcy
Clay	10^{-9}–10^{-6}	10^{-6}–10^{-3}
Silt, sandy silts, clayey sands, tills	10^{-6}–10^{-4}	10^{-3}–10^{-1}
Silty sands and fine sands	10^{-5}–10^{-3}	10^{-2}–1
Well-sorted sands	10^{-3}–10^{-1}	1–10^{2}
Well-sorted gravels	10^{-2}–1	10–10^{3}

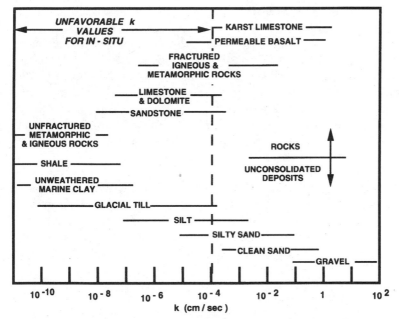

Figure 6.13 Range of hydraulic conductivity for various formations. (*U.S. EPA Bioremediation Workshop, Dec. 13–14, 1988, Athens, Ga.*)

$$1 \text{ cm/s} = 21{,}200 \text{ gal/day/ft}^2$$

$$1 \text{ cm/s} = 2835 \text{ ft/day}$$

Typical values for hydraulic conductivity and intrinsic permeability are given in Table 6.16 and a general range for various formations in Fig. 6-13. In situ bioremediation becomes difficult at hydraulic conductivity below 10^{-4} cm/s.

Intrinsic permeability is a direct function of the soil's particle size distribution and grain size and shape. Knowledge on the mean grain size allows one to estimate changes in intrinsic permeability within a site. Intrinsic permeability k is related to grain size by

$$k = Cd^2$$

where d is the mean grain size and C is a factor that depends upon the porosity of the soil, the range of and distribution of grain sizes, the shape of the grains, and their orientation.

Once the geological characteristics of a site are determined, one should evaluate data for potential zones of reduced permeability. It will be these zones that control the cleanup time of the site.

Hydrogeological characterization

In situ bioremediation requires a detailed understanding of the hydrogeology of a site. The degree of site characterization depends upon the heterogeneity of the aquifer, the composition of contaminants, and information on the health, ecological, and liability potential of a given situation. Subsurface hydraulics have a major effect on subsurface biodegradation rates, and an optimum hydraulic recycling rate is expected to exist for each site. Information that should be delineated is provided in Table 6.17. Of key importance for all remediation is establishing gradient and flow direction. For in situ bioremediation, drawdown testing to establish permeability and hydraulic conductivity is almost a must.

The geological characteristics of a site are usually established with core samples and well logs. These wells also can be used for monitoring the progress of bioremediation. Stratigraphic data and composition of geological units are useful in estimating porosity, permeability, and homogeneity within an aquifer. Lenses of dissimilar aquifer mate-

TABLE 6.17   Hydraulic Evaluations for Designing in Situ Bioremediation System

Piezometric measurements:
  Confined
  Unconfined
  Gradient
  Flow direction
Drawdown testing:
  Permeability
  Hydraulic conductivity
  Natural-gradient water velocity
Pumping test response:
  Leakiness
  Barriers
  Abnormalities
Tracer testing:
  Natural detention times
  Forced gradient detention times
  Percent capture of recycled water

rial, whether they are more (gravel) or less (clay) permeable, must be delineated to increase the degree of certainty in predicting hydraulics. The variations in aquifer material can cause heterogeneities in hydrodynamic dispersion and dramatically affect groundwater flow.

Geological data from well loggings and sample borings can be converted to a pictorial cross section of the subsurface (Fig. 6.14). The vertical columns in Fig. 6.14 illustrate actual position, depth, and geological findings for six loggings. By collecting data along the axis of the contamination plume and transverse to the plume, a three-dimensional characterization of the subsurface is developed (Fig. 6.15).

The major hydraulic properties of the aquifer can be established only through pump drawdown tests. These may be conducted over a duration of 30 min to several days. Establishing a plot of drawdown versus time for the extraction and observation wells provides information that can lead to the calculation of hydraulic conductivity, natural gradients, and groundwater velocity. Evaluation of data with modeling techniques can provide additional information on aquifer conditions. Such modeling aids in establishing the degree of leakiness of the aquifer system, transmissivity, storability, and the presence of directional anisotrophy. All local aquifers and their interactions must be established to properly locate injection and recovery wells. Any communication between an upper aquitard and a deep aquitard must be

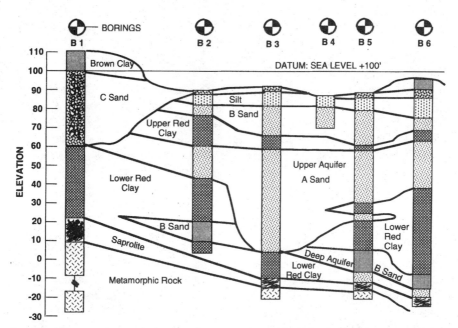

Figure 6.14 Geological characterization of a site from soil borings and well-drilling data.

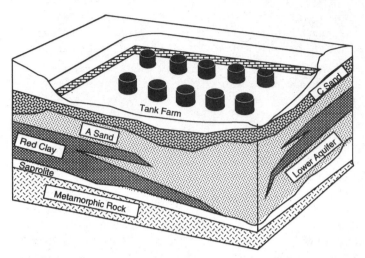

Figure 6.15 Three-dimensional characterization of the subsurface for improved conceptional design of in situ treatment.

recognized (Fig. 6.14, boring B3). The lack of boring B3 in Fig. 6.14 would have resulted in an incomplete characterization, and the assumption that the lower red clay unit maintained separation of the aquifers. However, a stream once cut through this clay unit. Hydraulic intervention for bioremediation in the upper aquifer could push contamination into the deep aquifer if designs did not recognize this communication. Since it is difficult to know how many and where to place borings, a tracer test to determine percent capture is recommended.

Depending upon the importance and environmental significance of a contaminated area, hydraulic modeling may be necessary as well as tracer tests to establish the degree of leakiness of an aquifer system. The advantage of using tracer tests is that one can gain a higher degree of confidence on the amount of injected fluid that is captured by the recovery wells. In addition, periodic application of tracers aids in establishing an aquifer's clogging or other changes in flow patterns as a result of bioremediation activities.

Confirming transport characteristics

Once the approximate location of recovery and injection wells has been established, it is wise to conduct a tracer test between these points to verify expected flow directions, velocities, and actual travel time for the injected agents. Tracer tests will give firsthand information on the transport characteristics within the actual zone for bioremediation. Tracer evaluations also provide for hydraulic transport model calibration.

Chloride and bromide are two salts that do not adsorb on soils or undergo transformation. Although chloride is used, its sensitivity is reduced as a result of natural background chloride. Thus bromide is usually a better selection. Bromide is usually injected as a pulse, the duration being a function of aquifer characteristics. Pulse duration of approximately 3 h has been used in field studies. Bromide injections have used concentrations of approximately 50 to 100 mg/L. The tracer study can be performed under natural gradient and various induced gradient conditions to characterize flow and nutrient transport in the treatment zone. A forced gradient is typically formed by continuous injection of groundwater and withdrawal at selected recovery wells. Multilevel sampling wells have been used to monitor the tracer. The salt is injected in the injection well and monitored at downgradient wells over time, or in the recovery well. The concentration history is plotted versus time.

A typical bromide trace study, as illustrated in Fig. 6.16, was conducted on a sandy freshwater aquifer located on Cape Cod, Massachusetts (Harvey, 1989). The aquifer sediments were deposited in layers as glacier outwash, contained little clay, and were composed largely of quartz and feldspar. Mean grain size, average porosity, and hydraulic conductivity were 0.5 mm, 0.38, and 0.1 cm/s, respectively. This field study also evaluated the transport of indigenous bacteria through the aquifer. A diverse mixture of indigenous bacteria was concentrated, marked, and injected. Transport of bacteria compared with bromide was rapid given the larger size of bacteria. Other field studies have demonstrated similar results, indicating that the use of a bromide tracer provides a good reference for estimating travel time for injected oxygen, nutrients, and microorganisms.

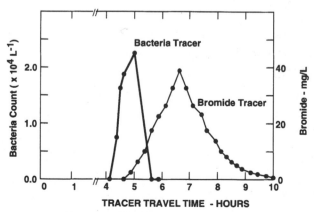

Figure 6.16 Concentration histories for bromide tracer study and a marked indigenous bacteria. (*Harvey, 1989.*)

For hydraulic abnormalities or sites with sensitive environmental surroundings, bromide tracer tests are suggested throughout the bioremediation program. Application of a periodic tracer test aids in establishing aquifer clogging or other changes in flow pattern as a result of bioremediation activities.

Modeling systems are available and have been applied to evaluate tracer results for in situ bioremediation. An example of one study is the in situ biodegradation of chlorinated hydrocarbons at the Moffett Naval Air Station, Mountainview, Calif. For this site a semianalytical model, RESSO, was utilized to simulate two-dimensional advective transport under induced injection and extraction rates as well as natural gradient conditions (Roberts, 1990). Mathematical simulation can provide a degree of confidence on the amount of injected fluid that is captured by the recovery well system. Typical information gained from mathematical simulation is presented in Table 6.18. The tracer studies can be used to calibrate the models, providing the ability to evaluate potential migration and risks of a specific well injection and recovery system.

An optimum injection and recovery rate exists for each site. Once well placements are established, optimization becomes a function of the zone of hydraulic influence, zone of biological influence, percent capture in the recovery wells, nutrient and electron acceptor transport rates, and changes in the aquifer conductivity. Optimizing these systems is difficult to project from modeling and must rely upon adequate field monitoring during the bioremediation program.

TABLE 6.18   Applications of Tracer Tests for Hydraulic Characterization

Verify flow direction
Verify flow velocities
Verify travel time
Verify nutrient transport
Verify capture percentage
Evaluate aquifer hydraulic conductivity changes
Establish operating injection and recovery rates

In Situ Bioremediation

Introduction

In situ treatment is the enhancement of in-place biodegradation of organic compounds within the subsurface or existing waste-holding lagoons. This chapter presents the principles for engineering subsurface in situ treatment systems. Lagoon treatment systems, those designed as surface reactors or those that are modified waste-holding ponds, are discussed in Chap. 8.

Although a concept of in situ bioremediation was patented by Suntech in 1974, in situ bioremediation did not receive significant attention until almost 15 years later. One of the first sites to receive engineering intervention for bioremediation was the U.S. Coast Guard Air Station in Traverse City, Mich. Over 25,000 gal of aviation gasoline, JP-4, and chlorinated ethenes were released into the groundwater. Studies at this site have evaluated various in situ techniques. Since these early developmental activities, many in situ bioremediation projects have been undertaken. These include aquifers contaminated with aviation gasoline, chlorinated organic compounds, benzenes, alcohols, acids, phenolic compounds, polyaromatic hydrocarbons, creosote, and pesticides.

In situ bioremediation has several advantages. One is that site disturbance is minimized. This is particularly important when the contaminated plume has moved under permanent structures. Although site characterization and monitoring requirements are more extensive and costly, the cost of removing contaminated material for surface treatment is frequently greater. The biggest limitation of in situ bioremediation has been the inability to deal effectively with metal contaminants mixed with organic compounds. However, new technologies such as electrokinetic techniques may prove effective when combined with in situ bioremediation in addressing this deficiency. In the past year, the combining of soil venting with in situ bioremediation, bioventing, has gained popularity.

With bioventing, soil venting operations are optimized to enhance biodegradation of the volatile organic compounds. This combination can significantly reduce the mass organic loading in vent gases, resulting in lower treatment costs for volatile emissions. Studies conducted by the U.S. Air Force have yielded volatile organic removal ratios of 80 percent by biodegradation and 20 percent by vent extraction. When air injection is possible over vacuum systems, vent gas treatment can be completely eliminated. Combining these processes significantly reduces the operating costs of vent gas treatment while satisfying the oxygen needs for bioremediation. Using combinations of these new techniques, sites contaminated with volatile compounds can be cleaned up in a shorter time at reduced cost.

The goal of in situ treatment is to manage and manipulate the subsurface environment to optimize microbial degradation. Engineering requires an understanding of the microbial processes that affect the biodegradation of the target compounds and the soil physical, chemical, and hydrological interactions. The major engineering challenge is the design of a delivery and recovery system that provides for a responsive control of the subsurface environment, and a monitoring system that provides data for process optimization. Biological processes require close monitoring of the microbial environment to maintain control of the overall degradation of target compounds.

Hydraulic Control

The first design stage of in situ bioremediation is selecting the technique for isolating and controlling the contaminated zone. Hydraulic control is necessary to either move or halt groundwater flow, raise or drop the water table, and control movement of the contaminated plume. The most frequent procedure for isolating the contaminated plume is through hydrological intervention. Hydrodynamic isolation systems are generally less costly than physical containment structures. They are also more flexible since pumping rates and flow patterns can be changed as needed over the period of treatment. The hydrological control system is designed so groundwater is centrally withdrawn and injected at select points. Ideally, the contaminated zone should be completely isolated (Fig. 7.1). Uncontrolled groundwater may enter the contamination plume, but contaminated groundwater cannot move from the bioremediation area.

It is never possible to achieve 100 percent control even in an ideal aquifer. However, if the hydraulic configuration is to open, the biore-

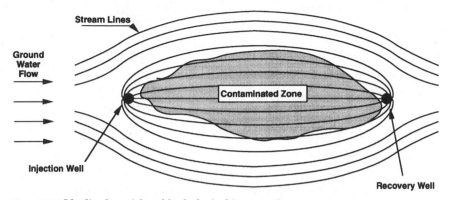

Figure 7.1 Idealized semiclosed hydrological intervention.

mediation may be unsuccessful or push contaminants into unpolluted zones. In some situations, hydraulic control can be improved when combined with physical barriers. Hydrological controls usually result in saturating the subsoil within the contaminated zone. For the unsaturated zone, in situ bioremediation uses bioventing procedures. Again, hydraulic control may be important for dropping the water table to expose more contaminants to the aerobic degradation process.

In addition to isolation of the contaminated area, the hydraulic system is the operator's means for process control. Fluid flow provides delivery of appropriate nutrients, substrates, and electron acceptors within the subsurface. The design goal is to provide a system of injection wells, recovery wells, and possibly barriers that allows control of mass transfer into and out of the contaminated area. A simplified in situ bioremediation system using recovery and injection wells is schematically illustrated in Fig. 7.2. The recharge system can use recovered groundwater that may be treated or water from an uncontaminated source. The injected water may be supplemented with nutrients, substrate, or electron acceptor or have additional microbial seed added before being reintroduced into the subsurface. It is through the proper engineering of mass transfer that the microbial processes are manipulated in the subsurface. Although the principle and equipment for in situ bioremediation are simple, achieving proper control is difficult for field applications owing to the heterogeneity of a site. A typical field injection system is presented in Fig. 7.3. Figure 7.3 shows several injection wells, some covered by protective covers and others open. Flexible tubing is used at this site to carry the oxygen and nutrients for forced injection. This in situ bioremediation is being applied to a tight aquifer, resulting in low injection rates. As a result, numerous injection points are required for adequate control.

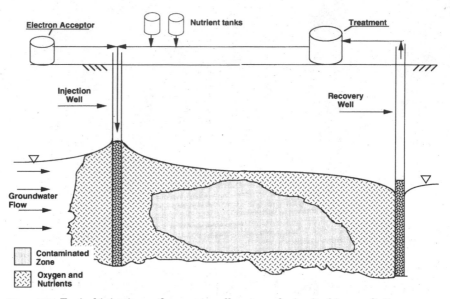

Figure 7.2   Typical injection and recovery well systems for in situ bioremediation.

Figure 7.3   Forced injection of water containing oxygen and nutrients for in situ bioremediation.

The design of hydrological controls requires an understanding of the hydrogeology of a site. This includes details on the groundwater flow, interaction of multiple aquifers, and other aquifer abnormalities. Aquifer test response plus tracer studies with model interpretation may reveal preferential flow paths or communication between aquifers or other abnormalities.

Injection rates influence biodegradation rates, and an optimum flow rate is expected to exist for each site. Flow rates impact on solute, gas transfer rates, and contact times. As infiltration or recirculation rate changes, so does the flow velocity and dispersion. With increased infiltration, the hydraulic detention time in the soil decreases, and zones of influence usually increase. Optimizing injection rate is difficult to predict from modeling techniques and must rely on adequate field monitoring during the bioremediation program.

Subsurface Delivery Systems

Subsurface delivery systems usually rely on the hydraulic system used to isolate the subsurface treatment area. The goal of the delivery system is to deliver the bioremediation control agents to the contaminated zone in the most efficient manner. This requires the minimization of losses, clogging, and promoting a uniform distribution within the desired microbial active zone.

Injection-recovery systems vary based on the extent and location of the contaminants as well as the hydrological characteristics of a site. These can consist of simple arrangements of recharge trenches, well-trench combinations, and well patterns.

Injection systems consist of gravity and force methods (Table 7.1). Required chemicals for the treatment can be applied directly to the contaminated soil or groundwater by flooding, spray irrigation, trenching, or one or more methods of infiltration. Forced delivery methods inject the necessary additives through a pressurized pipe.

TABLE 7.1 Subsurface Delivery Systems

Gravity systems:
  Flooding
  Ponding
  Trench
  Surface spraying
  Infiltration gallery
  Infiltration bed
Forced systems:
  Pump injection
  Air vacuum
  Air injection

Gravity injection

The application of gravity feed systems is limited, since few situations are suitable for this technique. Although gravity feed has been used for contaminants in the saturated zone, it functions better when contamination is located in the vadose zone. For contamination in the saturated zone, reagents may not mix adequately within the contamination area. Surface gravity delivery methods are suited for shallow waste deposits, sites of high permeability such as gravel and sand beds, and localized contamination.

Surface application methods can be considered for contamination within several feet of the surface. They are not appropriate where impermeable layers exist between the surface and the contamination zone. Some surface applications such as flooding, ponding, and spray irrigation may not be appropriate if the terrain has excess slopes. Slopes over 3 to 5 percent can result in runoff and reduced effectiveness of applied reagents.

Climate is important, since frost penetration depth and freezing conditions may hinder the application of liquids. The second consideration is precipitation rates and the ability of the soil to receive the applied liquid. Finally, consideration must be given to soil permeability and the organic load. Permeability determines the feasibility of getting to the contaminated zone by surface application within a reasonable time. Some field applications have found that nutrient and oxygenated water travel time should not exceed 4 to 6 weeks. If surface application is unfeasible, deep infiltration trenches can be used. These can be installed directly above the contaminated soil.

The most important consideration when evaluating the feasibility of gravity feed is the organic load. With gravity feed, the rate of electron acceptor and nutrient feed is controlled by the soil's percolation rate in the unsaturated zone, the hydraulic conductivity (vertical and horizontal), and the adsorption rate and decay rate during soil transport. If the organic load is too great, it may be impossible to deliver the electron acceptor and nutrients at a rate adequate for completing bioremediation within an acceptable time. Design of gravity delivery systems must evaluate the following:

1. Required delivery rate for electron acceptor and nutrients

2. Location of contamination, depth from surface

3. Sustained infiltration rate

4. Hydraulic conductivity of the site (vertical and horizontal)

5. Configuration of the contamination zone: aerial dimension, thickness of deposit

6. Depth to water table from the surface

7. Mounding height

8. Porosity and uniformity within the subsurface

9. Precipitation and frost penetration

10. Surface topography in area of the site

11. Aquifer thickness

The required amount of electron acceptor, nutrients, or substrates that must be delivered is calculated from the mass of contaminant or from respiration tests. Example calculations are provided in Appendix B. The rate of delivery for effective bioremediation is calculated from the rate of organic degradation, usually established during laboratory testing, or field pilot studies and the desired cleanup schedule. If the desired degradation rate is attainable by the selected infiltration system, then gravity feed is possible. If it is not possible to support the desired level of biological activity, force injection is necessary. For most gravity feed systems, the rate of bioremediation is limited by the low delivery rate of the electron acceptor. Although the capital and operating cost per month may be low, the ultimate cost can be greater than for other delivery techniques. Extending the treatment time extends both the supervision time and the monitoring program. The other alternative for shallow contamination is to excavate the contaminated soil and provide surface bioremediation as a solid or slurry system.

Surface applications. The simplest gravity injection methods are surface application such as flooding, spray irrigation, recharge basins, and trenches (Figs. 7.4 and 7.5). Although these systems have a low capital cost, they limit the ability to maintain process control. Other potential disadvantages are flooding and the need for runoff controls and a collection system. Flooding is avoided by establishing the cycle times between loading and drying. Extensive experience with surface gravity injection has been gained in years of recharge basin operation for groundwater recharge. Recharge basins constructed of natural soil without a grass or gravel layer have been the most effective (Caramagno, 1991).

Although gravel basins have an initial higher rate of percolation, they have been the least successful. The higher rates of recharge are short-lived as debris clogs the gravel beds.

The design and operation of gravity feed infiltration systems are dependent on the nature of the in situ treatment program. The first consideration is the location of the contamination—what mass of contaminant is located in the vadose zone versus the saturated zone. The gravity feed system may be designed to address the contamination in only the vadose zone, only the saturated zone, or both. The second consideration is how the electron acceptor will be carried, by

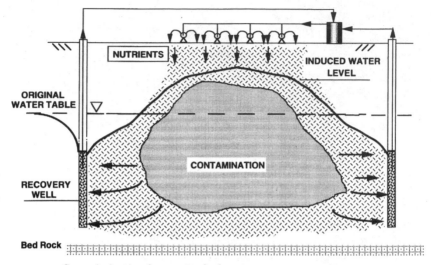

Figure 7.4   Spray irrigation for gravity feed system.

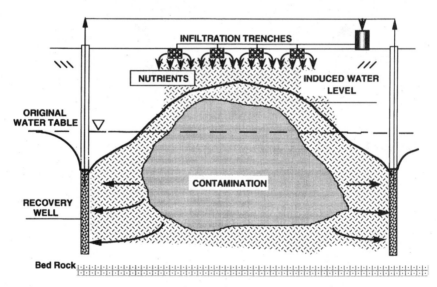

Figure 7.5   Infiltration trenches for gravity feed system.

water or air. If water carries only nutrients, and oxygen is supplied by air injection or extraction, then an unsaturated condition is desired to maintain a high gas permeability rate. If the electron acceptor will be delivered by water, then the goal is to maximize the rate of infiltration without flooding the surface from water mounding. These decisions determine if the gravity feed system will be designed and operated to maintain an unsaturated or saturated condition in the vadose zone. The rate of infiltration influences both the hydraulic detention time and the degree of saturation. A decreasing infiltration rate results in a corresponding decrease in saturation. This provides for higher air permeability and more surface area for oxygen transfer.

Hydraulic conductivity usually decreases after start-up and then stabilizes. Continuous injection results in reduced hydraulic conductivity. Maximum rates of infiltration are obtained when a cycle of injection and drying is used. The operational goal is to minimize drying time while maintaining hydraulic conductivity and microbial activity. The actual injection and drying cycle is usually optimized during field operations. The maximum flooding period for groundwater injection in Arizona was 14 to 21 days followed by a period of nonflooding. This yielded a hydraulic loading rate of 34 percent of the daily soil conductivity (Caramagno, 1991). An optimal injection-noninjection cycle will vary with regional precipitation, soil types, and the need to maintain biological activity.

Estimating mounding height.   Water mounding is undesirable when gas permeabilities are to be maintained or when a shallow aquifer is present. Surface flooding results in a loss of nutrients and electron acceptor as well as the potential for contaminated runoff. Controlled mounding is desirable if the goal is to saturate the contaminated area of the vadose zone. Mounding height is controlled by infiltration rate and can be estimated for an infiltration system (Fig. 7.6).

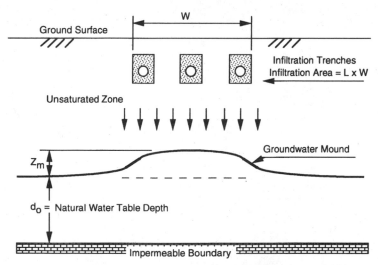

Figure 7.6 Groundwater mound height from gravity infiltration. (*Finnemore, 1993.*)

Accurate modeling of the rise in mound height requires complete representation of the saturated-unsaturated flow system. This, however, yields very complex calculations. The estimation of groundwater mounding from gravity infiltration is usually based on the method of Hantush (Finnemore, 1993). This procedure neglects unsaturated flow and uses the Dupuit-Forchheimer theory of unconfined flow that assumes horizontal streamlines. It is based on an initially near-horizontal saturated zone. Field observations under infiltration basins demonstrate good comparison with this simplified approach. This predictive model has been adopted by the North Coast Regional Water Quality Control Board of California (Finnemore, 1983) and used to predict mounding beneath on-site sewage disposal fields.

The original Hantush predictive model for mound height has been improved by Finnemore. Finnemore's model provides for a stronger theoretical base and extends the application to any rectangular infiltration field. This new procedure also extends the accuracy to periods of shorter duration for infiltration, making it applicable to bioremediation.

The rise of water mounding at the highest point, the center of the mound, is given by the equations (Fig. 7.6):

$$Z_m = \frac{Q}{4\pi\, KH} \, [3 - \gamma - J\,(r) - \ln \mu] \tag{7.1}$$

$$H = d_0 + 1/2\, Z_m \tag{7.2}$$

$$\mu = \frac{L^2 S}{16\, KHt} \, (1 + \frac{1}{r^2}) \tag{7.3}$$

and

$$r = \frac{L}{W} \tag{7.4}$$

where Q = average volume of water injected into the field (volume per time)

K = horizontal hydraulic conductivity of the aquifer

d_0 = natural depth of the saturated zone (the groundwater table elevation to the impermeable bottom boundary)

γ = Euler's constant = 0.577

S = specific yield of the aquifer

t = time since the beginning of water application

L = length of the infiltration field

W = width of the infiltration field. For multiple trenches the width is the total of all trenches (see Fig. 7.6)

The assumption made during the simplification of Eq. (7.1) requires that the time duration since the beginning of infiltration t be greater than the quantity

$$4.17 \frac{L^2 S}{KH} \tag{7.5}$$

for the results to be accurate. Thus a minimum t exists for the application of Eq. (7.1). However, even at t less than the minimum, the departure from other equations representing mounding height is insignificant.

If a desired value for Z_m is given, the time required to achieve this mounding height for various loading rates Q is given by

$$t = \frac{L^2 S}{16 \, KH} (1 + \frac{1}{r^2}) \exp [\frac{4 \pi K Z_m H}{Q} - 3 + \gamma + J(r)] \tag{7.6}$$

Estimation of the mounding height requires three aquifer characteristics: (1) the natural depth of the saturated zone d_o, (2) the horizontal hydraulic conductivity of the aquifer K, and (3) the specific yield of the aquifer.

The natural depth of the saturated zone is determined from water table elevations and the location of the impermeable bottom. Since the groundwater table varies with seasons, the calculation should be performed with the seasonal depth that is more pertinent to the design and operational goals. The value for d_o can be a mean normal saturated depth or the highest value if the maximum mounding height is not to be exceeded.

The hydraulic conductivity K of the aquifer should be determined for the zone that is affected by the groundwater mound. The necessary conductivity is the horizontal, and it can be determined from well pumping test, slug tests, or bail tests. The specific yield S of the aquifer is also obtained from conductivity tests. The specific yield of the aquifer is the volume fraction of the total aquifer that will drain freely. It must be less than the total porosity of the aquifer and can be estimated as the difference between the volumetric water contents above and below a falling water table (Finnemore, 1983). Gamma-ray and neutron-probe techniques are frequently used to measure water content. The typical range for specific yield is 0.10 to 0.25.

Solving for the mounding height is not straightforward, since it is included in both the H and μ equations [(7.2) and (7.3)]. Four different methods for solving Eq. (7.1) have been suggested (Finnemore, 1993). Three of these methods are presented below.

The most accurate method is by use of a computer or programmable scientific calculator that has an equation (root) solving capability. The equation is first stored and then values of each known variable are stored. Then the root solving capability is activated to solve for the unknown variable Z_m. This is the most convenient and accurate method for solving Eq. (7.1).

The second method is a graphical solution. The accuracy of this method is limited to reading from the graph. The graphical solution involves the grouping of terms into dimensionless quantities. These are:

$$f = \frac{Z_m}{d_0} \tag{7.7}$$

$$N = \frac{L^2 S}{16\, K d_0 t} \tag{7.8}$$

and

$$R = \frac{Q}{4\pi\, K d_0^{\,2}} \tag{7.9}$$

The resulting equation is

$$R = \frac{f(1 + 1/2\, f)}{3 - \gamma - J(r) - \ln\,\{N\,[1+(1/r^2)]/(1 + 1/2\, f)\}} \tag{7.10}$$

Since N and R contain all known quantities, a plot of f vs. R can be made for various values of r and N (Fig. 7.7).

The third method is an interactive procedure, starting with an assumed value for Z_m. The estimate Z_m is used to calculate H and μ, and the right side of Eq. (7.1). This yields an improved value of Z_m, and the process is repeated until the Z_m values agree within the desired accuracy. Using an estimate Z_m from the graphical plot (Fig. 7.7) will greatly reduce the number of iterations.

Mounding height is mainly influenced by three parameters, hydraulic conductivity, time duration since initiation of infiltration, and the natural depth of the saturated zone. The effect of specific yield of the aquifer is minor. The effect of hydraulic conductivity on mounding height and time is illustrated in Figs. 7.8 and 7.9. The influence of the natural saturated zone on mound height is illustrated in Fig. 7.10. These comparisons are based on an infiltration rate of 2500 gal/day over a square injection area of 100 ft².

Spray irrigation. Spray irrigation is a low-capital-cost system that can vary from soaker hose to irrigation delivery systems. Once the designed delivery rate for electron acceptor and nutrients is established, the design of a spray irrigation system is based on the necessary

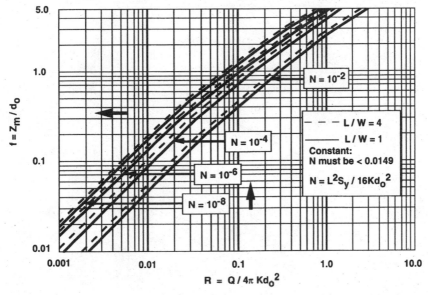

Figure 7.7 Graphical solution for the mounding height equation (7.1). (*Finnemore, 1993.*)

rate of discharge for a given surface area. The rate of discharge per unit length of a soaker hose should be measured with flowmeters under field operating pressure. Discharge rates vary with a hose, and actual field measurements are needed. Discharge can be controlled by automatic valves on a timed schedule. Soaker hose must be examined at intervals for deterioration as a result of sunlight and chemical reactions.

For a spray irrigation delivery system, the rate of discharge and the effective spray area must be determined. For a fixed nozzle there is a maximum distance for the spray's carry distance (Fig. 7.11). For a fixed rotating nozzle, the effective spray radius is a function of the nozzle head, its angle of discharge, and nozzle height above the ground. For a nozzle at ground level, the time duration that the liquid is airborne t is given by

$$t = 2\,t_h \qquad (7.11)$$

where t_h is the time for the water to reach its maximum height in the projectory. At this point, the vertical velocity component V_v is zero and

$$t_h = V\frac{\sin\alpha}{g} \qquad (7.12)$$

where V = discharge velocity at the nozzle

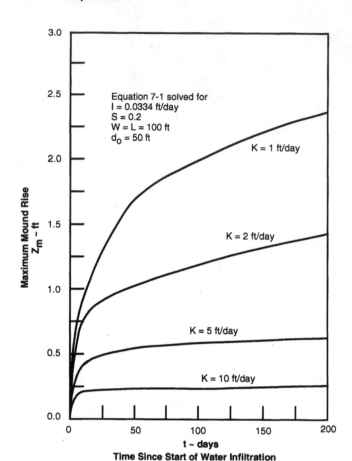

Figure 7.8 Effect of hydraulic conductivity on mounding height over short infiltration times. (*Finnemore, 1993.*)

α = angle of discharge

g = gravity constant

Since

$$V = C\sqrt{2\,g\,h} \tag{7.13}$$

where C is the nozzle discharge coefficient, then

$$t_h - C\sqrt{2h/g}\,\sin\alpha \tag{7.14}$$

If wind effects are neglected, the horizontal spray distance is

$$d = 2Vt_h\cos\alpha = 2\,C^2\,h\,\sin\alpha \tag{7.15}$$

For a rotating nozzle, d is the radius of the maximum spray pattern. Nozzle coefficients of discharge C vary from 0.85 to 0.92 (Fair, 1968).

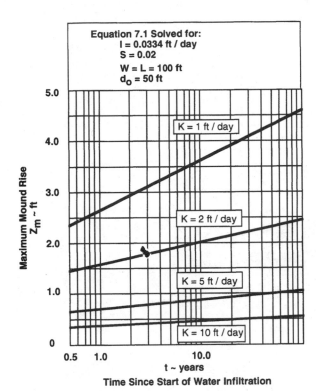

Figure 7.9 Effect of hydraulic conductivity on mounding height over long infiltration times. (*Finnemore, 1993.*)

The discharge rate Q is given by

$$Q = C(\Sigma a)\sqrt{2\,g\,h} \tag{7.16}$$

where Q = discharge rate, ft³/s
$\quad C$ = nozzle discharge coefficient
$\quad \Sigma a$ = sum of the individual area of nozzle openings
$\quad h$ = orifice head, ft
$\quad g$ = gravity constant

Subsurface applications. Trenches and subsurface drains are common methods for gravity injection. Subsurface drains are generally limited to shallow depths. The excavated area of pulled underground storage tanks also can be used as an infiltration gallery (Fig. 7.12). Appropriate sealing of sidewalls and the upgradient end of the excavated pit is necessary.

A trench drainage system is constructed by excavating a trench and laying perforated drainage pipe or a distribution medium such as large rock. The trench is then backfilled with gravel and sealed with

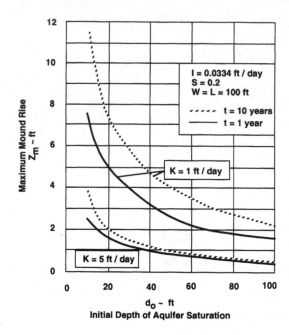

Figure 7.10 Influence of ground-water depth on mound growth. (*Finnemore, 1993.*)

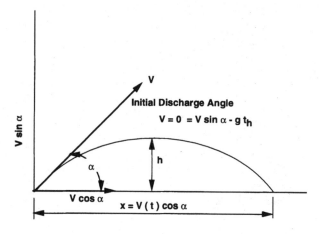

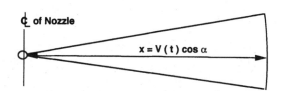

Figure 7.11 Design geometry and spray pattern of nozzles. (*Fair, Geyer, and Okun, 1968.*)

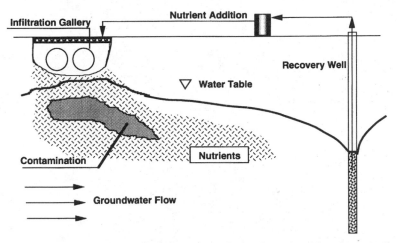

Figure 7.12 Infiltration gallery or excavated pit of a removed underground storage tank for gravity feed.

soil. Often the gravel is covered with fabric to prevent fine soil from entering the gravel from above and plugging the drain. An impermeable liner may be required on the upgradient end of a trench or on the surface to reduce rain infiltration.

Infiltration trenches have been placed near surface and at depths as great as 15 ft. Their lengths range from 10 ft to over 150 ft. A system of infiltration and extraction trenches is being used to bioremediate contaminated soil and a shallow aquifer at a large theme park in Texas (Piotrowski, 1993). This system is bioremediating a gasoline and diesel fuel plume approximately 400 ft long by 150 ft wide (Fig. 7.13). At this site the groundwater table ranges from 2 to 12 ft below the ground surface. Transmissivity values and percolation rates of 200 to 500 gal/day/ft and 0.5 to 2.6 in/h, respectively, exist for this shallow aquifer. An infiltration design rate of 5 gal/min was used after considering mounding effects. Groundwater modeling and site constraints due to structures resulted in a trench configuration consisting of three infiltration trenches and two extraction trenches (Fig. 7.13).

Design specifications for trenches vary, but it is typical to protect drains from clogging with filter fabric. The perforated pipe can be wrapped or the filter fabric can be placed on the gravel backfill. Shallow (1 to 3 ft) trench construction usually differs from deep (3 to 15 ft) trenches (Fig. 7.14). The discharge conduit is usually perforated pipe. Pipe of PVC, SCH8O, has been used. Orifice sizes of $\frac{1}{4}$ in or larger help reduce orifice clogging. Trenches are then backfilled with gravel and a layer of compacted soil. Shallow trenches may use an

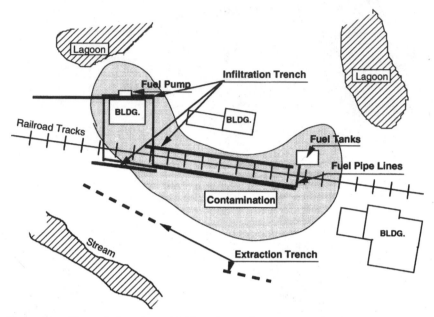

Figure 7.13 Example location of infiltration and extraction trenches relative to a fuel plume for in situ bioremediation. (*Piotrowski, 1993.*)

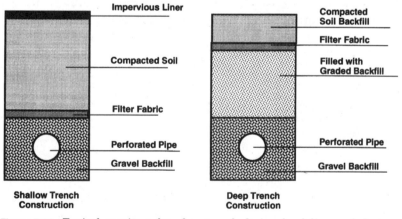

Figure 7.14 Typical gravity subsurface trench design for delivery of electron acceptor and nutrients.

impervious liner in areas of significant precipitation. Trenches tend to clog and design provision should provide a means to clear the trench of debris. Trench design can provide for pressurized back-flushing and points for pipe clean-out. For long trenches, riser pipes are installed at approximately 10- to 20-ft intervals (Figs. 7.15 and 7.16). Recovery trenches are usually equipped with groundwater pumps.

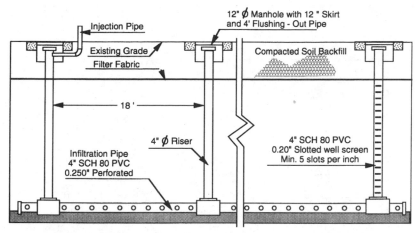

Figure 7.15 Typical design specifications for infiltration trenches. (*Piotrowski, 1993.*)

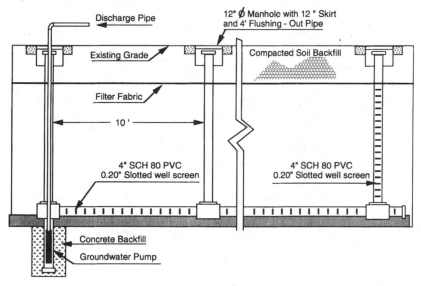

Figure 7.16 Typical design specifications for extraction trenches. (*Piotrowski, 1993.*)

An infiltration trench cannot be expected to deliver a uniform discharge over its length without proper design of the perforated pipe. Flow through a perforated pipe decreases at each opening. At least one bioremediation system has failed from this oversight. A pressurized recirculation system can be used to obtain superior distribution. The design of a perforated pipe distribution system requires flow to be discharged uniformly along the length of the pipe. This hydraulic

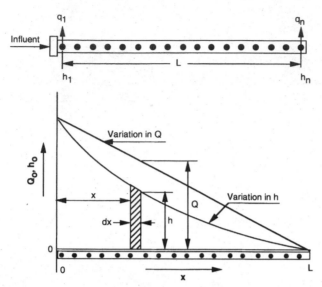

Figure 7.17 Variation of flow and head with uniformly decreasing flow. (*Fair, Geyer, and Okun, 1968.*)

equality is obtained by subjecting the dividing flow to equal fractional resistances. The flow is subdivided so that the discharge from the farthest orifice is near the value of the discharge of the first orifice in the pipe. The incoming flow to a distribution pipe Q is subdivided so that the discharge from an orifice is q_n (Fig. 7.17). The discharge from the first orifice is q_1 and the last is mq_1, where $m<1$.

If the outlets are orifices, the head loss across the first h_1 and last h_n orifice is given as

$$h_1 = Kq_1^2 \qquad (7.17)$$

and

$$h_n = K(mq_1)^2 \qquad (7.18)$$

combining these equations gives

$$h_n = m^2 h_1 \qquad (7.19)$$

The head loss from the entrance of the pipe at $x = 0$ to all points at each orifice exit must be the same (Fig. 7.17). Thus the head loss from the first orifice h_1 must equal the head loss in the last orifice h_n plus the loss in pipe length L between these orifices. If the head loss in a pipe of length L is given by h_L, then

$$h_L + h_n = h_1 \qquad (7.20)$$

and

$$h_n = m^2 h_1 \tag{7.21}$$

then

$$h_L = h_1 (1 - m^2) \tag{7.22}$$

For any desired flow mq_1 from the last orifice, the necessary head loss can be calculated.

For flow in a perforated pipe of constant diameter, the flow decreases in a stepwide manner at each orifice. If the outlets are evenly spaced, the flow decrease can be approximated by a uniform drop given by

$$Q_x = \frac{Q(L-x)}{L} \tag{7.23}$$

where Q_x = flow at any point of length x from the entrance to the pipe

Since head loss per unit length of pipe is proportional to the square of the flow velocity (Darcy-Weisbach equation),

$$\frac{h_0}{x} = K Q^2 \tag{7.24}$$

where K is a constant and h_0 is the head loss in a pipe.

The head loss from pipe friction at distance x from the inlet end of pipe L (Fig. 7.17) is

$$h_x = \int_0^L \frac{h_x}{x} \, dx \tag{7.25}$$

substituting and integrating yields

$$h_x = K Q^2 \left(x + \frac{x^2}{L} + \frac{x^3}{3L^2} \right) \tag{7.26}$$

Since

$$\frac{h_0}{L} = K Q^2 \tag{7.27}$$

one can express the head loss as

$$h_x = h_0 \left[\frac{x}{L} - \left(\frac{x}{L}\right)^2 + 1/3 \left(\frac{x}{L}\right)^3 \right] \tag{7.28}$$

If there was no orifice outlet except for the pipe end at L, then $x = L$ and

$$h_x = 1/3\, h_0 \qquad\qquad (7.29)$$

The head loss across a pipe with evenly spaced outlets is approximately equal to one-third of the head loss with no orifices.

To size a distribution pipe, the steps are as follows:

1. Given an orifice discharge coefficient, calculate the head loss h_1 across the orifice closest to the pipe's entrance [Eq. (7.17)].

2. Calculate the head loss across the length of pipe h_1 for a desired discharge from the farthest orifice [Eq. (7.22)].

3. Calculate the head loss h_0 that would occur in the pipe for the design delivery flow Q [Eq. (7.24) or (7.29)].

4. Use this head loss with the appropriate pipe flow equation to calculate a pipe diameter. The Hazen-Williams equation is typically used for full pipe flow.

Forced injection

Forced injection yields the greatest flexibility in process control, since the rate and location of delivery are controlled. The major limitation to forced injection is potential flooding when working with shallow water tables. Shallow injection points limit the amount of hydraulic head that can be applied to the injection system and limit the area of influence of injection wells.

Injection is achieved with wells screened over an interval corresponding to the zone to be treated. A concrete or grout seal usually is required to prevent water from flowing along the casing to the surface when the well is pressurized. Injection wells can be installed to any depth. In low-permeability soils, injection wells may be spaced closely to assure even, complete distribution of treatment chemicals. For silty aquifers, a larger number of wells will be required to provide the same treatment response that can be achieved in the more sandy-gravelly aquifers.

Permeability, aquifer thickness, and depth to water table all affect the rate of travel of injected water into a contaminated area. Each injection well has one or more zones of influence (Fig. 7.18). One zone of influence is the hydraulic zone that is dependent on the rate of injected water and the aquifer hydraulics. This zone can be established from pump testing. The second zone of influence is the travel distance of key bioremediation agents. The most important agent is the electron acceptor, usually oxygen. Other injected agents may include nutrients, a primary substrate, an electron donor, or other chemicals to optimize the biodegradation of the hazardous substances. Each of these chemicals has its own zone of influence as a

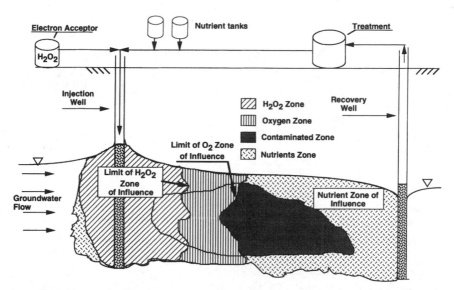

Figure 7.18 Zones of influence as established by transport and reaction rates during in situ bioremediation.

result of the concentration injected, adsorption on soil, and reaction or degradation potential. For example, the injection of hydrogen peroxide results in the following reactions:

Soil Catalyzed

$$H_2O_2 \rightarrow 1/2O_2 + H_2O$$

Microorganism Catalyzed

$$C_aH_b + O_2 \rightarrow aCO_2 + 1/2b\ H_2O$$

The zone of influence is a result of initial mass injection rates, travel time in the subsurface, and the rates of chemical and biological reaction. For bioremediation, the most important zone of influence is the one that supports the degradation of the contaminants. This zone is determined by the transport of necessary electron acceptor, substrates, and nutrients. The substance limiting the degradation rate establishes the effective zone of influence for biodegradation.

Since the degradation rate is usually oxygen-limited, the zone of influence for bioremediation is frequently established by the distance that oxygen can be carried to concentrations of at least 1 mg/L. This zone of oxygen influence determines the required well spacing and ultimate capital cost for force injection. Ideally, injection wells and any recovery well system should be at a distance that allows added nutrients to reach the area of contamination less than 6 weeks from time of injection. The travel time and reaction rates for injected chemicals are estimated from treatability studies and pump testing. These must be confirmed by field monitoring. The necessary studies for estimating zones of influence are discussed in Chap. 6 under assessments for in situ bioremediation and Chap. 10 on treatability studies.

Recovery systems

Recovery technologies also can be grouped under gravity and forced methods (Table 7.2). Gravity recovery depends upon intercepting the groundwater downgradient from the contamination zone. Fluid is normally collected in an intercepter system such as an open ditch or buried drain by simple gravity flow. Extraction trenches are designed like the injection trench system discussed above. The trenches should be designed with pressurized back-flushing and clean-out capabilities. Details for an extraction trench are provided in Fig. 7.16. Force recovery systems can use suction lifting and standard well pumps. Recovery by well points (suction lifting) is located downgradient from the contamination area (Fig. 7.19). Well points are best suited for shallow aquifers where excavation is not needed below more than about 22 ft. Beyond this depth, suction lifting is ineffective. Well pumps are used for greater depths and for greater hydraulic control. Recovery wells can be equipped with electronic water-level controllers to maintain a minimum drawdown for hydraulic control.

Injection and recovery well configurations

A variety of well patterns for injection and recovery can be used. Each site has its unique characteristics that influence well patterns. Patterns for well arrangements are a function of the variability of aquifer hydraulic conductivities, subsurface travel times, sustainable pumping rates, contaminated distribution, and subsurface as well as surface structures.

TABLE 7.2 Recovery Systems for in Situ Bioremediation

Gravity systems:
  Trench
  Buried drain
Forced systems:
  Well point
  Lysimeter
  Deep well pump

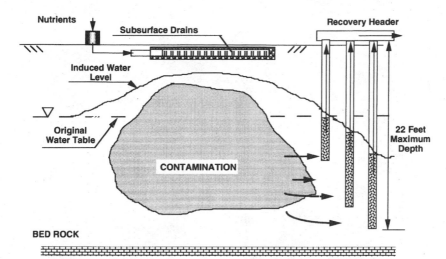

Figure 7.19 Recovery system using well points with subsurface drains.

The design of well systems can be aided by simulating groundwater flow patterns. Linear programs in combination with groundwater flow simulators have also been used to determine the best well-placement strategy. Other investigators use nonlinear programming combined with a groundwater flow model and an advective transport model. Pumping-injection patterns can be used to create stagnation zones and gradient barriers to pollution migration, control the trajectory of a contaminated plume, and intercept the trajectory of a contaminated plume.

Well patterns. Typical well patterns consist of (1) the injection-recovery pair (Fig. 7.20a), (2) a line of upgradient injection wells and downgradient recovery wells (Fig. 7.20b), (3) a series of injection wells with recovery wells around the boundary of the plume (Fig. 7.20c), and (4) a series of alternating injection and recovery wells that bisect the plume in the direction of flow with downgradient recovery wells (Fig. 7.20d). This well pattern is more efficient in mixing injected chemicals across a plume's gradient.

For tight aquifers (silty sand), adequate hydraulic control may require circulation of groundwater within the plume versus well patterns that are placed outside the plume. A triangular pattern has been used with one injection well surrounded by three extraction wells (Fig. 7.21a). One subsurface contamination in very tight soils was bioremediated by recycling water where natural groundwater flow was nonexistent (Hicks, 1992). A well pattern that has been used for tight aquifers is a five-spot pattern consisting of one injection well surrounded by four recovery wells (Fig. 7.21b)(Lee, 1988). Another

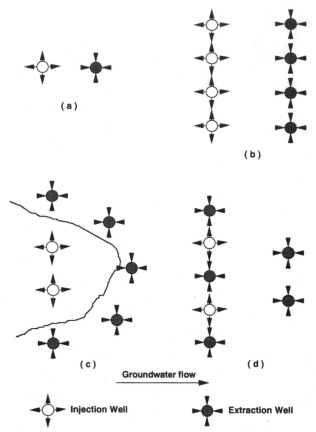

Figure 7.20 Typical injection and extraction well patterns used for in situ bioremediation.

used a nine-well pattern with one injection well surrounded by eight recovery wells (Fig. 7.21c).

A vertical hydraulic circulation system has been used for in situ remediation in Germany (Herrling, 1991). The system was originally developed for vacuum vaporization and called the UVB system (Unterdruck-Verdampfer-Brunnen). The vertical circulation is created at a single well by a specially designed well with two screen sections (Fig. 7.22). One screen is at the aquifer bottom and one near the groundwater surface. A well pump draws water in at the bottom screen and discharges at the upper screen. The resulting flow pattern is illustrated by streamlines in Fig. 7.23. Figure 7.23a illustrates the flow with zero natural groundwater flow and Fig. 7.23b with groundwater flow.

Other approaches combine forced injection and recovery with gravity feed systems. Each combination and configuration is site-specific. For

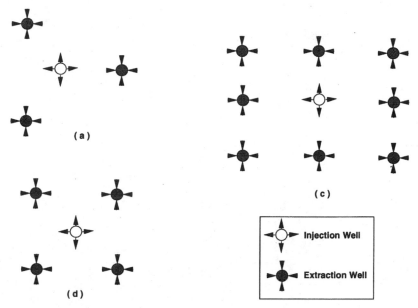

Figure 7.21 Injection and extraction well patterns for tight aquifers.

contamination above the groundwater table, oxygen can be supplied by air extraction and nutrients supplied by infiltration trenches. For contamination in both the vadose and saturated zone, the groundwater table can be raised, lowered, or remain unadjusted. If raised by water injection, oxygen is transported by water. Other systems have dropped the groundwater table by pumping and transport oxygen by applying a vacuum. Another approach is to provide oxygen by air extraction in the vadose zone and by oxygenated water in the saturated zone. Air sparging can supplement the oxygen carried by water injection.

Injection and recovery patterns may be determined by the need to establish specific microbial responses. Cometabolism, nitrate- or sulfate-reducing microorganisms, and multiple metabolic modes require designs that promote the proper microbial response. Mineralization of some compounds requires both anaerobic and aerobic metabolism. Providing a sequence of anaerobic and aerobic metabolic systems for in situ treatment is significantly more difficult to design and operate.

A two-zone in situ treatment approach has been proposed for the treatment of multiple contaminants (Vira and Fogel, 1991). Contaminants requiring anaerobic dehalogenation are first reduced in an anaerobic metabolic zone. These reduced products and other contaminants that are not mineralized by the anaerobic activity are then mineralized in a downgradient aerobic metabolism zone.

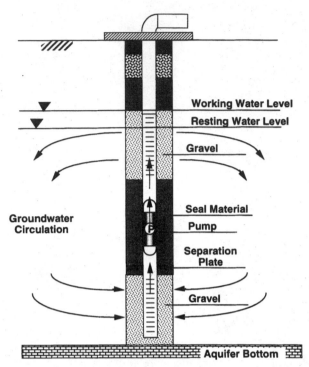

Working Water Level

Resting Water Level

Gravel

Groundwater
Circulation

Seal Material

Pump

Separation
Plate

Gravel

Aquifer Bottom

Figure 7.22 The UBV well circulation system for in situ bioremediation. (*Adopted from Herrling, 1991.*)

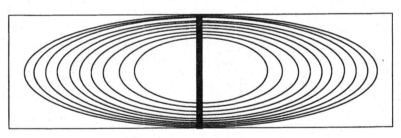

(a) Streamlines With No groundwater Flow

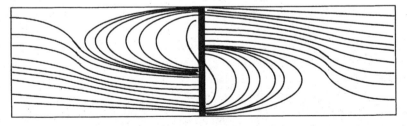

(b) Streamlines With Groundwater Flow

Figure 7.23 Flow patterns of the UBV well installation. (*Herrling, 1991.*)

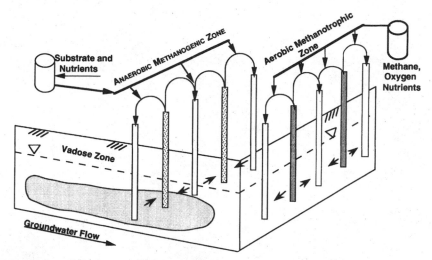

Figure 7.24 Injection-recovery systems for two-zone metabolic response. (*Vira, 1991.*)

To achieve this dual metabolic system a series of alternating injection and recovery wells are located in parallel rows to create the two-zone metabolism system (Fig. 7.24). The first row of injection-recovery wells is located immediately downgradient of the contamination plume. Nutrients and, if necessary, a primary substrate are injected to promote methanogens. This microbial metabolism mode promotes dehalogenation of chlorinated organic compounds.

The second row of injection-recovery wells is located at an appropriate distance downgradient of the first row to establish aerobic metabolism. This distance is a function of groundwater flow rates and the rate of dehalogenation. Significant dehalogenation is necessary before transformed compounds reach the second row of process control wells. The aerobic metabolism can be used to establish a consortium of organisms, or a specific group can be promoted. One potential candidate would be methanotrophic bacteria that can oxidize transformation products that are not completely dehalogenated.

Horizontal wells.   Forced injection has mainly employed vertical wells, but a few sites have used horizontal wells. The horizontal well has the advantage of delivering agents to the subsurface that cannot be reached by vertical wells because of surface structures or even river beds. Horizontal wells do not have to be drilled directly above a contaminant source to remove contaminants. Unlike vertical wells, horizontal boreholes are advanced in the direction of the plume for distances up to 500 ft.

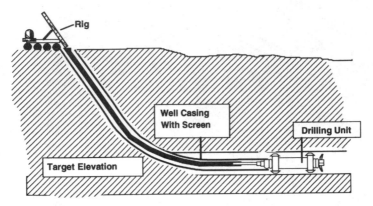

Figure 7.25 Horizontal wellbore system. (*Langseth, 1990.*)

Horizontal wells should provide superior performance to that available from a greater number of vertical wells even when pumping the horizontal wells at a lower total rate than the vertical wells (Langseth, 1990). Plume thickness is an important variable when comparing horizontal versus vertical well performance. Horizontal wells outperform vertical wells on thin plumes, and thick plumes respond better to vertical wells. The main disadvantage to vertical wells is that the cost can exceed $200 per foot. Costs are typically between $150 and $200 per foot. The Department of Energy is evaluating horizontal well application with in situ bioremediation at the Savannah River site.

The horizontal well is first straight-drilled, at an angle, to the required depth so a 100-ft-radius curve will reach the horizontal at the desired vertical depth. A curved drilling assembly is used to create the curve and horizontal portion of the well (Fig. 7.25). This drilling assembly is comprised of a dual drill string, a hydraulic downhole motor, an expanding drill bit, and a tool face indicator and inclination measurement device (Fig. 7.26) (Karlsson, 1990). The downhole motor converts the hydraulic energy of the pumped drilling fluid into mechanical energy that rotates the bit. The directional control on the motor is accomplished by stabilizer rings on the motor housing that place the motor in an eccentric position in relation to the hole axis. By orienting the direction of the bit offset, the hole can be steered.

Clogging Control

A potential operating problem is clogging of the aquifer and injection wells. Most clogging results from three mechanisms: suspended solids, biological growth, and aquifer chemistry. Wells, however, have

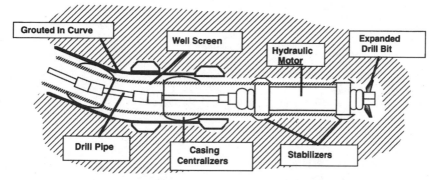

Figure 7.26 Horizontal drilling downhole equipment. (*Langseth, 1990.*)

been clogged from compaction of the gravel-sand pack around the well through surging during well development.

Suspended solids clogging

Suspended solids have rapidly clogged several aquifers (Caramagno, 1991). Suspended solids can be present since injected water is usually from a recovery system with or without surface treatment. The source of suspended solids can be clays from the recovery well, floc from poor surface treatment, or biological growth. Clogging an aquifer with suspended clay makes permeability recovery almost impossible. Suspended solids should be reduced to levels less than 2.0 mg/L in the injected water. Turbidity can be used as a control measurement and should be below 1 nephelometri turbidity unit (NTU) before water is injected.

Chemical or biological surface treatment of recovery well water can increase the suspended solids content. This is a function of the type of treatment provided. Biological growth and floc of biological and/or chemical origin are created during surface treatment. Dissolved alum or ferric hydroxides can cause clogging after injection because of pH and oxidation-reduction potential shifts in the aquifer. The use of alum in water treatment, even after flocculation and settling, will result in trace amounts of metal hydroxides in the injected water. The pH must be controlled for minimum solubility of these hydroxides prior to settling in a surface treatment system. A filter polishing unit is recommended.

If minimum solubility has not been achieved at the surface, a pH shift in the subsurface will result in metal hydroxide precipitation. The pH range for minimum solubilities for ferric and aluminum hydroxides is between 5 and 7. Once all precipitate has been settled in the surface treatment unit, and the supernatant removed, the pH can be adjusted to a level above pH 7 for countering the acidity produced by biodegradation in the subsurface. The response of surface-

TABLE 7.3   Rates of Microorganism Transport in
Soil

Soil type	Distance traveled, m/week
Fine sand	0.7
Sand and sandy clay	1.3
Sand and pea gravel	150
Coarse gravel	200
Crystalline bedrock	160–200

SOURCE: Hagedorn, 1984.

treated water to pH and redox potential changes as a result of mixing with groundwater should be evaluated before injection.

Growing seed organisms in surface reactors and injection can enhance specific microbial response and provide shorter acclimation periods. Microorganisms can be transported in the subsurface. Movement mainly depends on the hydrological conditions, and transport can be more rapid than an inorganic tracer (see Fig. 6.16). However, the injection of undispersed microbial flocs should be avoided to prevent clogging. The rates of microorganism transport in various soil types are presented in Table 7.3.

Biofouling

Injection screens.   A major operating problem with in situ bioremediation is microbial clogging at the injection wells. The addition of oxygen and nutrients stimulates biological growth at the well screens and the surrounding aquifer media. To reduce clogging potential at the injection well, a pulse injection procedure can be utilized. The electron acceptors and nutrients are separately pulsed at specified intervals. The separate injection of agents eliminates optimum conditions at the injection point, discouraging biological growth.

Pulse injection requires information on the transport phenomenon in the aquifer, since a uniform distribution in the contaminated zone is achieved by hydrodynamic dispersion. Pulsing at the adequate frequency and duration allows for natural dispersion to create a uniform distribution by the time the agents reach the contaminated zone. The pulse injection can be cycled with periods of bactericide treatment at the well screen and boreholes. Hydrogen peroxide has been used at concentrations of 200 to 500 mg/L as a bactericide.

It is easier to control injection well clogging by staggered injection when more than one substrate is necessary for supporting biological growth, as with cometabolism. Under this metabolism mode a primary substrate must be injected as well as the electron acceptor and pos-

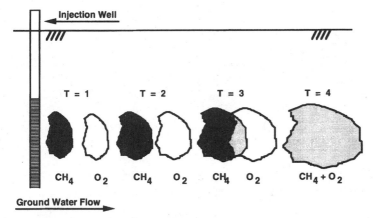

Figure 7.27 Clogging prevention by pulse injection of electron acceptor and primary substrate. (*Semprini, 1988.*)

sibly nutrients. The secondary substrate is the contaminant. The primary substrate and electron acceptor are injected as separate pulses (Fig. 7.27). The pulse injection of two essential agents for in situ treatment has been used for the dehalogenation of chlorinated organic compounds.

The primary substrate, methane, is injected with the appropriate electron acceptor, oxygen, to support cometabolism for dehalogenation with methanotrophic bacteria. The alternating time pulse injection of these two agents separately prevents excessive biological growth at the injection point. This principle was applied to the pilot in situ bioremediation of chlorinated hydrocarbons at the Moffett Naval Air Station site, Mountain View, Calif. Methane and oxygen were injected at alternating time pulses. The alternating time pulses were controlled so the methane and oxygen mixed at the contaminated zone as a result of hydrodynamic dispersion during transport in the aquifer. The methanotrophic bacteria are then simulated at the point of contamination.

To design a dual pulse injection system for methanotrophs, it is necessary to select:

1. The ratio of the individual pulses of methane and oxygen
2. The overall pulse length
3. The cycle frequency

The ratio of methane to oxygen for the individual pulses is calculated from knowledge on the stoichiometry of methane oxidation as discussed in Appendix B. Based upon the oxidation-reduction reaction, 2 moles of oxygen are required per mole of methane, corresponding to a mass ratio of 4 g of dissolved oxygen per g of methane.

The overall pulse length and frequency depend on the aquifer transport characteristics and the hydrodynamic dispersion of the aquifer system. Tracer testing between the injection and recovery wells provides data for establishing these characteristics. The pulse length can be established by employing a transport model that provides for periodic input of alternating agents (Valocchi and Roberts, 1983). Valocchi and Roberts modeled the effects of convection, dispersion, and adsorption on transport and mixing under conditions of uniform flow. The model was correlated with pulse injection of a tracer and the adjustment of dispersion coefficient to match field measurements. The model is then used to provide the operating sequence for pulsing methane and oxygen individually into the aquifer.

Iron bacteria. A difficult biofouling problem to deal with results from the growth of iron bacteria. Iron-related biofouling is a common problem in water-well maintenance and surface-treatment systems. Even waters that have little measurable iron have yielded iron biofouling of fixed-film bioreactors.

Several bacteria have been implicated in water-well biofouling problems. The two predominant species are *Gallionella ferruginea* and *Lepothrix* spp. (Tuhela, 1993). *Gallionella* is the predominant biofouling organism, since it is widely distributed in the environment. The organism is capable of deriving carbon from CO_2 or from organic compounds (Hallbeck, 1991). The bacterium lives as a free-swimming, flagellated cell and produces a stalk when exposed to ferrous iron. The organism is capable of oxidizing ferrous iron on the stalk or sheath surface:

$$4\,Fe^{2+} + O_2 + 4\,H^+ \rightarrow 4\,Fe^{3+} + 2\,H_2O$$

The ferric ions yield ferric oxides such as ferrihydrite ($Fe_5HO_8 \cdot 4H_2O$) that have low solubilities (Tuhela, 1993):

$$5\,Fe^{3+} + 12\,H_2O \leftrightarrows Fe_5HO_8 \cdot 4H_2O + 15H^+ \qquad K_{sp} = 10^{-40}$$

Chemical clogging

The final clogging mechanism, chemical precipitation, is difficult to predict because of the interactions of nutrients, substrate, oxygen, and biological activity with aquifer chemistry. Treatability studies and computer models on geochemical equilibrium can be applied to evaluate the potential effect of aquifer chemistry on the injected water mixture. Chemical reactions include oxidation and reduction, precipitation, and ion exchange. For most aquifers, ion exchange does not have a significant effect on clogging potential (Caramagno, 1991). The most prevalent problems result from changes in the redox potential and the addition of nutrients.

Redox changes. During the injection of oxygenated water, reducing conditions are changed to oxidizing conditions. Redox reactions involving iron, sulfate, and ammonium will occur during bioremediation. Many groundwaters contain sufficient iron, resulting in precipitation.

The oxidation of reduced iron minerals, such as pyrite, produces acid. Acids are also produced by microbial activity. This acidity can be neutralized by the carbonate buffer system in the aquifer, or the acidity may remain if the aquifer is naturally acidic, such as the Magothy aquifer in New York. The ultimate state of the iron, sulfur, and other ions depends on the pH and pE, electron activity, within the aquifer after injection.

The graphical representation of redox equilibria, called pE-pC diagrams, is useful for evaluating potential clogging reactions. Procedures for the development of these diagrams are provided by Snoeyink, 1980. Although these diagrams assume equilibria and may not accurately represent the natural system, they provide significant information on the likely chemical changes from injected water.

The solubility of iron in relation to pH and pE is diagramed in Fig. 7.28 (Snoeyink, 1980). The diagram has been developed for a total inorganic carbon content ($C_{T,CO3}$) of 10^{-3}M and a total inorganic sulfur content ($C_{T,S}$) of 10^{-4}M. The multiple boundaries between the solid phase and solution phase (Fe^{2+} or Fe^{3+}) represent different solution concentrations of iron. The iron concentrations range from 0.57 mg/L (10^{-5} M) to 570 g/L (10 M).

If one visions the upper portion of the diagram in the oxygenated zone of an aquifer and the bottom portion of the diagram in the anaerobic zone of the aquifer, the iron species and concentration can be estimated for a well sampled at the three indicated depths. Each depth is labeled as 1, 2, and 3 on the diagram. At well sampling location 1, the water oxygen is in equilibrium with the atmospheric oxygen. The iron content of the water is governed by the equilibrium:

$$Fe\ (OH)_3\ solid + 3H^+ \leftrightharpoons Fe^{3+} + 3\ H_2O \qquad K = 10^3$$

Since the pH is 6 to 7, the iron content of the water is low, less than 1 μg Fe^{3+}/L. Actual groundwater may contain iron concentrations that are higher owing to organic complexes, but concentrations are usually less than 0.5 mg/L.

Within the anaerobic zone at sampling location 3, the iron is in the form of ferrous iron and the sulfur is in the sulfide form. Iron disulfide has precipitated and the well water iron concentration is governed by

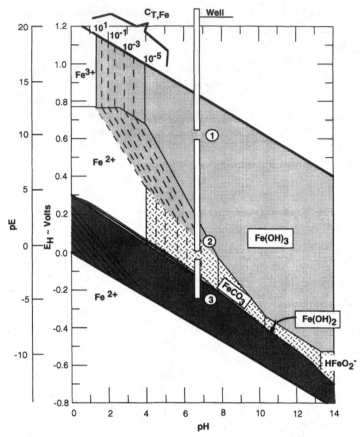

Figure 7.28 Iron solubility in relation to pH and pE at 25°C. (*Snoeyink, 1980.*)

$$\text{FeS}_2 \text{ solid} \rightleftharpoons \text{Fe}^{2+} + \text{S}_2^{2-} \qquad K = 10^{-26}$$

Water at sampling location 3 also will have low soluble iron concentrations and contain some hydrogen sulfide.

At sampling location 2, the iron concentration is controlled by ferrous carbonate (FeCO_3):

$$\text{FeCO}_3 \text{ solid} + \text{H}^+ \rightleftharpoons \text{Fe}^{2+} + \text{HCO}_3^- \qquad K = 10^{-4.4}$$

For a bicarbonate (HCO_3^-) concentration of 10^{-3}M, the iron concentration would be approximately 20 mg/L.

The injection of oxygenated water into well location 2 or 3 will change the pE, resulting in an equilibrium shift. As Fe^{2+} is oxidized to Fe^{3+}, more Fe^{2+} is released by the equilibrium shift, causing dissolution of FeS_2. The additional Fe^{2+} will ultimately be oxidized to Fe^{3+}, resulting in more Fe^{3+} and eventually more $Fe(OH)_3$.

Nutrients. Phosphate is another chemical element that has caused aquifer clogging. Orthophosphate has been responsible for some aquifer clogging during bioremediation (Aggarwal, 1991). The development of phosphate precipitates may not be immediate because of the need for a precursor formation. For example, the precipitation of hydroxyapatite or chloroapatite is preceded by the formation of a more soluble octacalcium phosphate or brushite. These precursors are transformed upon aging to the stable precipitate (Aggarwal, 1991).

Orthophosphate concentrations below 10 mg/L should minimize precipitation in most geochemical environments (Aggarwal, 1991). This concentration, however, may not be adequate for bioremediation given the difficulty of subsurface transport of phosphate. Phosphate should not be injected in calcareous soils. For such soils, much of the added phosphate will be adsorbed and precipitated. Alternate phosphate sources have been used for in situ bioremediation. These include sodium tripolyphosphate and other polyphosphates. Possible species include pyrophosphate $[P_2O_7^{-4}]$, and trimetaphosphate $[P_3O_9^{-3}]$.

The use of computerized geochemical models to calculate chemical species and saturation levels for given pH, pE, and element concentrations provides greater insight on potential clogging situations. Several models are listed below:

- U.S. EPA MINTEQA2 (EPA 600/3-87/012)
- SOLMINEQ.88 of the U.S. Geological Survey (Kharak, 1988)
- PHREEQE Model (Parkhurst, 1980; Plummer, 1990)

Treatability tests should be performed to evaluate chemical responses in actual soil samples prior to any subsurface injection. The addition of any chemical to the subsurface must be evaluated, and it is always best to inject chemicals separately rather than as a mixture. Separate chemical mixing and storage tanks should be used for each supplement.

In Situ Oxygen Sources

For those systems requiring aerobic microbial processes, oxygen is generally the limiting factor for in situ treatment. Oxygen can be added by forcing molecular oxygen into the system or by chemical reaction. Molecular oxygen can be carried in water or air. With water as the carrier, oxygen is dissolved from air or pure oxygen and injected or infiltrated into the contaminated subsurface. With air, forced air injection or a vacuum is applied to the subsurface. Chemical oxygen supply involves the addition of a chemical that can be converted to oxygen such as hydrogen peroxide.

In selecting an oxygen supply system, the following criteria must be evaluated:

1. Oxygen demand of the contaminant load and its location

2. Oxygen demand of natural organic deposits

3. Design rate of degradation, which establishes the oxygen delivery rate

4. Oxygen transfer rate supplied by each method

5. Ease of transportation and utilization of oxygen source

6. Cost of the oxygen source

The ultimate oxygen demand, coupled with the rate of degradation, establishes the rate at which oxygen must be delivered. Oxygen supply should be balanced with oxygen demand to optimize the cleanup time for bioremediation. Too low an oxygen supply rate relative to the contaminant load results in extended remediation times. Too high a rate of oxygen supply results in elevated remedial cost and potential for soil gas binding.

Treatability studies on hydrocarbon degradation usually follow a pseudo-first-order rate of degradation and oxygen demand. However, this is seldom demonstrated in field applications, since oxygen is usually limiting the rate of degradation. For PAHs, the limiting rate can be that of organic desorption from the soil matrix. If a field design attempts to provide the required rate of oxygen for optimum biodegradation, the delivery system may be oversized after several months of treatment. The cost-effectiveness of a system's delivery capacity must be balanced with the bioremediation cleanup time.

The oxygen demand rate and contaminant location determine the feasibility of various oxygen supply systems. Oxygen can be delivered by:

1. Aeration of a well bore

2. Injection of oxygenated water

3. Injection of hydrogen peroxide

4. Venting

5. Injection of air

Aeration of a well bore

The simplest system with the lowest equipment cost is diffusing air into a well bore through a porous bubbler device. Sparging of well

TABLE 7.4   Oxygen Supplied by Aeration of a Well Bore

Hydraulic conductivity, gal/day/ft^2	Hydraulic gradient, ft/ft					
	High 0.1		Medium 0.001		Low 0.001	
	Air, lb/day	Pure oxygen, lb/day	Air, lb/day	Pure oxygen, lb/day	Air, lb/day	Pure oxygen, lb/day
10^4 (gravel)	6	30	0.6	0.3	0.06	0.3
10^2 (medium sand)	0.006	0.3	6×10^{-3}	3×10^{-2}	6×10^{-4}	3×10^{-3}
10^{-1} (silt)	6×10^{-5}	3×10^{-4}	6×10^{-6}	3×10^{-5}	6×10^{-7}	3×10^{-6}

SOURCE: Brown and Crosbie, 1989.

water for aeration has used carborundum diffusers, porous stone, cinc-tured metal, fitted glass, or silicon carbide diffusers, and Du Pont Viaflo tubing (Lee, 1988; Brown, 1989). Air spargers of silicon carbide can defuse atmospheric air into groundwater at a rate of up to 10 ft^3 of air per minute (Brown, 1989). The oxygen-saturated water diffuses out into the subsurface. The amount of oxygen delivered to the contamina-tion zone is a function of the rate of water flow from the well bore. This, in turn, is a function of the hydraulic conductivity and the gradi-ent in the subsurface area of the formation affected by contamination.

Table 7.4 provides the pounds of oxygen per day per well from aera-tion of a well bore for different hydraulic conductivities and gradients (Brown, 1989). These calculations assume a 30-ft saturated thickness and that the lateral influence of the well is 3 ft. The rate of oxygen supplied is significantly affected by the hydraulic conductivity. For the same hydraulic gradient, it varies from a low of 6×10^{-5} lb/day to a high of 6 lb/day. Depending upon the temperature of the groundwa-ter, only 8 to 10 mg/L of oxygen can usually be dissolved by air sparg-ing. Pure oxygen will increase the amount of oxygen delivered by 5 times, but with increasing operating cost.

Injection of oxygenated water

The most common method of delivering oxygen is by pumping oxygen-saturated water into a contaminated aquifer. The pounds per day of oxygen supplied is a function of injection rate and the concentration of oxygen in the water (Table 7.5).

Both aeration of a well bore and pumping of oxygenated water into the subsurface have limitations. Biofouling often inhibits flow of air outward into the well bore. Biofouling occurs at the well's screen and in the adjacent soil, decreasing flow and diffusion into and out of the soil zone adjacent to the injection well. Remedying these problems generally entails the pulling of the spargers for cleaning and treat-ment of wells with an appropriate chemical such as hydrogen perox-

TABLE 7.5 Oxygen Supplied by Oxygenated
Water Injection

Injection rate, gal/min*	Oxygen supplied, lb/day, single well	
	Aerated water, 10 mg/L DO	Oxygenated water,* 50 mg/L DO
1	0.12	0.60
10	1.2	6.0
50	6.0	30.0
100	12.0	60.0
150	18.0	90.0

SOURCE: Based on Brown and Crosbie, 1989.
*Other injection rates and DO concentrations can
be determined by direct proportion.

TABLE 7.6 Mass Requirements to Deliver Oxygen or Nitrate for
Mineralization of Hydrocarbon

Electron acceptor	Concentration and carrier medium	Mass requirements, kg carrier per kg hydrocarbon
Oxygen (in air)	8 mg/L in water	400,000
Oxygen (pure)	40 mg/L in water	80,000
Oxygen (in H_2O_2)	100 mg/L H_2O_2 in water	65,000
Oxygen (in H_2O_2)	500 mg/L H_2O_2 in water	13,000
Nitrate	50 mg/L in water	90,000
Oxygen (in air)	20.9% in air	13

SOURCE: Hinchee, 1991.

ide (H_2O_2). The frequency at which biofouling occurs often limits the
application of these aeration techniques.

The low rate of oxygen delivery is a major disadvantage of these
methods. Soil contaminated with 1000 gal of hydrocarbons will
require the delivery of 200 million gal of water at 10 mg/L of dissolved
oxygen to satisfy the biodegradation demand for oxygen. For these
reasons, hydrogen peroxide and soil venting have become popular
options. The mass requirements to deliver oxygen or nitrate for bioremediation of hydrocarbons are compared in Table 7.6.

Injection of hydrogen peroxide

Oxygen can be chemically supplied by hydrogen peroxide. Hydrogen
peroxide is highly soluble in water and decomposes to give water and
oxygen:

$$H_2O_2 \rightarrow H_2O + 1/2\ O_2$$

One part of hydrogen peroxide supplies 0.47 part of oxygen. In the presence of soil, 500 mg/L of hydrogen peroxide produces 235 mg/L of dissolved oxygen. Pumping a solution of this strength at 100 gal/min would deliver 280 lb/day of oxygen (Table 7.5).

A hydrogen peroxide delivery system needs a saturated subsurface to prevent significant hydrogen peroxide decomposition. Forced injection is usually necessary, since gravity feed of hydrogen peroxide yields poor results. With gravity feed, the hydrogen peroxide does not mix adequately with the groundwater. The major problem in using hydrogen peroxide is controlling the rate of its decomposition.

Hydrogen peroxide decomposes because of soil catalyzed reactions. Soils containing high concentrations of natural organic compounds, iron, nickel, and copper catalyze this decomposition. Ferric iron will catalytically decompose H_2O_2 (Lawes, 1991). An alkaline pH is also known to accelerate its decomposition. Soil degradation can limit the usable hydrogen peroxide concentration from 500 mg/L for clean sands to as little as 100 mg/L for other soils (Brubaker, 1989). Gradual increases in hydrogen peroxide concentrations have been continued up to a level of 1000 ppm of H_2O_2 (Staps, 1989). Concentrations of 100 to 500 mg/L are frequently used for most in situ applications. At 100 mg/L or above, hydrogen peroxide keeps the well free of heavy biological growth. This prevents fouling, allowing more equal and quicker transmission of oxygen to the treatment area. At concentrations of 1200 to 2000 mg/L, hydrogen peroxide has been found to be toxic.

The bacterial tolerance to hydrogen peroxide varies and may be very low for some species. Of particular importance is the report that hydrogen peroxide partially inhibits methanotrophic activity (Mayer, 1988). Lethal concentrations of hydrogen peroxide have been established at 30 to 50 mg/L for soil microorganisms.

Lawes (1991) measured the first-order rate constant for the decay of H_2O_2 in various soils. The decay rate to the base e varied from -1 per hour for soil in northwest Vermont to -69 per hour for soil from Stockton, Calif. The decay rate decreases with additional applications of H_2O_2, dropping by one-third the initial rate after the fourth treatment. Thus the potential exists to carry H_2O_2 greater distances in the subsurface after repeated applications. A significant portion of the decay of H_2O_2 is related to bacterial activity.

Because of soil-induced decomposition, H_2O_2 is not practical for unsaturated zone bioremediation. The bulk of the oxygen is lost and not utilized for bioremediation. The oxygen utilization efficiency E in percent was evaluated for in situ bioremediation of a sandy unsatu-

rated soil at Eglin Air Force Base (Hinchee, 1988). Utilization efficiency was calculated by

$$E = \frac{100 \, k_2(O_2)}{k_1(H_2O_2)}$$

where k_1 = oxygen generation rate from H_2O_2 decomposition
      k_2 = oxygen utilization rate for degradation
  (H_2O_2) = hydrogen peroxide concentration
    (O_2) = dissolved oxygen concentration

Field observations of k_2 were 0.06 to 0.006 per hour, and k_1 of 6.0 to
0.6 per hour. For the best case and H_2O_2 and O_2 concentrations of 300
mg/L and 40 mg/L, respectively, the utilization efficiency was only
1.3 percent. Thus a significant portion of the oxygen is lost. Since the
H_2O_2 was delivered with infiltration galleries and spray irrigation,
most of the oxygen was probably lost as a gas from the infiltration
galleries. In the saturated zone, a higher percent utilization occurs
because some of the oxygen is dissolved in water and some gas phase
is consumed in the vadose zone.

Rapid hydrogen peroxide decomposition can supersaturate the
groundwater with oxygen before reaching the desired location. Excess
oxygen has the potential to cause gas blockage of the formation. This
phenomenon limits the concentration at which hydrogen peroxide can
be used. The overall result is a loss of oxygen and poor general transport of dissolved oxygen to the contamination zone.

If hydrogen peroxide is the selected oxygen source, consideration
should be given to the selection of nutrient species that help prevent
its decomposition. Hydrogen peroxide can be carried greater distances
when stabilized with phosphate. To reduce phosphate adsorption to
the soil, a combination of simple and complex phosphate salts has
been used. The injection of phosphate can be twofold, (1) as a nutrient
and (2) to stabilize hydrogen peroxide. Potassium monophosphate will
help stabilize hydrogen peroxide solutions (Lee, 1988), and orthophosphate will block most of the iron-induced decomposition of H_2O_2
(Lawes, 1991).

In Situ Cometabolism

Most in situ bioremediation projects are designed to utilize the contaminant as the primary substrate. When contamination consists of
chlorinated aliphatic compounds, a cometabolism system is usually
desired. Cometabolism usually requires the injection of a primary
substrate. When the contamination is mixed with both petroleum

hydrocarbons and chlorinated organic compounds, the petroleum hydrocarbons can serve as the energy and carbon source for cometabolism of the chlorinated aliphatic compounds.

This approach has been applied in situ by the U.S. EPA at Traverse City, Mich. At this site, hydrogen peroxide was injected to stimulate the aerobic transformations of the petroleum hydrocarbons. After oxygen depletion, the in situ system supported anaerobic dehalogenation of the chlorinated hydrocarbons. The transformation products from the aerobic activity and one of the original contaminants, toluene, served as the energy and carbon source during this anaerobic dehalogenation. The reductive dehalogenation of tetrachloroethylene, trichloroethylene, and cis-dichloroethylene was complete, with ethene remaining as an end product. Ethene is considered an acceptable end product.

Although in situ bioremediation projects have mainly used oxygen as the electron acceptor, in situ studies have evaluated nitrate. Field demonstrations have been conducted in Germany, by U.S. EPA at Traverse City, Mich., and by Stanford University. Indigenous denitrifying organisms can be rapidly stimulated when an energy source and nitrate are present.

Microorganisms that use nitrate as an electron acceptor have been stimulated in situ by injecting acetate as a primary substrate and nitrate as the electron acceptor (Semprini, 1991). This in situ study evaluated the ability to stimulate the cometabolism of chlorinated aliphatic compounds. Sulfate-reducing bacteria also are stimulated by acetate injections. Thus a consortium of microorganisms developed when both nitrate and sulfate were available as electron acceptors. A transitory buildup of nitrate has been observed within the first 60 h of acetate injection. The in situ response yielded a ratio of nitrate to acetate consumption of 1:1, which is lower than the stoichiometric conversion of nitrate to nitrogen gas (Semprini, 1991). This lower ratio is expected when considering the biomass cellular growth. Acetate was pulse fed to prevent clogging problems at the injection well screens. Semprini (1991) used a pulse cycle of 320 mg/L of acetate for 1 h out of a 13-h period. Nitrate was injected continuously.

Most injected nitrate and acetate is consumed within a short distance of the injection well. However, degradation of carbon tetrachloride occurred in a zone farther removed from the presence of nitrate and acetate. A second microbial population that is stimulated by the denitrifiers has been proposed as the dehalogenator.

At the Traverse City site, nitrate was injected without acetate as a primary substrate. The object was to utilize the contaminant, JP-4 jet fuel, as the substrate. Since these compounds are more difficult to degrade than acetate, a longer acclimation period should be anticipat-

TABLE 7.7   Average Removal Efficiencies for TCE

Concentration of phenol injected, mg/L	TCE concentration, μg/L	Removal %*	Transformation yield, g of TCE/g of phenol
12.5	62	60–89	0.0044
12.5	125	68–87	0.0087
12.5	250	70–88	0.018
12.5	500	68–88	0.035
12.5	1000	68–77	0.062
19	1000	75–85	0.045
25	1000	75–90	0.036

SOURCE: Hopkins, 1993.
*Range represents the average numbers for three monitoring wells.

ed. The U.S. EPA experienced a 60-day acclimation period before nitrate uptake abruptly increased (U.S. EPA, 1991). Nitrate was injected at concentrations below the drinking water standard of 10 mg/L. The in situ microbial activity decreased the nitrate concentration to 0.1 to 0.5 mg/L.

For the BTEX components of jet fuel, benzene was removed below drinking water standards and no toluene could be detected in monitoring wells. The removal of ethylbenzene and xylenes was varied. Some monitoring wells were below detection limits and a few still contained levels of 4 to 56 μg/L (U.S. EPA, 1991). Oxygen was present in the injected water and measurable in monitoring wells. Thus it is difficult to establish the portion of BTEX oxidization coupled with NO_3 versus O_2 at the Traverse City site.

The cometabolism of trichloroethylene (TCE) by injection of phenol has been evaluated during in situ pilot studies (Hopkins, 1993a). At the Moffett Naval Air Station, Mountain View, Calif., the efficiency of TCE dehalogenation increased with increasing phenol concentrations (Table 7.7). Phenol-oxidizing bacteria were responsible for the degradation of TCE. The responding growth in microbial population caused oxygen demands to increase from 1.5 g of oxygen/g of phenol to 2.0 g of oxygen/g of phenol after 1000 h of phenol injection. The stoichiometric oxygen demand for phenol mineralization is 2.4 g of oxygen/g of phenol.

Electron Acceptor Delivery Rate

Acclimation for biological response is necessary in the field as typical of laboratory systems. The demands for electron acceptors will not maximize until the microbial population has developed. Field operations are usually started with lower delivery rates than the expected operational level. After the indigenous population has increased, the delivery rate is increased in increments. As with any process operation, this rate of increase is based on process monitoring data. The original estimate for delivery rates is based on laboratory studies or in situ respiration tests. A predictive procedure using treatability test data is provided in Appendix B.

Initially, the electron acceptor will be used by microorganisms in the vicinity of the infiltration point. When these contaminants have been degraded, the transport of oxygen or other electron acceptor occurs over longer distances. As remediation at the site progresses, the electron acceptor must be carried increasingly longer distances until it eventually breaks through at the recovery well.

The acclimation and microbial population response can be a considerable time, and pumping rates should be controlled to minimize the loss of excessive electron acceptor, nutrients, or primary substrate. When developing a population of methanotrophs in situ, the breakthrough of methane in recovery wells is wasteful. After breakthrough occurs at recovery wells, fluid injection can be stopped until the methanotrophic growth is substantial. This is established by monitoring for methanotrophic activity. Indirect methods include monitoring for decreases in oxygen and methane, and the transformation of chlorinate compounds. The injection should again commence to support the developed methanotrophic population. The fluid residence time in an aquifer with appropriate primary substrate and oxygen for significant methanotrophic growth to incur is approximately 100 days (McCarty, 1991).

Bioventing

In situ bioremediation was originally applied to contaminants in the saturated zone, since hydraulic control is necessary for optimization of the process. Bioremediation of the unsaturated zone has used water injection to raise the water table for hydraulic control. However, it is now recognized that significant bioremediation can be achieved in the unsaturated zone by forcing air into the soil through soil venting or air injection. Optimizing bioremediation of the unsaturated zone with air is called bioventing (Fig. 7.29).

One of the first projects to control soil venting for enhanced biodegradation was at the Hill Air Force Base in Utah. During field cleanup activities using soil vapor extraction for remediation of JP-4 fuel, removal also was occurring from biodegradation. Bioremediation in the unsaturated zone has been reported for gasoline-contaminated soil piles (Conner, 1988), toluene-contaminated soil (Ostendorf, 1989), and other fuels (Ely, 1988). During the U.S. Air Force treatment of a site

Figure 7.29 Enhancing biodegradation through air extraction (bioventing) at the Fallon Naval Air Station.

contaminated with JP-4 jet fuel, removals of 15 to 80 percent could be attributed to bioremediation (Dupont, 1992).

Air venting provides an effective means to remove volatile organic compounds from the subsurface and induce oxygen. Air venting provides for multiple treatment modes. First, oxygen is supplied for biological remediation. Second, air venting removes significant levels of volatile organic compounds by volatilization from the unsaturated zone. When a vacuum is applied, the amount of oxygen supplied is a simple function of the airflow. Given atmospheric air at 20 percent oxygen, the pounds per day of oxygen supplied for specific airflow rates are given in Table 7.8.

TABLE 7.8   Oxygen Supplied by Venting Systems (Unsaturated Soils)

Gas flow (surface ft³/min), lb/day	Single well oxygen delivered
1	23
5	117
10	233
20	467
50	1170
100	2330

SOURCE: Brown and Crosbie, 1989.

Application

Bioventing is applicable to the unsaturated zones that have good gas permeabilities. Gaseous permeability is related to hydraulic conductivity. However, soil moisture significantly affects gaseous permeability. As soil moisture increases, gaseous permeability drops. At 55 percent water saturation, gaseous permeability is decreased by approximately 80 percent.

Bioventing is applicable to contaminants that are degraded through aerobic metabolism and have vapor pressures less than 1 atm. Highly volatile organic compounds will volatilize rather than biodegrade. The relative suitability of remediating important contaminants by bioventing is given in Table 7.9. For example, tetrachloroethylene is not amenable to bioventing since it is not degraded under aerobic conditions. Vinyl chloride which is easily degraded under aerobic conditions has been characterized as moderately suitable because of its high volatility. Significant portions may be volatilized rather than biodegraded. The BTEX compounds are good candidates for bioventing. They degrade readily under aerobic conditions and have a low enough vapor pressure that biodegradation is the main removal mechanism.

When feasible, bioventing provides a low-cost installation with low operation and maintenance. Oxygen can be delivered to the subsur-

TABLE 7.9   Applicability of Bioventing to Organic Contaminants

Rank*	Compound	Applicability†
1	Trichloroethylene	M
2	Toluene	G
3	Benzene	G
4	PCBs	P
5	Chloroform	M
6	Tetrachloroethylene	P
7	Phenol	G
8	1,1,1-Trichloroethane	M
9	Ethylbenzene	G
10	Xylene	G
11	Methylene chloride	M
12	trans-1,2-Dichloroethylene	M
13	Vinyl chloride	M
14	1,2-Dichloroethane	M
15	Chlorobenzene	M

SOURCE: INET, 1993.
*Rank is the U.S. EPA listing for contaminants most frequently reported at Superfund sites.
†G, M, and P refer to good, moderate, and poorly suited for bioventing.

TABLE 7.10   Cost Comparison of Bioventing versus Soil Vapor Extraction for 2500 yd^3

	Bioventing	SVE
Pilot Study		
Soil gas permeability test	$10,000	$10,000
In situ respiration test	10,000	0
Subtotal	$20,000	$10,000
Design and Installation		
Design	$10,000	$ 10,000
Vent well installation	25,000	7,000
Monitoring point installation	5,000	5,000
Blower acquisition and installation	7,000	20,000
Catalytic incinerator	0	80,000
Subtotal	$47,000	$122,000
Monitoring		
Pretreatment soil sampling	$ 5,000	$ 0
Flux monitoring	8,000	0
In situ respiration testing	25,000	0
Posttreatment soil testing	10,000	10,000
Air emissions	0	10,000
Subtotal	$48,000	$20,000
Operation and Maintenance		
Makeup fuel	$ 0	$ 15,000
Power	2,000	5,000
Maintenance	1,000	10,000
Subtotal	$ 3,000	$ 30,000
Total	$118,000	$182,000
Cost per yd^3	$ 47.20	$ 72.80

SOURCE: Hinchee, 1993b.

face with significantly less energy than using water as a carrier. It also provides cleanup of heavier hydrocarbons that are not amenable to soil vapor extraction. Its limitations are that contaminants must be degradable by aerobic metabolism and the remediation of volatile compounds is slower than with soil vapor extraction. However, the cost of gas treatment can represent 50 percent of soil vapor extraction costs. Bioventing significantly reduces this cost or eliminates it. Bioventing has been estimated to cost 65 percent less than soil vapor extraction (Table 7.10). The treatment of 2500 yd^3 is projected at $47.20 per yd^3 compared with $72.80 per yd^3 for soil vapor extraction (Hinchee, 1993b).

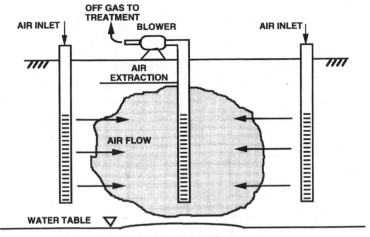

Figure 7.30 Schematic of soil venting operated as a bioventing process.

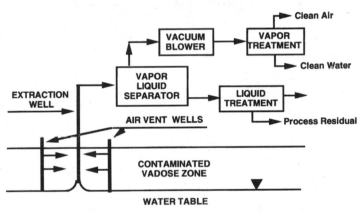

Figure 7.31 Typical vapor extraction and off gas treatment schematic.

A typical soil vapor extraction system that has been modified for enhanced biodegradation is illustrated in Fig. 7.30. A vacuum is applied in the heart of the contaminated zone. The rate of air transport is reduced over the operational rates for soil vapor extraction to reduce volatilization over biodegradation. Although this configuration enhances biodegradation, it does not optimize bioremediation over vaporization. Off-gas treatment is usually required. A typical process scheme is illustrated in Fig. 7.31.

An improved configuration for bioventing is illustrated in Fig. 7.32, where the vacuum is applied outside of the contaminated zone. This configuration expands the in situ bioreactor as vapors are drawn into clean soil, but a greater percent of the organic compounds is biodegraded. If designed properly, off-gas treatment may be eliminated.

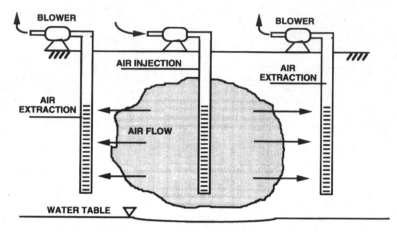

Figure 7.32 Basic bioventing process schematic.

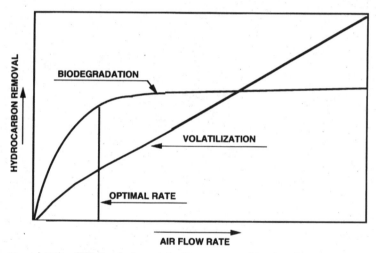

Figure 7.33 Effect of air extraction rate on bioremediation versus volatilization. (*Hinchee, 1993b.*)

With air extraction, the amount volatilized versus biodegradation is a function of extracted airflow rate (Fig. 7.33). There is a point at which the amount biodegraded is optimized.

An improved approach for maximizing biodegradation is air injection in the contaminated zone with no vacuum wells (Fig. 7.34). For this configuration, the amount bioremediated versus volatilization is illustrated in Fig. 7.35 (Hinchee, 1993b). A properly designed system results in complete degradation of the organic compounds before they can exit the surface. Air injection should not be used if a potential exists for the discharge of vapors into other structures.

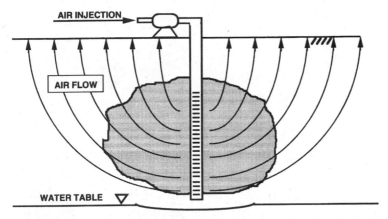

Figure 7.34 Providing oxygen by air injection for reduced emission of volatile compounds.

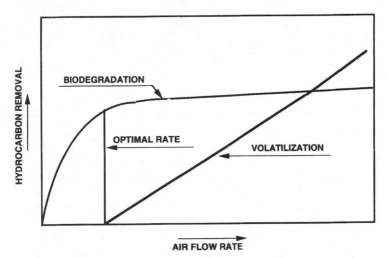

Figure 7.35 Effect of air injection rate on bioremediation versus volatilization. (*Hinchee, 1993b.*)

Design

The design of bioventing systems, like any bioremediation program, is based on the site's demand for oxygen for cleanup and the ability to deliver this oxygen. The radius of influence that oxygen can be carried must be estimated and then monitored during field operations.

The radius of influence is an estimate based on the soil gas pressures, oxygen concentration, and gas flow. Significant gas flows can occur well beyond the measurable radius of influence in highly permeable gravel.

TABLE 7.11   Recommended Spacing for Monitoring Points

Soil type	Depth to top of vent well screen, ft*	Spacing interval, ft†
Coarse sand	5	5–10–20
	10	10–20–40
	>15	20–30–60
Medium sand	5	10–20–30
	10	15–25–40
	>15	20–40–60
Fine sand	5	10–20–40
	10	15–30–60
	>15	20–40–80
Silts	5	10–20–40
	10	15–30–60
	>15	20–40–80
Clays	5	10–20–30
	10	10–20–40
	>15	15–30–60

*Assuming 10 ft of vent well screen: if more screen is used, the 15-ft spacing will be used.

†Note that monitoring point intervals are based on a venting flow range of 1 ft³/min/ft screened interval for clays to 3 ft³/min/ft screened interval for coarse sands.

The radius of influence is established by conducting air permeability tests. Air is withdrawn or injected for approximately 8 h at vent wells located in the contaminated soils (Department of the Air Force, 1992). Pressure changes are then monitored at a number of points to determine air permeability. Recommended spacing for monitoring points is provided in Table 7.11. Monitoring points can be constructed from small-diameter ¼-in nylon or polyethylene tubing. Each point contains a screen approximately 6 in long by ½ to 1 in in diameter. A typical construction detail is illustrated in Fig. 7.36.

For air extraction the soil gas permeability for steady state is given by

$$K = \frac{Q\mu \ln(R_w/R_I)}{H\pi P_W [1 - (P_{atm}/P_W)^2]} \tag{7.30}$$

where Q = flow rate, cm³/s
H = length of screen
R_I = radius of influence, cm (this is an estimated value from data)
R_W = radius of vent well
μ = 1.8×10^{-4} dynamic viscosity, g/cm-s at 18°C
P_W = absolute pressure at vent well, g/cm-s²

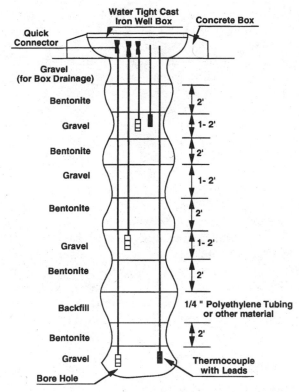

Figure 7.36 Construction detail for monitoring points.
(*U.S. Air Force, 1992.*)

P_{atm} = atmosphere pressure, $g/\text{cm-s}^2$
P_{atm} = 1.013×10^6 $g/\text{cm-s}^2$ at sea level

The absolute pressure at the vent well is the atmospheric pressure, 14.7 lb/in^2, plus or minus the measured gauge pressure or vacuum. One lb/in^2 equals 6.9×10^{-4} $g/\text{cm-s}^2$.

If air is being injected, then the steady-state permeability equation is

$$K = \frac{Q\mu \, \ln(R_w/R_I)}{H\pi \, P_{\text{atm}} \, [1 - (P_W/P_{\text{atm}})^2]} \qquad (7.31)$$

The value of R_I can be taken as the outer limit of vacuum or pressure influence under steady-state conditions. This can be measured in the field or extrapolated from field data by plotting the vacuum/pressure at each monitoring point versus the log of its radial distance from the vent well. The value R_I is then taken as a straight-line extrapolation to zero vacuum or pressure.

The rate of oxygen uptake is determined by in situ respiration testing (Chap. 6). Values for the in situ rate of hydrocarbon degradation are illustrated in Fig. 6.3. The rate varies significantly with soil characteristics and must be established for each site. Given the rate of oxygen uptake per day, the volume of air per mass of soil as based on soil porosity, and the mass ratio of hydrocarbon to oxygen for mineralization, a degradation rate in units of milligrams of hydrocarbon per kilogram of soil per day can be calculated. Given that the bioventing system maintains an adequate oxygen concentration in the soil, the time required for cleanup can be projected from the calculated degradation rate.

Vent wells are typically 2 to 4 in in diameter and constructed of schedule 40 polyvinyl chloride (PVC). Typical construction specifications are provided in Fig. 7.37 (U.S. Air Force, 1992). Blowers are selected according to their performance curves and the airflow required. Blowers should be designed to meet only necessary flows. An overdesigned blower results in excessive volatilization and energy consumption. Typically, blowers of 1 to 5 hp are adequate. The U.S. Air Force Center for Environmental Excellence has recommended blower specifications for use during site evaluations (Table 7.12).

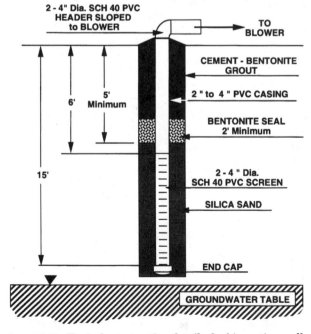

Figure 7.37 Typical construction detail of a bioventing well. (*U.S. Air Force, 1992.*)

**TABLE 7.12   Typical Blower Specifications for
Bioventing Evaluations**

For sandy soils:
    Explosion-proof regenerative blower
    20 to 90 surface ft³/min at 20 to 100 in H₂O, respectively
    3-hp explosion-proof motor
    Single-phase 230-V power source
For silt and clay soils:
    Explosion-proof pneumatic blower
    50 surface ft³/min at 130 in H₂O
    5-hp explosion-proof motor
    Single-phase 230-V power source

SOURCE: Department of the Air Force, 1992.

Other considerations in the design of bioventing systems include

1. Potential manipulation of the water table
2. Monitoring stations for measuring soil gas composition
3. Monitoring vent gas composition
4. Design for variable vent gas flow rates to optimize bioremediation and reduce volatilization
5. Potential treatment for surface emissions
6. Need for moisture and nutrient addition

Nutrients and moisture addition may be necessary and can be added by trenching or other subsurface gravity feed methods (Fig. 7.38).

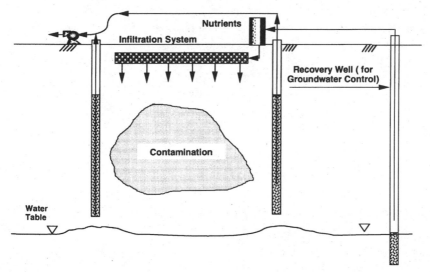

Figure 7.38   Combined nutrient addition and bioventing for oxygen delivery.

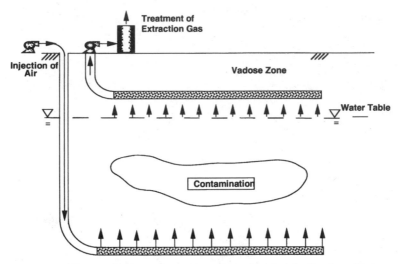

Figure 7.39 Bioremediation using soil venting with air injection by a horizontal well. (*Looney, 1992.*)

Bioventing can be applied to soil that has been saturated after adjustment of the water table. However, dropping the water table can yield significant quantities of contaminated water for surface treatment.

Bioventing with air sparging in the saturated zone is being evaluated with horizontal wells. Significant rates of oxygen can be transported into a contaminated aquifer by forced air. When air is sparged in a well screened below the water table, the air migrates upward through the soil, creating a two-phase system in the saturated zone. Volatile compounds are transferred to the gas phase by diffusion and oxygen is transferred to the water phase by diffusion. The gas phase carries some volatile contaminants into the vadose zone. Proper design provides for degradation before they are discharged as surface emissions, or they can be collected in a vacuum well.

A potential configuration for horizontal well application is illustrated in Fig. 7.39. The need for collecting gases above the water table is a function of the volatility of the contaminated airflow rates and the degree of bioremediation that occurs in the unsaturated zone. This zone has the potential of acting as a biological vapor trap if adequate biological activity is maintained.

Costs and Selection of the Oxygen Source

The most important factor in selecting the oxygen source is the bioremediation demand rate for oxygen. For most sites, one is limited to the use of oxygenated water if the cleanup time is to be minimized.

TABLE 7.13   Cost Comparison of Electron Acceptors for in Situ Bioremediation

Oxygen source	Hydrogen peroxide	Nitrate	Bioventing
Construction	$ 35.0	$ 90.0	$20.0
Labor and monitoring	$ 55.0	$ 74.0	$31.0
Chemicals	$385.0	$ 23.0	$ 0.3
Electricity	$ 19.0	$ 9.0	$ 5.0
Total costs per cubic yard	$494.0	$196.0	$56.3

SOURCE: U.S. EPA Bioremediation in the Field (EPA 540/2-91-018).

The next factor is cost of the delivery system. However, a low delivery cost may be more than offset by the cost of extending the cleanup time. The costs for the injection of hydrogen peroxide, nitrate, and bioventing are compared in Table 7.13. The final factors are ease of transport and maintenance of the oxygen supply system.

For air sparging, the most significant operating cost is maintenance of the air sparger. Maintenance includes the compressor and the diffuser and well screen. Injection of oxygenated water is also a low-cost operation. The main cost of operation is the control of biofouling of the injection system. With air sparging and injection of oxygenated water, the primary capital costs are the pumps needed to drive the system. Maintenance is simple and power consumption is minimal.

Hydrogen peroxide is usually a low-capital, easy-to-maintain system. Operation cost is higher owing to the hydrogen peroxide use rate. The main factor influencing operation cost is how quickly hydrogen peroxide decomposes in any given soil matrix. Laboratory treatability tests must include an evaluation on the stability of hydrogen peroxide for a given soil matrix. During 1993, hydrogen peroxide cost ranged between $0.70 and $1 per pound on a 100 percent H_2O_2 basis.

For unsaturated subsurface conditions, bioventing is frequently more trouble-free than air sparging or injection of oxygenated water. However, the air discharge may require treatment. Treating the vapor discharge could be the largest potential cost for the bioventing system, particularly if the bioremediation process is not optimized.

Off-Gas Treatment

Treatment of the exhaust gases from bioventing frequently uses one of the following:

1. Activated carbon

2. Catalytic oxidation

3. Internal combustion engines

4. Packed bed thermal process

5. Biofilters

Activated carbon is not an ideal sorbent for short-chain volatile compounds, but it does have a high affinity for aromatic hydrocarbons such as the BTEX group. It has a low capacity for halogenated aliphatic compounds, resulting in rapid bed exhaustion and significant regeneration costs. However, activated carbon has been a popular choice for small systems that can use carbon drums.

Catalytic oxidation utilizes a combustion unit with a catalyst that is a metallic mesh, ceramic honeycomb, or packed bed of impregnated pellets. The catalyst allows combustion of hazardous compounds at lower temperatures than straight thermal combustion. Natural gas or propane has been used as fuel supplements.

The internal combustion engine uses an automotive or industrial engine. The carburetor is modified for vapors rather than liquid fuel. Propane is usually used to supplement the hydrocarbon vapors.

The packed bed thermal process uses ceramic beads heated to 1800°F. The excellent heat transfer of packed beds and minimum short circuiting results in almost complete destruction. This system performs well on chlorinated aliphatic compounds.

Biofilters have a high potential for exhaust gas cleanup, but as with many biological processes, their design requirements have not been fully appreciated. Therefore, mixed results have been obtained. Soil biofilters have been used extensively in Europe for the control of odorous gases. Promising biofilters for vent gas cleanup are biologically active activated carbon beds. The beds are operated to optimize degradation of specific chemicals.

A true cost comparison for oxygen supply systems can be made only after data are obtained from laboratory treatability studies, site characterization has been completed, and the rate of cleanup has been established. Based on capital cost, the least costly is air sparging followed by bioventing and hydrogen peroxide by water injection. The least total project cost is not always related to the least capital cost system. Operational maintenance and site monitoring costs make total project cost highly dependent on the cleanup time.

Cleanup Time

The cleanup time for one system may be as much as 10 times greater than for another. Thus cost comparisons are very site-specific and vary with such factors as:

Site heterogeneity

Soil characteristics

Contamination location and distribution

Soil moisture content

Contamination level

Desired cleanup time

Air emissions

The greatest performance benefit is usually obtained with design dollars applied first to the oxygen delivery system and second to the control of the hydraulic system. The performance obtained, however, can be optimized and documented only by an adequate monitoring program.

The cleanup time for in situ bioremediation is theoretically a function of:

- Degradation rate of the organic compounds
- Availability of the organic compounds
- Volume of site under active cleanup
- Required cleanup limits
- Soil characteristics
- Hydrological characteristics
- Rate of delivery for electron acceptors and other agents
- Site heterogeneity

From a practical standpoint only three of the above usually control cleanup time. The first is what portion of the site is under actual cleanup. Larger sites are frequently divided into sections that are scheduled for a sequential cleanup. The second is the design rate for delivering the necessary chemicals. The third, and probably the most important, is site heterogeneity.

In situ cleanup time is controlled by the ability to deliver the electron acceptor and other necessary agents to the contaminated zone. This rate of delivery is based on more than the ability to inject these chemicals. The spatial distribution of hydraulic conductivity and distribution of organic pollutants are keys to estimating cleanup times. Most heterogeneity configurations that have been evaluated in modeling schemes increase cleanup time (Schafer, 1991). Even for aquifers with only moderate heterogeneity, the cleanup time is projected to be twice the time for homogeneous conditions. Several reported cleanup times are provided in Table 7.14.

Site Monitoring

Monitoring programs for in situ bioremediation must be designed for multiple goals. Data are collected for the evaluation of environmental

TABLE 7.14   Example in Situ Bioremediation Cleanup Times

Spill type	Time, months	Reference
Gasoline	10–18	Lee, 1988
BTEX	15	Lee, 1988
Trichloroethane	20	Mahaffey, 1990
Diesel	$1\frac{1}{2}$–2	Brown, 1990
Gasoline (10,000 yd³)	24	Harding-Lawson, 1990

impacts, process operation, and documentation of cleanup level. These monitoring requirements are more costly to perform than monitoring for surface treatment systems. Data acquisition is more difficult and elaborate because of the subsurface conditions. Operators do not have the benefit of visual observations of a process performance as with surface reactors. The monitoring program must monitor for contaminant migration from the site. Both groundwater and soil sampling are usually necessary. Some projects also require monitoring for air emissions. The net result is a more extensive and complicated data acquisition program. Bioremediation usually requires months or years, depending on the degree and size of the contaminated site.

To optimize process conditions for successful in situ bioremediation, the injection of necessary agents must be controlled according to the biological response. For this reason, the data acquisition program must measure the principal process control parameters at regular intervals. An automated sampling and analysis program will provide frequent data acquisition at significant cost savings versus manual programs.

The first consideration in the design of the automated sampling program is establishing the process control parameters. The second step is selecting analytical instruments that can operate with minimum attention under field conditions. An automated system developed by Stanford University in cooperation with the U.S. EPA and the Public Works Department at Moffett Naval Air Station is summarized in Table 7.15, and illustrated in Fig. 7.40. In situ bioremediation was evaluated for a sand and gravel aquifer contaminated with chlorinated aliphatic compounds. The microbial process used cometabolism to degrade vinyl chloride, cis-1,2-dichloroethylene, trans-1,2-dichloroethylene, and trichloroethylene. The process control parameters for optimizing the biological aerobic dehalogenation of these solvents included:

The primary substrate—methane

The electron acceptor—dissolved oxygen

TABLE 7.15 Moffett Field Automated in Situ Monitoring System for Process Control

1. A programmed microcomputer selects the next well sample location

2. A multihead peristaltic pump is activated flushing the sample manifold line with rinse water

3. The selected sample location is purged by the multihead peristaltic pump to obtain a representative sample

4. The pumped sample is split into two streams. One flow is directed to the DO probe and the second to the pH probe

5. The microcomputer stops the sampling pump

6. The microcomputer collects the DO meter and the pH meter readings

7. The microcomputer opens the sample manifold solenoid and a multichannel technician pump draws sample for processing into one or more gas chromatographs

8. The outputs from the GC detectors (ECD and Hall) are processed

9. The GC data are stored in the system's database as integrated peak areas and as computed concentrations

10. Upon completion of the analysis, the microcomputer automatically collects the next sample

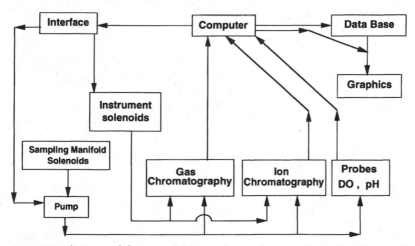

Figure 7.40 Automated data acquisition and control system for in situ bioremediation. (*Semprini, 1988.*)

The secondary substrate—halogenated compounds

Environmental factors—pH

Monitoring wells should be installed to provide adequate coverage of the subsurface biological response. They are usually grouped in clusters of three to six, depending on the depth and hydrogeology. To prevent volatilization and adsorption of chemical constituents, the fluid sampling lines should be fabricated with minimum storage volume. Small-diameter stainless-steel tubes minimize volume within the sampling line and provide for rapid line flushing. Well construction has used 6-mm stainless-steel tubing for each well. Tubing for each well cluster contains discrete 40×40 mesh stainless-steel screens spaced about 3 to 6 ft apart. Each screen is separated with bentonite seals to prevent channeling. A typical monitoring well design is illustrated in Fig. 7.36.

Wells within each cluster are located both within and below the contaminated zone to monitor biodegradation and potential contaminant transport. Piping to each well screen is numbered or color coded according to depth. Nylon tubing has been used for some installations, since it comes in a variety of colors. In cold climates, well lines should be placed below land surface and routed to an analysis building or trailer.

Water samples can be withdrawn from the subsurface with multihead peristaltic pumps. For wells too deep for peristaltic pumps, compact stainless-steel monitoring pumps are available (11 in long by 1.8 in in diameter). These monitoring pumps provide automatic purging followed by sampling at lower flow rates. Sampling flows are controlled to provide recommended rates for volatile compound sampling.

Another approach is to place monitoring probes directly in the wells. Compact probe units are available that measure five or more parameters, including pH, dissolved oxygen, redox potential, temperature, and conductivity. These probes will transmit digital signals to a personal computer.

The data acquisition system for the Moffett Field program was driven by a 6-MHz microcomputer that included a Techmar megafunction board, CGA composite monitor, and a 20-Mbyte hard disk and modem. The program controlling the analytical system was written and compiled with Microsoft's Quick Basic. The computer program could be operated in the automatic mode or manual mode. The manual mode allows selection of individual sample locations, analysis of stored data, and calibration of instruments. During the manual mode, the operator can validate the chromatography peaks and have the computer perform recalculations when necessary.

In the automatic mode, the computer program selects the next scheduled well for sampling and activates the sampling pump. The sampling lines are flushed with rinse water, and then the selected well is pumped for an adequate time to obtain a representative sample. Line flushing and sampling can be performed by electrically activated valves if the pump is not designed to perform this automatically. The pumped water sample can then be split and sent to multiple analyzers. The computer program collects all data results and stores these for later review.

At the Moffett Field site, methane was measured with a gas chromatograph equipped with a flame ionization detector. The chlorinated solvents were measured by a gas chromatographic instrument equipped with Hall and ECD detectors. A spectra physics 4270 integrator processed the output from the GC. All data are stored in the system's database as integrated peak areas and as computed concentrations.

Calibrations were performed in manual operation mode using external standards. The data acquisition system at Moffett Field provided for reliable operation and oversight capabilities that could not have been achieved in a manual mode. During 1987, the system collected data over 85 percent of the year and processed approximately 9300 samples. Sending this volume of samples to an outside laboratory would have easily cost four times the analytical and computer equipment costs for this program. The cost savings in manual field sampling add another major benefit to automated data systems.

Solid- and
Slurry-Phase
Bioremediation

Introduction

The biological treatment of municipal waste solids and sludges is not a new concept. It has been practiced worldwide for at least two centuries and for thousands of years in China. During the treatment technology advances of the 1960s and early 1970s, microbial systems received only secondary attention compared with physical-chemical systems. By the late 1970s, however, interest was renewed in biological treatment of solid wastes. Physical-chemical processes were judged by some as environmentally unacceptable or too expensive.

The need to clean up soil and lagoon sludges contaminated with hazardous substances has rekindled solid-phase biological treatment. This chapter presents treatment concepts for aboveground degradation of hazardous chemicals in solids and slurries. There is no clear distinction between slurry and liquid wastes. Slurries are sometimes referenced as solid wastes that have been suspended in water to enhance handling and treatment without any regard to water content. Others consider wastes with water contents of 95 percent or more as liquids. For the discussion of treatment modes here, slurries are defined as water suspensions that are impractical to treat by the liquid reactor technologies discussed in Chap. 9. These typically are mixtures with less than 90 to 98 percent water by weight. A 98 percent water content is typical of unthickened sludges from liquid biological processes.

The solid-phase treatment processes for hazardous compounds consist of:

Land farming

Composting

   Windrow system

   Static pile system

   In-vessel system

Slurry reactors

Land farming is usually conducted on the surface in an open environment. Composting is conducted as both an open surface operation and with enclosed vessels. The slurry-phase systems are conducted in existing sludge lagoons or in designed bioreactors. Solid-phase bioremediation has been used to treat soils contaminated with hydrocarbons and pesticides. Success has been achieved with petroleum fuel spills and for the pesticides of 2,4-dichlorophenoxyacetic acid (2,4-D) and 4-chloro-2-methylphenoxy-acetic acid (MCPA) (Borrow, 1989). Several polynuclear aromatic hydrocarbons, including phenanthrene, fluoranthene, and pyrene, have been treated to required cleanup levels. Chlorinated aromatic compounds such as pentachlorophenol and several aroclors have responded well to solid-phase treatment.

Design Considerations

A typical design for solid-phase bioremediation starts with consideration to runoff control and groundwater protection. Specifications for site grading, berm construction for directing surface water flow, and a leachate collection system are usually required. The solid-phase treatment system consists of some form of containment. This can be a completely enclosed vessel or a simple lined-berm drainage pad.

Typical leachate collection consists of lateral perforated drainage pipe placed on top of an impervious surface, such as a clay liner or a synthetic liner. High-density 80-mm liners with heat-welded seams have been used. Clay liners are not suggested for highly soluble compounds or the halogenated aliphatic compounds. Sand has been used to protect the liner and leachate drainage pipe. When specifying sand depths, consideration should be given to the loss of surface sand during the batch treatment cycles. Typically 6 to 12 in can be lost during the removal of treated soil, depending on the type and size of the materials handling equipment.

Overhead spray irrigation is a frequent procedure for applying nutrients and water addition. Local precipitation, potential air emissions control, and need for enhanced heating influence considerations for total enclosure. The collected leachate can be recycled for nutrient and microbial inocula or treated in a surface reactor.

TABLE 8.1   Significant Features of Solid and Slurry
Processes That Differ from Liquid Processes

1. Materials handling techniques
2. Mixing and pumping limitations
3. Heterogeneity:
    Chemically
    Biologically
4. Lower mass transfer rates for chemical solutes and gases
5. Degradation rates limited by desorption rates
6. Higher effective toxicity levels
7. Different microbial species
8. Analytical monitoring deficiencies

Mass transfer

The treatment of solids and slurries presents features that are significantly different from liquid systems (Table 8.1). The first major difference is that solids and some slurries cannot be easily pumped or circulated. The head loss in pumping slurries may be 2 to 10 times greater than that of water. Head loss is not simply a function of solids concentration but is also influenced by a sludge's organic content and plasticity. Most pumps used for slurries are nonclog design centrifugal pumps. The more popular centrifugal pumps are torque-flow since the size of particles that can be handled is dependent on the diameter of the pump's suction and discharge openings. Cantilever centrifugal pumps have been successfully used in the bioremediation of sludge and soil in lagoons. The cantilever design provides for direct placement of the pump blades in the lagoon from surface floats. This eliminates the need for seals that are difficult to maintain in high-organic-waste mixtures. Other pump types include the diaphragm, progressive cavity, and plunger.

Most differences between solid- or slurry-phase and liquid-phase treatment result from the heterogeneity of solids and the distribution of contaminants. This heterogeneity limits the overall efficiency of bioremediation. The heterogeneity problem of solid systems is first realized when characterizing the chemical contaminants and estimating the total contaminant content. Establishing the total mass of contamination in soil is difficult because of the heterogeneous distribution. One is fortunate to obtain estimates of ± 25 percent accuracy, and field results often illustrate that original estimates are in error by as much as 100 percent. This heterogeneity exists because of limited transport within soil systems and the adsorption affinity of specific compounds for soil.

The predominant soil contaminants are those that have a high affinity for the soil matrix. This affinity is influenced by the chemical properties of water solubility, octanol-water partition coefficient, vapor pressure, and the organic composition and chemistry of the soil. As a contaminated site ages, the more soluble contaminants are transported from the site by water and the volatile contaminants are vaporized. The remaining residuals have high soil affinities and have probably moved little from their original spill location. Thus the distribution can be very heterogeneous and related to the spill type, release location, and time sequence of occurrence. At a given site, it is not uncommon to be treating soil piles that contain significant levels of different compounds according to soil location on the site. This heterogeneity makes it extremely difficult to establish mass contaminant loadings unless an extensive number of soil samples are analyzed. The sampling job may become cost-prohibitive.

As with a heterogeneous chemical distribution, so is the microbial species distribution and therefore their response to bioremediation. Microbial identifications in soil treatment systems have confirmed the heterogeneity of microbial populations. For example, PCB-degrading organisms can be found in one soil pile while absent from a second soil pile of similar material from the site. Engineers and operators of soil treatment systems must be cognizant of these potential heterogeneity problems and address corrective solutions. Cross inoculation of soils and the use of seed organisms are frequently advantageous for solid-phase treatment. One major advantage of slurry systems over solid-phase is that they significantly reduce the heterogeneity of both the chemical and microbial distribution.

For solid-phase treatment, polynuclear aromatic hydrocarbons (PAHs) usually control the cleanup time because of their low rate of degradation. The affinity of a chemical contaminant for the soil matrix and its rate of desorption controls the availability of the compound as a microbial substrate. It has been demonstrated that the microbial degradation of several PAHs is controlled by the PAH desorption rate (Mihelcic, 1988). The rate of desorption is dependent on the soil-water ratio. The formation of a stable emulsion of PAHs greatly enhances biodegradation rates (Bleam, 1988). The solubility variances between different PAHs are illustrated in Table 8.2.

The half-lives for PAH degradation in solid-phase systems have been reviewed and the 95 percent confidence intervals established for specific PAHs (Fig. 8.1). These data have been grouped regardless of the soil nature, test conditions, or microbial communities. Given these variations, projected rates of degradation based on first-order kinetics are presented in Fig. 8.2. In general, the rate of degradation decreases with increasing number of aromatic rings.

Another significant feature that separates solid- and liquid-phase treatment response is the greatly reduced mass transfer rates within solid systems. Chemicals are immobilized by physical entrapment

TABLE 8.2   Solubility of Some PAHs in
Water

Chemical	µg/L
Two rings:	
Naphthalene	31,700
Three rings:	
Anthracene	73
Four rings:	
Pyrene	135
Benzo[a]anthracene	14
Five rings:	
Benzo[a]pyrene	3.8
Dibenzo[a,h]anthracene	2.4
Perylene	0.4
Six rings:	
Benzo[g,h,i]perylene	0.2

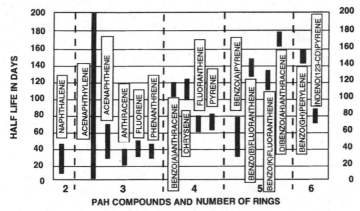

Figure 8.1 Half-lives 95 percent confidence interval for polynuclear
aromatic hydrocarbons. (*Sherman, 1988.*)

within the soil structure. Transfer rates of organic compounds are
reduced because of both sorption equilibria and lower diffusion rates.
First, intraparticle diffusion as well as surface boundary layer diffusion is involved. Second, transfer rates depend on the amount of
interstitial space and whether water or gas fills this space. Solids and
slurries are more difficult to mix or circulate. Thus mixing and mass
transfer become more limiting than in liquid-treatment processes.
This results in decreased degradation rates that are usually limited
by mass transfer rather than microbial degradation.

The rate of degradation for soil treatment is usually a mixed-order
reaction. Zero, first, and between zero and first orders are frequently
experienced. In soil systems, biodegradation rates can vary consider-

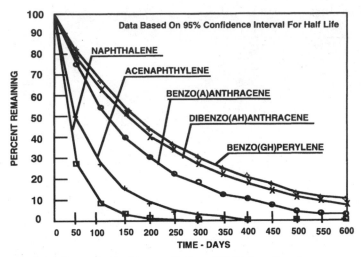

Figure 8.2 Rates for the bioremediation of PAHs in soil systems.

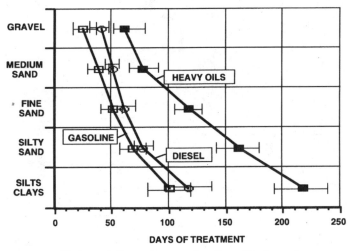

Figure 8.3 Biodegradation rates as a function of hydrocarbon source and soil type. (*Hicks, 1993.*)

ably over small distances. This is probably a direct result of the heterogeneity of bacteria, chemicals, and mass transfer limitations. The influence of soil type and nature of the contaminant on cleanup time is illustrated in Fig. 8.3.

Liquid-phase processes are usually mixed to a degree that eliminates significant heterogeneity of both chemical and microbial distributions. Even the addition of water to soil to yield a liquid-like solution for enhanced mixing will not eliminate all heterogeneity. Once attached to soil, many compounds dissociate little. Slurry systems, however, can result in improved rates of degradation over solid-phase systems. The heterogeneity of a soil matrix presents one of the more challenging aspects of solid-phase treatment.

An advantage of these reduced mass transfer rates is that many contaminated mixtures can be biologically treated at concentrations that would be toxic in liquid-phase processes. The adsorbed state of the chemical reduces the effective toxicity level that microorganisms are exposed to. Contaminated soil and industrial sludges have been successfully bioremediated even in the presence of aromatic compounds and metals that would be toxic to liquid-phase systems. A second advantage is that solid-phase systems are capable of supporting organisms with enzymes that have the ability to oxidize complex compounds. Examples are fungi that do not develop significant populations in liquid processes.

Materials handling

Materials handling is the main on-site operation for treating contaminated soil and sludges. Screening operations and washing techniques may be the first process of the treatment system. Equipment used to move materials within a site includes front-end loaders, back hoes, trucks, flails, dredges, belt conveyors, screw conveyors, drag chain conveyors, and pumps. When possible, in situ treatment provides for greatly simplified operations.

Alternative pathways for material flow should be considered when designing a facility that involves the movement of large volumes of material. Design should consider alternative conveyance routes and the potential for bypasses or temporary storage. The failure of one piece of equipment should not paralyze the complete treatment facility. Smaller multiple units provide greater flexibility than large single units. For example, the system can plan for the use of front-end loaders if a conveyor should fail. Operations for material handling can be made flexible by providing surge bins within feed systems prior to major processes such as a bioreactor. This also provides for the continuation of preprocessing of a solid or sludge if the receiving bioreactor is unavailable for reloading.

Design and process of soil treatment systems vary significantly in degree of sophistication. They include systems that are no more than till and mix operations and those with state-of-the-art materials handling and process control instrumentation. Some have been seeded with select organisms for enhanced dechlorination of persistence compounds while most use indigenous microorganisms. The level of sophistication in process design is a function of the level of hazard, the potential for off-site transport, the sensitivity of exposure pathways for both ecology and human health, and the magnitude and expected duration of the cleanup activities. At a minimum the need for the following facilities should be addressed during the design stage:

- Facilities for groundwater protection
- Facilities for surface runoff collection
- Facilities for treating runoff or leachate
- Facilities for the minimization of volatilization
- Facilities for the mixing and storage of process additives
- Facilities for enclosure of the process
- Facilities for spill containment
- Facilities for site security
- Facilities for decontamination:
  - Heavy equipment
  - Personnel
- Facilities for the control of dust
- Facilities for environmental monitoring

Typical costs for soil treatment are provided in Table 8.3.

Spillage problems are associated with almost all equipment, and containment for potential spills must be a part of site design. Impervious surfaces, equipment wash-down facilities, and collection facilities are important factors of a site's safety program. Conveyor spillage is a frequent problem, and the amount of operator time to clean up can be reduced with catch pans and the proper selection of the conveyor. Decontamination facilities are necessary for site equipment as well as personnel.

TABLE 8.3   Relative Cost Comparison for Containment Pads for Land Farming or Soil Piles

Containment system	Relative cost per yard capacity
Graded compacted soil	<$ 5
Visqueen	<$10
Geomembrane	<$30
Concrete	<$70

SOURCE: Hicks, 1993.

Wet materials adhere to rubberized conveyor belts and do not drop or transfer completely at their destination. Installation of belt scrapers reduces the amount of spillage. For sludges, the adherence problem can be reduced by placing a layer of dry material on the belt before adding the wet material. Dusting with sawdust has significantly reduced the spillage of sludges (EPA, 1989). Sand, shredded bark, wood chips, shredded agricultural crop waste, or even rock can be used. Rock and wood chips can be recovered by screening. Belts shaped into troughs are better for containing the materials. Other choices are drag chain, cleated belt conveyors, and screw conveyors.

Both belt and screw conveyors are limited to straight-line paths unless multiple units are used. Such systems require long inclined runs. Drag chain conveyors have experienced high maintenance cost owing to jamming and breakage at composting facilities (EPA, 1989, 625/8-89/016). The cleated belt conveyer has folds (or cleats) sewn into the belt fabric. These folds create a series of buckets that transport the material up steeper inclines.

A part of any process operation is the monitoring of materials flow. Process mass balances must be monitored and daily records kept. For the excavation of soils, truck scales and other measuring devices should be provided to track the flow of materials. Measuring devices include flowmeters for liquids and scales or measuring bins for solids. Scales can be incorporated into conveying systems and data retrieved by computer.

Operators need immediate notification of soil or sludge stoppage in the system. Sensors should be provided at all key transfer points to detect plugging. These sensors can be connected to audio alarms or even computerized so that upstream conveyors are automatically shut down.

Material handling may require screening and other separation techniques. Gravel can be damaging to equipment used in the treatment processes. Soil washing and screening can be cost-effective for soil containing significant amounts of gravel. A high percent of chemical contaminants are distributed with the fines because of the larger surface area for adsorption and the low adsorption affinity of silica surfaces versus clays and organic mixtures. Washing techniques make it possible to separate clean gravel from the bulk of the contaminated soil. Wash waters can be recycled and mixed with the fines for bioremediation.

Dust control during excavation and material transport must be considered. If moisture addition is not detrimental to the selected treatment modes, wetting systems can be considered. Other approaches use enclosures or plastic hoods around the materials handling operations. Some composting facilities have installed vacuum systems with filtering of discharge air for dust and biological aerosol control.

Land Farming

Land farming is a bioremediation process that is performed in the upper soil zone or in biotreatment cells. The process consists of the control degradation of contaminated materials in an above-ground system or with in situ treatment near the soil's surface. Land farming is widely used and has been applied to many waste types (Table 8.4). This technique has been successfully used for years in the management disposal of oily sludge and other petroleum refinery wastes. In situ systems have been used to treat near surface soil contamination for hydrocarbons and pesticides. Equipment requirements are typical of agricultural operations (Fig. 8.4). These land farming activities cultivate and enhance microbial degradation of hazardous compounds in place. This may be suitable when only the upper 12 in of soil is contaminated and

TABLE 8.4   Land Farming Waste Applications

Waste Sources:*
  Municipal sludges
  Chemical plant residues
  Creosote sludges
  Food processing sludges
  Petroleum refinery sludges
  Petroleum hazardous wastes
  Textile mill sludges
  Wood preservation wastes
  Paper process sludges
Hazardous chemicals:†
  Benzene, toluene, ethylbenzene, and xylene
  Fuels of gasoline, diesel, and oil
  Pentachlorophenol
  Polychlorinated biphenyls
  Polynuclear aromatic hydrocarbons:
    Acenaphthene
    Anthracene
    Benzo[a]anthracene
    Dibenzofuran
    Fluoranthene
    Fluorene
    2-Methylnaphthalene
    Naphthalene
    Pentachlorophenol
    Phenanthrene
    Pyrene
Pesticides:
    2,4-Dichlorophenoxyacetic acid
    4-Chloro-2-methylphenoxy-acetic acid

*Source: Sherman, 1988.
†Source: Barnhart, 1989; Block, 1989; Hutzler, 1989; Borow, 1989; Focht, 1985; Strou, 1988.

Figure 8.4 Land farming using agricultural equipment to enhance biodegradation of contaminants in soil. (*Courtesy of VFL Technology Corporation.*)

chemical migration is unlikely. Deep in situ techniques are possible with specialized drilling equipment equipped with mixing blades.

Design and operation

Land treatment is designed to optimize degradation by application of an aerobic biological process while using the soil as inoculum and support medium for biological growth. Biodegradation is optimized by aerating the soil and working in nutrients with agricultural tilling techniques. Tilling procedures serve to condition the soil to enhance biodegradation of the contaminants. General guidelines for soil preparation are (Hicks, 1993):

1. Condition soils to obtain a relatively uniform mix by removing larger rocks and other debris.

2. Apply all additives using mixing techniques that achieve at least 90 percent contact between soils and additives.

3. Inorganic nutrients generally benefit solid-phase bioremediation. They are generally added in mass ratios of 100:10:1 with respect to hydrocarbon:nitrogen:phosphorus. Trace amounts of potassium may be beneficial. Single-dose applications should be limited to 3 lb of nutrients per cubic yard of soil to prevent undesirable osmotic effects. Higher nutrient requirements should use time-release nutrients or repeated applications.

4. Increase the porosity of the soil as much as possible. Either increase porosity by mixing during earth-moving activities or add a bulking agent. Avoid compaction during soil placement.

5. Soils with a high clay content usually benefit from a bulking agent such as sand, sawdust, or wood chips. Gypsum can be used to reduce moisture content of clayey soils. Additions typically range from 10 to 30 percent by volume.

6. The soil's pH can be adjusted with either lime, alum, or phosphoric acid as required.

For most soil contamination below 1 ft, the soil is excavated and treated on a prepared bed. In a prepared bed system, the contaminated soil may be physically removed to the prepared area or temporarily stored while the excavated surface area is prepared for the treatment (Fig. 8.5). Although process facilities are usually not capital-intense, significant costs are associated with materials handling and controls to prevent migration of contaminants.

Bed preparation consists of measures to prevent contaminant transport from the site. The soil subsurface is protected by the placement of a clay or a plastic impermeable liner. Plastic liners are usually a high-density polyethylene of about 80 mil. The installation of a liner for one of two 1-acre land treatment cells is illustrated in Fig. 8.6. This soil treatment system is being used to treat contaminated soil from wood-treating operations in Libby, Mont. Relative cost of containment pad systems is provided in Table 8.5. Multiple liner systems with leak detection and leachate collection systems may be required. The scope of facilities to control contaminant release is a

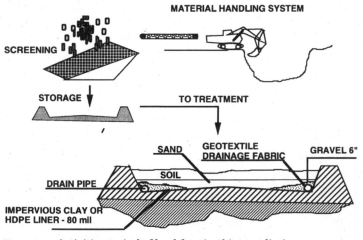

Figure 8.5 Activities typical of land farming bioremediation.

Figure 8.6   Construction of a land farming cell for the treatment of soil contaminated with PAHs and creosote.

TABLE 8.5   Typical Cost for Soil Treatment

Activity	Costs, dollars per cubic yard
Construction	$30
Contaminant*	$10
Soil conditioning*	$25
Soil disposal*	$10
Operation	$20–$40
Total	$50–$115

SOURCE: Adapted from Hicks, 1993.
*Optional activities.

function of the potential hazard and environmental health risk, environmental sensitive areas, and the projected time duration of the cleanup effort.

Beds for land farming are usually prepared with a base of sand or soil media with or without a gravel subbase. The base is used to protect the clay or plastic liner from equipment (Fig. 8.7). The sand or soil cover varies from 2 to 4 ft in thickness. The minimum thickness is usually 1 to 2 ft, since some surface sand is lost during treated soil removal with front-end loaders. Construction lasers have been used to control cutting depths. A thicker initial bed provides for more applications of soil for treatment before base renewal is necessary. A

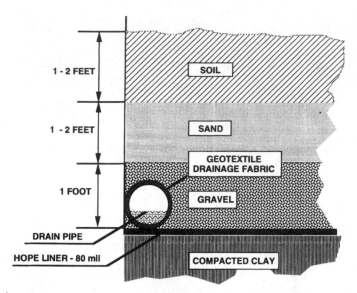

1 - 2 FEET

1 - 2 FEET

1 FOOT

SOIL

SAND

GEOTEXTILE
DRAINAGE FABRIC

GRAVEL

DRAIN PIPE

HOPE LINER - 80 mil

COMPACTED CLAY

Figure 8.7   Bed preparation for land farming bioremediation.

drainage system can be incorporated into the sand cover for collection of leachate. The leachate is recycled back to the soil being treated or sent to a surface treatment system.

The facility can be enclosed or covered by a roof depending on local weather conditions and emission controls. A cover reduces precipitation loading and can be cost-effective when balanced with the reduced leachate volume for treatment. Enclosed facilities are also desirable for operations during winter months. Excessive soil moisture content, above 70 percent by weight, will hinder gas transfer, causing excess anaerobic metabolism that may not be desired. Soil systems are never 100 percent aerobic, since some anaerobic metabolism takes place in the inner portions of soil or sludge particles.

Contaminated soil is excavated, screened if necessary, and placed on the sand bed. Soil is spread to a depth of 1 to 1.5 ft. However, the depth is a function of the tilling equipment's capabilities. The pH of the soil should be adjusted for optimum degradation and adequate nutrients added. Agricultural lime is typically used to raise pH levels. The initial amount to be added is based on the development of a liming curve. A soil slurry in distilled water is mixed with incremental additions of lime and the resulting pH values are recorded. The pH will drift for soil slurries because of the low transfer rates. Adequate time between intervals of lime addition must be provided for pH stabilization. Lime is added as necessary according to field monitoring of pH drop during bioremediation. Care must be taken to prevent excess lime, since high pH levels will hinder microbial growth. The potential degree of pH drop is a function of the original amount of organic load and the system's buffer capacity.

Nutrients are usually added by using a standard agriculture fertilizer mixture. These may be applied in a dry form and worked in by tilling or first dissolved and sprayed as a water solution. The best approach is based on ease of handling and the need for moisture addition. The initial nutrient requirement is based on stoichiometry calculations, treatability studies, or the need to control specific microbial response as with *Phanerochaete chrysosporium*. Soil samples should be analyzed at intervals to maintain optimum process parameters, and adjustments should be made during the operation.

For the more difficult to degrade organic compounds, a microbial seed may be used to shorten acclimation, enhance degradation rates, or provide a specific organism known to degrade the compounds. Examples include organisms that have known ability to degrade chlorinated compounds. Sewage sludge and cow manure are frequently used to promote microbial seeding and serve as an additional energy source.

After all additives have been applied, the contaminated soil is tilled, disked, or plowed. These tilling operations mix the soil to increase homogeneity, distribute the added chemicals and microbial seed, and increase oxygen transfer by promoting atmospheric diffusion. For aerobic systems, the redox potential should be maintained above 800 mV. Some operations provide tilling every day. Although the frequency required for plowing or disking is basically a function of the oxygen uptake, tilling will optimize other degradation variables. Laboratory studies illustrate that stimulated oxygen uptake occurs as a result of the mixing (Fig. 8.8) (Harmsen, 1991). Tilling not only introduces oxygen but redistributes nutrients, contaminants, and microorganisms.

Process performance for land farming is established by a program of scheduled monitoring. Degradation of contaminants in soil systems is established by monitoring their disappearance in soil through time. Leachate samples must be monitored to document the loss by water transport versus true degradation.

Bioremediation applications

Soil contaminated with pentachlorophenol (PCP) from wood preservation has been remediated by land farming. Approximately 5600 yd^3 of sandy soil with an average PCP concentration of 100 mg/kg soil was treated to 5 mg/kg within 4 months (Hutzler, 1989). A half-life of approximately 25 days was experienced. The land farm facility was

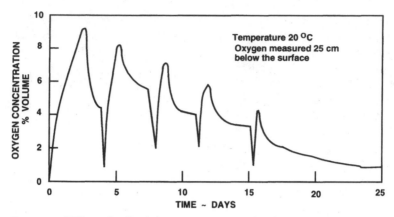

Figure 8.8 Effect of soil mixing on oxygen uptake during the solid-phase treatment of oil-contaminated soil. (*Harmsen, 1991.*)

designed to treat 100 yd^3 per year. The treatment consisted of excavation of contaminated soil and spreading to a depth of 1.25 ft. Three inches of cow manure and sewage sludge was added as a nutrient supplement. Bed preparation consisted of a high-density polyethylene membrane (80 mil), covered with 6 in of gravel over a leachate collection system. The soil was placed on top of the gravel bed, and the entire facility was covered by a roof.

Others have reported on the degradation of pentachlorophenol from 100 mg/kg soil to 20 mg/kg within a period of 10 to 20 weeks (Harmsen, 1991). Recent field studies using fungi inoculums have been successful for solid-phase degradation of pentachlorophenol. Pentachlorophenol-contaminated soil containing up to 6730 µg/g has been bioremediated with the fungi *Phanerochaete chrysosporium* and *P. sordida* (Lamar, 1990). PCP contamination resulted from a tank storage facility that held PCP in an 84 percent mineral spirits solution. Contaminated soil was an alkaline (pH 9.6) gravelly sand that was almost devoid of organic matter and deficient in nitrogen and potassium.

To enhance PCP degradation, the soil was amended by seeding with fungi grown on wood chips. Organic substrate was increased by adding peat moss having a pH of 4.0. Peat was applied at a rate of 1.93 percent (dry weight basis). The fungi seed was prepared by growing *P. chrysosporium* and *P. sordida* on aspen (*Populus tremuloides*) wood chips. The moisture content of the chips was adjusted to 60 percent, and they were inoculated from fungi grown up on 2 percent malt agar. Chips were separately inoculated with each fungus and incubated at 39 and 30°C for 6 weeks for *P. chrysosporium* and *P. sordida*, respectively.

The seed wood chips were applied at a rate of 3.35 percent (based on dry weight of soil). The chips, peat moss, and nutrients were mixed with a rototill into the upper 30 cm of soil. The soil-water potential was adjusted to -0.05 MPa and covered with polyethylene covers to prevent excessive moisture loss. The moisture level was monitored by in-place tensiometers.

The two fungi, *P. chrysosporium* and *P. sordida,* demonstrated similar rates of PCP degradation. An overall decrease of 88 to 91 percent of PCP occurred in 6.5 weeks with field temperatures ranging from 22 to 10°C. Most of the PCP was converted to nonextractable soil-bound products. The nature of these compounds is not known, but naturally occurring phenolic compounds are incorporated into soil humus.

Composting

The first systematized process approach to composting is reported to have occurred in India in the early 1900s. Early applications were anaerobic metabolic processes. Since then, composting has developed into many process configurations. These include in-ground trenches, rotating drums, circular tanks, open bins, silos, windrows, and open piles. In the last 25 years, composting has become one of the main processes for stabilizing municipal sewage sludge.

Process description

Most modern municipal sludge plants use aerobic rather than anaerobic decomposition. The disadvantage of anaerobic systems is the generation of odorous compounds such as hydrogen sulfide, mercaptans, and disulfides. The high organic content and available sulfur in sewage sludge can yield 10 lb per day of odorous compounds per ton of compost (Cookson, 1990). Some municipal sludge is treated in enclosed anaerobic digesters prior to composting. Aerobic composting provides for a greater degree of degradation for most sewage sludge compounds.

For hazardous waste treatment, the use of anaerobic processes can have advantages. As discussed in Chaps. 4 and 5, many halogenated and complex chemicals are treated more successfully under an anaerobic than under an aerobic mode. However, the degradation of PAHs under anaerobic conditions is not supported by most findings (Bleam, 1988). Thus treatment protocols should not be applied without consideration to the metabolism modes necessary for the target compounds.

The composting of contaminated soil has significant differences from composting of sewage sludge. Municipal sewage sludge has a high organic carbon and energy source. Typical municipal sludges are 70 to 80 percent organic solids. It is uncommon to have this high an energy source when applying composting to soils contaminated with hazardous compounds. This high energy content provides for the microbial heat generation during composting, achieving temperatures of 60 to 70°C. Temperatures of 55°C are required for 3 or more continuous days to achieve kill of pathogenic organisms. If no pathogenic organisms are associated with the contaminated soil and sludges, elevated temperatures are not necessary. Temperatures of 60°C and above are detrimental to the rate of degradation of many compounds (Cookson, 1990b). Composting of hazardous compounds has been successfully pilot tested at ambient temperatures. The term soil heap bioremediation has been applied in some instances to illustrate that the process is low-temperature composting.

Most composting systems utilize a bulking agent. Its purpose is to increase porosity of the media to be treated and decrease moisture levels. For composting, moisture levels should not exceed 60 percent or gas transfer rates decrease significantly, resulting in decreased rates of degradation (Cookson, 1990b). The optimum moisture for the degradation of many hazardous chemicals is about 50 percent. For hydrocarbons, 60 percent is recommended.

The effect of water content on the degradation rates for oil-contaminated soils has been evaluated in laboratory respiration studies (Stegmann, 1991). The compost mix was composed of 8 parts soil to 1 part compost obtained from a windrow composting plant. This mixture had a maximum water capacity of 48 percent by weight. The maximum oxygen uptake occurred at a moisture content of 60 percent of the total water capacity. The rate of degradation decreased significantly at moistures less than 60 percent and greater than 80 percent.

For treating soils, the use of a bulking agent is mainly to increase porosity. Bulking agents consist of wood chips, shredded bark, sawdust, leaves, agricultural crop wastes, and shredded rubber tires. The bulking agent, except tires, will serve as an additional carbon and energy source. Some bulking agents such as wood chips can influence pH. The pH of the complete mix should be optimized as well as the moisture content.

Screening of the compost mix is a common practice to recover the bulking agent for recycling. Screening equipment has included vibratory decks, rotary screens, and trammels. Except for the handling of raw materials, screening is a major source of dust and aerosol suspensions. The potential release of biological aerosols and dust containing nonbiodegradable hazardous substances should be evaluated. Compost facilities have used enclosed screening systems with air filtration equipment. The capital cost of screening must be compared with the value of the lost bulking agent if it is not recovered for recycling. This evaluation is dependent on the expected life of the cleanup operation, the treatment required, and the final deposition of the treated soil or sludge. The recycling of bulking agents also provides a microbial seed source for incoming soil.

Municipal sludge composting has been divided into two stages, a first stage that has high-rate degradation and a second stage with low-rate degradation. The second stage is used for polishing or, as it is called in the composting industry, curing. The first stage has the highest rate of biological activity and therefore the highest rate of oxygen demand and degradation. For aerobic systems, this first stage requires close attention to adequate oxygen supply. Oxygen is supplied by frequent mixing or forced aeration. In this stage high temperatures are attained and there is a high potential for odor production. The second stage may or may not have forced aeration or pile mixing. Oxygen is usually provided by natural convection with frequent zones of anaerobic biological activity. Since the level of biological activity is greatly reduced, as a result of significantly lower energy supply, objectional odors or temperature buildup do not occur.

For municipal sludge, the curing stage is usually conducted in a second paved area. The compost pile is usually broken down with front-end loaders and transported to the curing pad, where it is again piled. This operation provides some degree of mixing that may aid degradation of incompletely composted portions. For the composting of hazardous compounds, this stage has the disadvantage of additional materials handling. Each time material is handled, the potential for spills and employee exposure exists. If aeration pad space is available, it is best to perform the complete composting of hazardous substances with minimum movement.

Modern composting systems are usually divided into three process configurations:

- Windrow system
- Static pile system
- In-vessel system

The windrow system is the simplest to operate and least capital-intense (Table 8.6). However, it requires greater space and it is impossible to control volatile emissions. The in-vessel system provides for excellent capture of emissions and requires less space but is greatly limited in its flexibility.

Windrow system

The contaminated soil or sludge is mixed with the bulking agent, and the mixture is distributed in long rows (Fig. 8.9). Maximum dimen-

TABLE 8.6   Comparison of Three Composting Systems

Control level	Windrow system	Static pile system	In-vessel system
Operational skill level	Low	Moderate	High
Process flexibility	High	Medium	Low
Material load flexibility	High	Medium	Medium
Process control	Low	Medium	High
Moisture control	Low	Medium	High
Air emission control	Low	Medium	High
Runoff control	Medium	Medium	High
Space requirement	High	Medium	Low
Pathogen destruction	Medium	High	High
Climatic dependency	High	Medium	Low
Capital cost	Low	Medium	High
Maintenance cost	Low	Medium	High

Figure 8.9 Windrow composting of contaminated sludges and soils. (*Courtesy of VFL Technology Corporation.*)

sions for a windrow are 4 to 5 ft in height and 10 to 12 ft in width. These rows are mechanically mixed to maintain aerobic conditions by convective air movement and diffusion. Windrow piles are turned or mixed daily to maintain an aerobic condition. However, for high-oxygen-demand waste, zones of anaerobic activity will develop.

Mechanical mixing is performed with either a front-end loader or specially designed equipment (Fig. 8.10). Front-end loaders are one of the least expensive mixing approaches, but the quality of the mix depends on the amount of time spent by the operator. An impervious

Figure 8.10 Mechanical mixer for bioremediation of contaminated soil. (*Courtesy of VFL Technology Corporation.*)

bed is constructed on which a layer of bulking agent is placed. The soil or sludge is placed on top, and the two layers are mixed. After mixing, some operations use a manure spreader to increase the mixture's porosity (EPA/625/4-85/014). Equipment designed for compost mixing, such as the Cobey and SCARAB, also provides a mixture of good porosity. Other mixing devices include agricultural rototiller and feed mixers.

The windrow system is very flexible. It can handle large variations in material volumes and allow for process loading variations. The capital cost is low, with major requirements being the bed for pile construction, a windrow machine, and a front-end loader.

Static pile system

The static pile system uses forced air to maintain aerobic decomposition in a much larger pile mass than is possible with the windrow (Fig. 8.11). Enclosed operations improve the control of moisture from rainfall and dust emission. Bins have been used to improve space utilization. Figure 8.12 illustrates an enclosed static pile composting facility. Compost piles are constructed to heights of 20 ft. Height is usually limited by the on-site capabilities of the front-end loaders. The aeration system consists of a series of perforated pipes running underneath each pile and connected to blowers that draw air through the pile. The perforated pipe is placed on an impervious surface designed for collecting all runoff. The pipes are covered with the bulking agent, usually wood chips, to act as a manifold for air distribution.

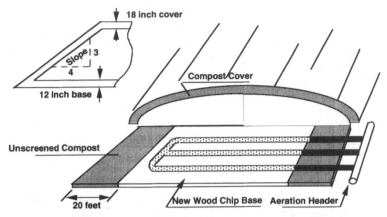

Figure 8.11 Schematic of a static pile compost system. (*Albrecht, 1983.*)

Figure 8.12 Enclosed bins for static pile composting.

The mixture to be composted is placed on the bulking agent. The compost mixture is usually covered with a bulking agent or other material to reduce dust emissions. If elevated temperatures in the compost pile are desired, a blanket of finished compost can be used as insulation.

Single or multiple blowers are used for aeration. Facilities frequently use multiple blower units of 1 to 5 hp. Flexibility is important, and several smaller units are preferred over a single unit. Variable-speed blowers are required. Aeration rates should not be

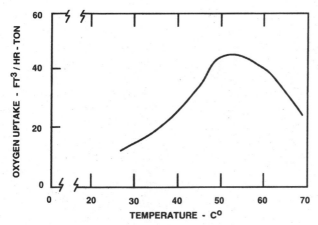

Figure 8.13 Effect of composting temperature on rate of degradation as measured by oxygen uptake. (*Cookson, 1990b.*)

controlled by on and off cycling periods for aerobic systems. Even short periods of no aeration, 15 min, result in significant zones of anaerobic activity (Cookson, 1990b). A vacuum aeration system is preferred since off-gases can be controlled and treated.

Aeration rates for the composting of sewage sludge are based on the need to control pile temperature and drying. Temperatures above 60°C result in excessive generation of odors. In addition, the diversity of microorganisms decreases, resulting in low degradation rates (Cookson, 1990b). The rate of degradation as measured by oxygen uptake per weight of compost mix can decrease by 50 percent at temperatures of 65°C versus 55°C (Fig. 8.13) (Cookson, 1990b). Excessive temperature levels should not be a problem with contaminated soils, since the energy source is much lower per unit volume of mix. The energy loading can be controlled during preparation of the mix.

In-vessel composting

In process theory, in-vessel composting is identical to the other configurations. However, the in-vessel operations do not allow the degree of flexibility of open systems. For example, if there is a material handling problem, such as compaction of the mix in the vessel, correction by remixing with a front-end loader is not an alternative. Thus most in-vessel systems use sophisticated mixing equipment. Pug-mill and plow-blade mixers have been used to create a more homogeneous mix (EPA/625/8-89/016). Generally, in-vessel systems use belt conveyors, screw conveyors, cleated belt conveyors, and drag conveyors for material transport. The use of conveyors to transport material is a major difference between in-vessel and the other composting configurations. They also represent one of the major operator maintenance problems.

For in-vessel composting the high-rate stage is always conducted in a closed reactor. The second stage, curing, may be conducted in a reactor or an exterior pile. In-vessel composting reactors are of two general types: plug flow reactors (vertical and horizontal) and agitated-bed reactors. In the plug flow reactor the mix moves as a plug either from top to bottom or horizontally through the reactor chamber. Most vertical plug flow reactors use a screw for material discharge (Fig. 8.14). In horizontal plug flow reactors, the material is transported by a moving floor or a hydraulic door (Fig. 8.15). The agitated-bed reactors use mechanical mixing to mix the compost either in place or as it moves through the reactor (Fig. 8.16).

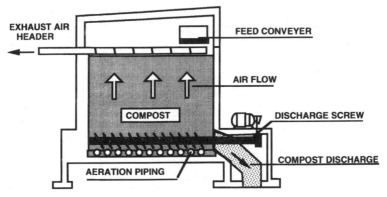

Figure 8.14   In-vessel compost reactor—vertical flow. (*U.S. EPA, EPA/625/8-89/016.*)

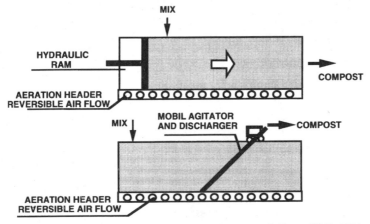

Figure 8.15   In-vessel compost reactors—horizontal flow. (*U.S. EPA, EPA/625/8-89/016.*)

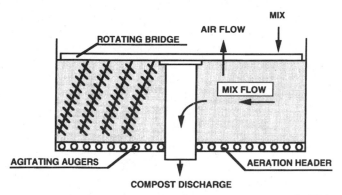

Figure 8.16 In-vessel compost reactor—agitated bed. (*U.S. EPA, EPA/625/8-89/016.*)

Process control

For sewage sludge composting, process control is based on pile temperature response. Temperature-controlled aeration rates are varied day to day as based on pile temperature. As a result, aeration rates for aerobic biological activity are less than the requirement for heat removal. Three important factors are related to pile temperature. First, if sewage sludge is used as an energy source and seed, an adequate temperature must be maintained for pathogenic organism destruction. This is readily achieved for high-energy-source mixes. Second, temperatures above 60°C result in significant odor production. Odor control is a major operational concern, and ineffective systems have resulted in facility closures. Third, temperatures above 60°C result in significantly decreased rates of biological degradation.

In Chap. 3 the rate of enzyme reactions was noted to increase exponentially with temperature. However, for biological systems there is an upper temperature limit that surpasses the optimum for various microbial species. At temperatures above this optimum, biological growth abruptly falls. As temperatures increase in the composting operation, the microbial diversity decreases, resulting in a corresponding decrease in degradation rates. The rate of degradation of hazardous compounds could decrease even more sharply than that for sewage sludge, since mixed microbial consortiums are usually necessary.

These relationships and the need to reduce odor production have made temperature the prime process control parameter. Other process control parameters are moisture content, pH of mix, material balances, and biological oxygen needs during stages when heat removal is not controlling the aeration rate.

For the treatment of hazardous compounds, the control parameters should be selected for assuring the degradation of the target compounds. For hazardous compounds, the main process control may not be based on temperature. Operating a process for pathogenic organism destruction is not necessary if the source of carbon and energy isn't sewage sludge. Second, the microbial heat-generation rate is a function of the organic load in the mix, pile size, and configuration. Mixes with lower organic content than municipal sludge will generate proportionally less heat. The smaller the pile, the greater the heat loss by convection. Thus aeration rates may be based not on temperature as with sewage sludge but on the rate of aerobic metabolism.

For low-energy mixes, the rate of aeration should be balanced to the needs for aerobic degradation and good distribution in the pile. Good air distribution may control the aeration rate rather than biological needs. Improved air distribution is achieved by limiting the distance for diffusion and dead zones. An improved design over that for sewage sludge composting provides for aeration piping, usually slotted PVC, throughout the pile (Fig. 8.17). The radius of influence around each pipe is based on the soil mix composition and must be determined for each mix type. Vacuum pressures and air velocities can be measured for each slotted piping run. Vacuums can be measured using a Magnehelic gauge and air velocity with a thermal anemometer (Hicks, 1993).

The aeration system is adjusted by individual pipe valves as well as the use of variable-speed blowers to yield equal vacuum at all piping runs. Valve adjustment alone can result in overheating blower motors. Vacuum monitoring is recommended at least once a month after the system is initially balanced. Other design recommendations are (Hicks, 1993):

1. Provide a minimum of 5 ft^3/min of airflow per piping run.

2. The vacuum system should be sized to remove at least 15 soil pore volumes per day.

3. Soil moisture should be frequently monitored. An agricultural tensiometer has been successfully used. Optimal soil moisture is a function of soil properties and the chemicals. Typical ranges are given in Table 8.7.

4. Moisture can be added by a drip irrigation hose placed on top of a soil pile or with spray discharges.

The proper composition of the compost mix, both organic energy source and bulking agent, must be established in treatability studies. Laboratory studies have been used to optimize the mix as based on oxygen uptake rates. Using compost from a windrow facility and oil-contaminated soil, ratios of 2:1, 4:1, 8:1, and 16:1 of soil:compost (dry weight) have been evaluated. The ratio of 2:1 yielded the maximum

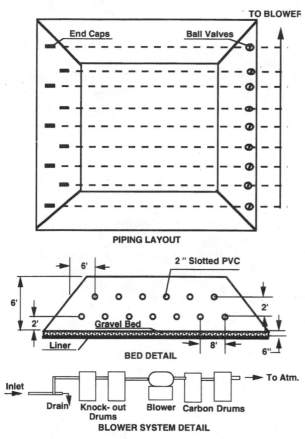

Figure 8.17 Design concepts for soil pile bioremediation. (*Hicks, 1993.*)

TABLE 8.7 Typical Soil Moisture Ranges for Soil Heap Composting

Soil type	Moisture range, centibar*
Sandy soils	20–40
Medium-textured soils	40–60
Clay soils	60–80

SOURCE: Hicks, 1993.
*1 centibar = 0.01 atm, measured by agricultural tensiometer.

rate of oxygen uptake and the lowest final hydrocarbon concentration (Stegmann, 1991).

The compost age also influences the rate of hydrocarbon degradation (Stegmann, 1991). Older compost (6 weeks) when mixed with diesel-fuel-contaminated soil was significantly better than fresher compost (2 weeks). Although the 2-week-old compost had a higher amount of biomass, the hydrocarbon degradation was twice as rapid for older compost. The abundance of other organic matter in fresh compost may hinder diesel fuel degradation. Higher natural organic matter has also hindered the degradation of pesticides. The need for supplemental nutrients also will depend on the energy available in the source material and the resulting nutrient level of the final mix.

A treatability study must confirm the suitability of mix rates and the significant process control parameters. For a compost mixture, documenting degradation of the target compounds is significantly more difficult than that of soil or liquid-phase systems. This results from the dilution effect of mixing contaminated soil with a bulking agent or sludge and the resulting decrease in analytical detection sensitivity. Both dilution and sorption on the bulking agent will occur. For sewage sludge the typical mix ratios are approximately 3 parts sludge to 2 parts wood chips by volume. Depending on the water content of the soil or sludge and the nature of the bulking agent, the dilution and sorption effect can be significant. Since dilution has never been acceptable as treatment, the monitoring and regulatory program must formulate an acceptable treatment level and how successful treatment will be established.

Bioremediation applications

The application of composting techniques to the treatment of contaminated soils has not been extensive. One limitation is the mixing of noncontaminated material with contaminated soil, and the potential that an unsuccessful operation may result in a greater quantity of contaminated material. Pilot testing is strongly recommended for any composting of hazardous wastes. Composting applications for hazardous waste include pentachlorophenol (Sikora et al., 1982), refinery sludges (Deever and White, 1978), insecticides (diazocin, parathion, and dieldrin) composted with cannery wastes (Rose and Mercer, 1986), and ethylene glycol with landfill sludges (Mays et al., 1989). Although full-scale applications are few, there have been many successful pilot studies.

Pilot composting studies have demonstrated effective degradation of PAHs and Aroclor 1232 (Hogan, 1988). Sewage wastewater sludge containing aliphatic hydrocarbons, PAHs, and Aroclor 1232 was composted over 35 days. The compost mix contained 1-octadecene, 2,6,10,15,19,23-hexamethyltetracosane, phenanthrene, fluoranthene, pyrene, and chrysene at concentrations of 1 g/kg of mixture. The PCB, Aroclor 1232, was present at 11 mg/kg of mixture. Equal parts of belt filter press sewage sludge were mixed with a stabilized sludge-derived compost as a bulking agent. A microbial seed consisting of material composted for 1.2 to 7 months in the presence of test compounds was added to comprise 1 percent of the mixture. All hydrocarbons were successfully biodegraded within 35 days of composting except chrysene (Table 8.8). The higher degradation rate at 35°C versus 50°C reflects the greater microbial population diversity, as discussed above.

Effective removal of PAHs by composting has been performed at ambient temperature of 4°C (Berg, 1991). Composting was applied to the treatment of PAH-contaminated soil in northern Norway. A coke works contaminated 20,000 tons of soil with PAHs averaging 500 mg/kg. The treatment goal was 20 mg/kg. Soil was sorted, crushed, and mixed with pine bark in a ratio of 1:1 on a volume basis. The soil was sandy and had very little capacity to retain moisture. Moisture was maintained by watering at intervals, varying from 2 to 6 weeks. Several oxygenation methods were evaluated: pile turning, forced aeration, and hydrogen peroxide addition by water. Forced aeration was superior, reducing PAHs from 500 to 20 mg/kg total PAHs in 7 weeks (Table 8.9). Nitrogen and phosphorus addition significantly improved the rate of degradation.

TABLE 8.8 Removal After 35 Days' Composting

Compound*	35°C	50°C
Octadecene	98.2	95.5
Hexamethyltetracosane	75.1	50.9
Phenanthrene	99.9	97.5
Fluoranthene	94.6	90.4
Pyrene	93.1	87.4
Aroclor 1232		81

*Initial concentrations were 1 g/kg of compost mixture for all except Aroclor 1232, which was 11 mg/kg of mixture.

TABLE 8.9 Effect of Aeration on Composting Rate for PAHs

Aeration method	Weeks to achieve 20 mg/kg of PAHs*
Forced aeration	7
Hydrogen peroxide	14
Turning	12

SOURCE: Berg, 1991.
*Initial PAHs 500 mg/kg.

Slurry Treatment

Slurry-phase reactors have been successfully applied to the decontamination of solids and sludges. Hazardous substances treated by slurry-phase systems include polynuclear aromatic hydrocarbons, pesticides, petroleum hydrocarbons, and heterocyclic and chlorinated aromatics associated with wood preservation facilities. The ability to handle a broad range of compounds is illustrated by the response of a slurry-lagoon bioremediation project in Harris County, Texas (Table 8.10). This bioremediation project was initiated after the U.S. EPA indicated that they planned to issue a record of decision for incinera-

TABLE 8.10   Concentrations of Hazardous Substances Bioremediated by Solid- and Slurry-Phase Systems

	Lagoon sludge and subsoil concentrations	
Chemical	Sludge, ppb	Subsoil, ppb*
Semivolatile Organic Compounds		
Phenol	32,000	0
1,2-Dichlorobenzene	4,700	0
bis(2-Chloroisopropyl) ether	6,100	0
n-Nitroso-di-n-propylamine	23,600	0
Isophorone	4,500	0
2,4-Diethylphenol	2,300	0
2,4-Dichlorophenol	1,200	0
Naphthalene	2,652,500	318,000
Acenaphthylene	582,000	47,000
Acenaphthene	233,000	22,000
4-Nitrophenol	9,300	500
2,4-Dinitrotoluene	8,100	1,200
2,6-Dinitrotoluene	160,000	1,800
Fluorene	560,000	67,000
Pentachlorophenol	23,000	400
Phenanthrene	900,000	118,000
Anthracene	246,000	35,000
Fluoranthene	206,000	16,500
Pyrene	269,000	21,600
n-Nitrosodiphenylamine	260,000	14,000
Butylbenzylphthalate	17,600	50
Benzo[a]anthracene	68,000	4,600
bis(2-Ethylhexyl)phthalate	12,800	3,500
Chrysene	80,500	5,100
Di-n-Octyl phthalate	20,000	0
Benzo[b]fluoranthene	51,500	300
Benzo[k]fluoranthene	84,000	300
Benzo[a]pyrene	37,000	500
Indeno[1,2,3-cd]pyrene	23,000	0
Dibenz[a,h]anthracene	23,600	90
3,3-Dichlorobenzidine	2,400	0

TABLE 8.10   Concentrations of Hazardous Substances Bioremediated by
Solid- and Slurry-Phase Systems   (*Continued*)

Chemical	Lagoon sludge and subsoil concentrations	
	Sludge, ppb	Subsoil, ppb*
Metals		
Arsenic	2,600	220
Beryllium	90	70
Cadmium	1,500	40
Chromium	112,000	8,300
Copper	546,000	14,500
Mercury	2,000	50
Nickel	283,000	3,000
Lead	143,000	11,000
Selenium	200	20
Zinc	794,000	34,000
Pesticides		
Dieldrin		11
PC13-1232	428,000	56,000
Volatile Organic Compounds		
Vinyl chloride	600,000	17,000
Methylene chloride	8,600	2,500
1,1-Dichloroethene	5,200	700
1,1-Dichloroethane	318,000	14,500
trans-1,2-Dichloroethane	674,000	503,000
Chloroform	16,600	400
1,2-Dichloroethane	1,230,000	24,400
1,1,1-Trichloroethane	500	0
1,2-Dichloropropane	153,000	10,700
Trichloroethene	25,700	2,400
1,1,2-Trichloroethane	78,000	3,000
Benzene	299,000	19,000
Tetrachloroethane	6,000	400
Toluene	285,000	53,000
Chlorobenzene	16,000	5,000
Ethylbenzene	89,400	345,000

*2 to 4 feet below lagoon bottom.

tion of the contaminated sludge and soil. Incineration was projected
to cost $120 million (French Limited Task Group, 1988). The bioreme-
diation pilot study not only proved successful, but the field bioremediation program for the sludge and soil was completed at a cost of $55
million. In situ slurry-lagoon bioremediation achieved the cleanup criteria for vinyl chloride (<5 ppm) and total PCBs, benzene, and
benzo[a]pyrene at concentrations less than 10 ppm.

TABLE 8.11   Advantages of Slurry-Phase Bioremediation

1. Greater and more uniform process control
2. Enhanced solubilization of organic chemicals
3. Physical breaking of soil-sludge particles
4. Increased contact between microorganisms and contaminants
5. Ability to enhance solubilities of contaminants with surfactant applications
6. Improved distribution of nutrients, electron acceptors, or primary substrates
7. Faster biodegradation rates

Advantages over solid phase

Field bioremediation projects have demonstrated several advantages for slurry reactor systems over solid-phase systems (Table 8.11). Slurry reactor systems increase degradation rates compared with rates observed for solid systems. The half-life of phenanthrene has been reported at 8 days in slurry-based systems compared with 32 days in solid-phase systems (Yare, 1991). Similar enhanced rates have been demonstrated for other compounds that have a high soil adsorption affinity (Table 8.12).

TABLE 8.12   Comparison of Degradation Half-Life for Solid-Phase and Slurry-Phase Bioremediation

Compound group	Solid phase		Slurry phase	
	Initial concentration, mg/kg	Half-life, days	Initial concentration, mg/kg	Half-life, days
Total volatiles*	1,065	7.6	498	1.2
Total PAHs*	55	27.3	57	7.1
Phenanthrene	36	32.4	47	8.0
Total organic carbon	21,900	311	15,800	21.7

*Total volatiles consisted of:
Benzene
Chlorobenzene
Chloroform
1,1-Dichloroethane
1,2-Dichloroethane
1,1-Dichloroethylene
1,2-trans-Dichloroethylene
Ethylbenzene

Methylene chloride
1,1,2,2-Tetrachloroethane
Tetrachloroethylene
Toluene
1,1,2-Trichloroethane
Trichloroethylene
Vinyl chloride

Total PAHs consisted of:
Anthracene
Benzo(a)anthracene
bis(2-Chloroethyl) ether
1,2-Dichlorobenzene
1,3-Dichlorobenzene
1,4-Dichlorobenzene
Hexachlorobenzene

Hexachlorobutadiene
Di-n-Octyl phthalate
Fluoranthene
Fluorene
Naphthalene
Phenanthrene
Pyrene

source: Yare, 1991.

TABLE 8.13   Disadvantages of Slurry-Phase Bioremediation

1. Additional energy requirements
2. Increased material handling
3. Possible physical separation process for stones and rubble
4. Liquid and solids separation processes
5. Increased water handling and treatment costs

The enhanced rates for degradation are a direct result of improved contact between the microorganisms and the hazardous compounds. The water phase provides for a larger amount of soluble compounds and improved homogeneity. Typical applications include difficult-to-degrade organic compounds such as PAHs, pesticides, and TNT, viscous and oily sludges, clay and silty soils, and halogenated hydrocarbons.

Slurry-phase systems do have several disadvantages (Table 8.13). These are all related to additional process requirements and materials handling, causing higher costs for facilities. Slurry-phase treatment is more costly than land farming and composting. It is, however, usually less costly than incineration, solvent extraction, and thermal desorption (Brox, 1993). Significant quantities of wastewater can result from solids separation and dewatering after slurry treatment. This wastewater may require treatment before discharge. However, these disadvantages are often offset by the improved performance of slurry bioremediation over solid-phase and the growing list of successful site closures.

Soil contaminated with 2,4-dichlorophenoxy acetic acid (2,4-D), 4-chloro-2-methyl-phenoxyacetic acid (MCPA), alachlor, trifuralin, and carbofuran have been reduced from levels of 800 to 20 mg/kg within 2 weeks. The residuals were further treated through land application (U.S EPA /540/2-90/016). Slurry treatment has met closure on the treatment of wood preserving contaminated sludges, achieving greater than 90 percent removal for pentachlorophenol and PAH compounds. Results for several field bioremediation projects are combined in Table 8.14. Typical cost ranges for slurry treatment are $80 to $150 per cubic yard of soil. Cost breakdowns for one slurry treatment program of PAH-contaminated soils are provided in Table 8.15. Others have confirmed rapid degradations, reporting 2,4-D degradation from 1500 ppm to less than 10 ppm in 13 days (Ross, 1988).

The use of nutrients and microbial augmentation usually improves degradation rates for slurry treatment. The addition of nitrogen and phosphorus has been reported to result in final concentrations 50 to 90 percent lower for PAH-contaminated soils (Strou, 1988).

Augmentation with specific microorganisms and cometabolism has been evaluated in pilot studies. Slurry treatment of soils contaminated with chlorinated compounds can be enhanced by augmentation

TABLE 8.14   Slurry Bioremediation of Wood Preserving Contaminated Sludges

Compound	Initial concentration, mg/kg	Final concentration, mg/kg	Percent removal
Phenol	1.4	<0.1	92.8
Pentachlorophenol	64	0.8	92.8
Naphthalene	343	1.6	99.5
Phenanthrene and anthracene	2870	13.7	99.5
Fluoranthene	511	4.6	99.1
Carbazole	139	0.3	99.8
2-Chlorophenol	2	<0.01	99.5
Acenaphthylene	999	1.4	99.8
Chrysene and benzo(a)anthracene	519	1.4	99.7
Benzo(a)pyrene	83	0.1	99.9
Benzo(b)fluoranthene	519	<0.03	99.9

SOURCE: U.S. EPA/540/2-90/016.

TABLE 8.15   Cost Breakdown for Slurry
Bioremediation of PAH-Contaminated
Soil

Activity	Cost, dollars per ton
Treatment unit:	
Slurry preparation	$50–$60
Biological treatment	$40–$50
Dewatering process	$20–$30

Design, project administration, and treatability studies can add another 40 percent to the
final cost per ton.
SOURCE: Jerger, 1993.

with organisms having degrading capabilities for halogenated hydrocarbons. Augmentation with dichloromethane-degrading organisms of
Hyphomicrobium GJ21 or *Methylobacterium DM4* decreased degradation half-lives by one-fifth and two-thirds, respectively, over that of
the indigenous organisms (Janssen, 1991). The addition of methane
to slurry reactors containing 45 percent solids by dry weight
improved cometabolism degradation of trans,-1,2-dichloroethylene
(Janssen, 1991).

Slurry bioreactors

Slurry-phase treatment is usually handled as a batch process. The
liquid-solids reactors include lagoons, open vessels, and closed systems. The first step includes excavation, screening to remove stones

longer than $\frac{1}{10}$ to $\frac{1}{2}$ in in diameter, and creation of an aqueous slurry. The slurry phase can be 60 to 95 percent water by weight, depending on the nature of the biological reactor. The reactor can be a designed containment unit or an existing lagoon. The treatment facility typically requires a settling tank or thickener and solids dewatering. One process flow diagram is illustrated in Fig. 8.18.

Various slurry reactor designs are available that differ in the mechanics of oxygenation and mixing of the solid suspension. Aeration and mixing is provided by floating direct-drive mixers with draft tubes, turbine mixers, spargers, and rotating drums. Some systems have better solids-handling capabilities and they vary significantly in energy requirements. Since hydraulic retention times are often long, reactors that use less energy become more cost-effective.

A typical draft tube reactor that uses a floating aerator is illustrated in Fig. 8.19. These systems have lower initial capital cost, but they frequently result in poor mixing near the reactor bottom. The direct-drive mixer is subject to significant abrasion. Air is provided by spargers. Some spargers are reported to have significant clogging problems.

Another reactor design uses a turbine mixer with spargers (Fig. 8.20). Two or more impellers are mounted on a shaft to provide improved mixing and maintain suspension of the higher-specific-gravity solids near the tank's bottom. Bottom impellers may be high shear–air dispersion types, and upper impellers are axial pumping types (Brox, 1993).

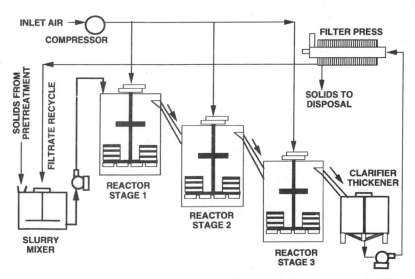

Figure 8.18 Typical process flow sequence for slurry-phase treatment. (*Brox, 1993.*)

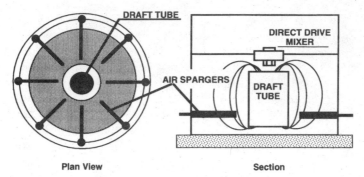

Figure 8.19 Draft tube liquid-solids contact reactor for slurry bioremediation. (*Courtesy of Eimco Process Equipment Co., Salt Lake City, Utah.*)

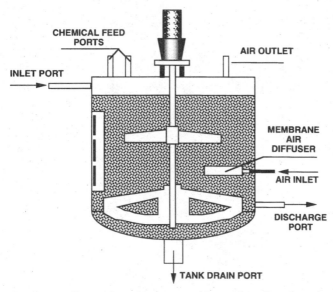

Figure 8.20 Slurry-phase bioreactor with two impellers for mixing. (*Brox, 1993.*)

A dual-injection reactor has been designed in the Netherlands to treat coarse and fine particles by providing separation based on settling velocity. The first stage of the "dual injected turbulent separation reactor" separates the coarser material (greater than 100 μm) by simultaneous upflow of air and injected slurry. Particles up to 4 mm in diameter can be treated (Brox, 1993). The fine (smaller than 100 μm) flow to the other reactor stages where they are kept in suspension by aeration. The largest reactor used to date has been 1000 gal in the Netherlands. The main disadvantages are high energy input to maintain suspension and the potential for volatilization.

Figure 8.21   Rotating drum bioreactor for the bioremediation of soil slurries. [*Courtesy of G. Brox (Brox, 1993)*.]

A rotating drum reactor has been designed in Germany to handle high solids levels of 65 to 75 percent by weight (Fig. 8.21). This system is applicable to waste with low oxygen demands or anaerobic systems. There is limited oxygen transfer capability, since oxygen is supplied by natural ventilation of head space gases. It is also limited to slurries that do not have clumping potential.

An air lift enclosed reactor using slurry agitators typical of the cement industry is available from Eimco (Fig. 8.22). A rotating rake maintains solids mixing below the air diffusers. These bottom solids are recirculated to the reactor's top with air lifts. The main aeration is provided by an air sparging system of chemically resistant rubber membranes. This reactor design has a relatively low energy input (0.15 to 0.25 hp/1000 gal) and operates at solids concentrations of 35 to 45 percent by weight (Brox, 1993).

Many industrial facilities have used open lagoons as holding ponds for waste sludges. They vary from small lagoons of several hundred square feet to lagoons of several acres. These existing waste lagoons are frequently the source of hazardous chemicals and can be modified to serve as a slurry bioremediation facility. For many remediation activities, minimizing materials handling reduces the associated risks to ecology and humans. In situ treatment is frequently a sound environmental approach as well as the least costly. Engineering in situ lagoon bioremediation, however, provides challenges that are significantly more difficult than those found for surface bioreactors.

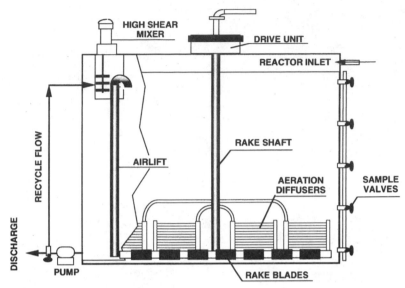

Figure 8.22 Airlift slurry-phase reactor for treating contaminated soil. (*Courtesy of Eimco Process Equipment Co., Salt Lake City, Utah.*)

Slurry lagoon systems

Slurry treatment systems can be constructed as a lagoon facility incorporating all of the latest controls and safeguards, such as double liners with leak detection underdrains. However, this is not the typical bioremediation lagoon facility. The typical lagoon system is an in situ bioremediation project. The lagoon has probably existed for several decades, receiving a mixture of organic and inorganic hazardous substances. Many have resulted in both soil and groundwater contamination. Cleanup usually requires a combined treatment approach.

The first design task is to establish the volume of contaminated liquid, sludge, and subsoil and, second, the nature and concentration of contaminants. Establishing the volume of liquid, sludge, and subsoil to be treated in a lagoon is similar to collecting soil borings for soil contamination. Instead of developing the usual grid pattern, lagoon sections are characterized to establish contours (Fig. 8.23). Soundings for water and sludge depths and soil borings are collected at intervals for each section to develop a profile of the cross section (Fig. 8.24). From these cross sections the volume of contaminated water, sludge, and subsoil is estimated.

Sludge samples can be collected with a piston closing tube that allows for the piston to be withdrawn as the tube is lowered into the sludge. The piston is at the lower end of a stainless-steel or PVC tube. The tube is lowered vertically into the sludge as the piston is raised

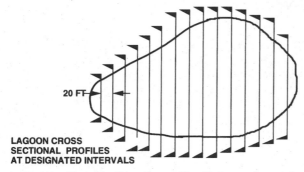

Figure 8.23 Characterizing the profile of a lagoon by establishing cross sections for data collection.

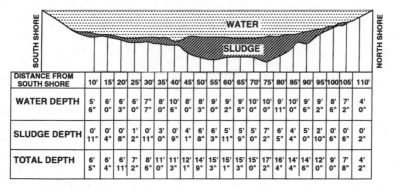

DISTANCE FROM SOUTH SHORE	10'	15'	20'	25'	30'	35'	40'	45'	50'	55'	60'	65'	70'	75'	80'	85'	90'	95'	100'	105'	110'
WATER DEPTH	5'6"	6'0"	6'3"	6'0"	7'7"	8'0"	10'6"	8'0"	8'3"	9'0"	9'2"	9'6"	10'0"	10'0"	9'11"	10'0"	9'6"	9'2"	8'6"	7'2"	4'0"
SLUDGE DEPTH	0'11"	0'4"	0'8"	1'2"	0'11"	3'0"	0'9"	4'1"	6'8"	6'3"	5'11"	5'9"	5'0"	7'2"	6'5"	4'4"	5'0"	2'10"	0'6"	0'6"	0'2"
TOTAL DEPTH	6'5"	6'4"	6'11"	7'2"	8'6"	11'0"	11'3"	12'1"	14'9"	15'3"	15'1"	15'3"	15'0"	17'2"	16'4"	14'4"	14'6"	12'0"	9'0"	7'8"	4'2"

Figure 8.24 Establishing a lagoon's cross-sectional areas of contaminated liquid and sludge. (*French Limit Task Group, 1988.*)

according to the rate of tube descent. A vertical coring device using a split PVC pipe contained within a 3-in-diameter by 5-ft-long Shelby tube also has been used. The sampler is driven into the sludge and subsoil. After removal, the split PVC pipe is removed from the Shelby tube and laid open without disturbing the core sample.

The second engineering decision is establishing that the waste is biodegradable and that the mixture does not present toxic or enzyme-inhibiting conditions. Unlike contaminated soil and groundwater, sludge lagoons can contain a broader chemical mixture. In addition, the chemical nature of sludges at one end or depth of a lagoon is different at another location. This occurs because of several factors:

1. Industrial discharges to lagoons have usually occurred over many years with significant changes in the process discharge characteristics. These discharges seldom mix but remain as pockets of different chemicals.

2. Transport mechanisms that are active in soil and groundwater are not present to partition or cause segregation of chemicals. Thus

metals and the high-molecular-weight PAHs are found with the soluble aromatic compounds.

3. Waste lagoons tend to be self-sealing as a result of viscous sludges and tars. This condition helps retain the more soluble compounds since they are not easily transported from the lagoon.

4. The viscous sludges and lighter hydrocarbons provide for lower volatilization rates of solvents.

5. Lagoon surfaces may contain a floating scum layer that reduces volatilization.

One example of a lagoon's content is the characterization of the French Limited Site located in Crosby, Tex. This 7.3-acre lagoon was an abandoned industrial waste management facility (Fig. 8.25). Contaminated media consisted of a lagoon liquid zone, a lagoon sludge zone, a lagoon subsoil, and groundwater zones. Chemical compounds consisted of a semivolatile fraction, a metals fraction, and a volatile fraction (Table 8.10).

Establishing that a lagoon's content can be biodegraded and that toxicity does not exist is not a straightforward evaluation, nor can one perform a simple toxicity test. First, it is unlikely that all compounds are biodegradable. For example, bioremediation will not reduce the metals listed in Table 8.10. Combined treatment approaches are fre-

Figure 8.25 Aerial view of bioremediation activities on the French Limited site. (*Courtesy of ENSR Consulting and Engineering, Houston, Tex.*)

quently necessary, since bioremediation will not complete the cleanup effort. For metals, a stabilization process can be used after bioremediation. Bioremediation, however, will significantly reduce the amount of soils that require stabilization. Volume reductions of 80 percent and more are possible.

A simple toxicity test on the mixture will not provide adequate information for operating a bioremediation process. Toxicity of lagoon sludges is a function of mixing and other techniques to enhance solubilization. Since contaminants will exist at different concentrations in the two or three lagoon waste phases (liquid, sludge, and subsoil), the design and operational question is to what degree can these phases be mixed without toxicity, and how does the operator control the level of mixing for maximum degradation. A second treatability question is how does mixing influence volatilization and can this be minimized?

Toxic compounds will partition between the soluble and adsorbed phase on soil and sludge solids. The degree of mixing sludge bottoms from the lagoon into the biological liquid zone can be controlled to prevent toxic concentrations. Oxygen uptake rates measured as milligrams per liter per minute have been used to monitor toxicity and control the degree of mixing within lagoons. Unless demonstrated otherwise by the treatability tests, the use of solubilizing and emulsifiers should be limited to the later stages of a bioremediation project when the potential for toxic loading conditions has been significantly reduced. By controlling the degree of mixing, a microbial population of 1×10^7 to 4×10^8 colony-forming units was maintained in the lagoon waste presented by Table 8.10.

The next engineering decision is selection of the microbial metabolism mode for degradation of the hazardous compounds. Because of the typical chemical variety, it is unlikely that one system will achieve complete treatment. If an aerobic system is selected, attention must be given to the stripping of volatile compounds. Volatile emissions have been a serious problem at some lagoon treatment facilities.

For waste containing volatile chlorinated hydrocarbons the application of an anaerobic metabolism mode followed by an aerobic phase has several advantages:

1. Anaerobic conditions can be used to dehalogenate the chlorinated volatile compounds.

2. Air stripping of the chlorinated volatile compounds would not occur.

3. Greater detoxification of chlorinated aromatic compounds would occur.

4. Operating cost is reduced, since oxygen is not required during anaerobic treatment.

5. After adequate anaerobic dehalogenation of volatile compounds, aeration can be provided for aerobic degradation of the remaining hydrocarbons and transformation products.

Disadvantages of anaerobic treatment would be the production of odors and the overall slower rate of degradation under anaerobic conditions. Odor capture and treatment can add significant costs to the project.

Monitoring for process control should be performed at a higher frequency for slurry reactors. The higher loading rates of soils and organic compounds coupled with the dynamics of mixing and degradation can cause rapid changes in pH, microbial populations, nutrient needs, and electron acceptors. Microbial demand for these supplements is a function of organic loading. Unlike liquid bioreactors, the organic loading of slurry reactors is a function of mixing. Enhancing slurry mixing with centrifugal pumps has approximately doubled chemical oxygen demands in hazardous waste mixed liquor. The degree of mixing and its effectiveness are not easily controlled in lagoon systems.

Since the required biological requirements are provided, mixing is probably the most important process variable in slurry systems. Mixing must accomplish the following:

1. Maintain suspended the biomass and soil and sludge particles.

2. Dispense water-soluble compounds from soils and sludges to the mixed liquor.

3. Break insoluble soil and sludge particles that would settle into smaller particles that can be maintained in suspension.

4. Aid in emulsification of oils, tars, and viscous substances.

5. Lift soils from the lagoon bottom.

6. Provide for variation in mixing degree to prevent increase in toxicity, but maintain maximum degradation rates.

Because of the need to provide a realistic solids distribution in treatability reactors (zones of increased solids concentration with depth), small reactors are unsuitable. Treatability studies have used 55-gal drums and larger tanks as reactor vessels. Mixing becomes the limiting factor for the rate of bioremediation. Mixing power per unit volume of liquid for the treatability test must represent practical field levels. The availability of hazardous compounds for biological degradation is frequently limited by the available mixing and solubilization procedures.

Several techniques have been used for slurry mixing. These include air spargers, pumping, mechanical, and dredging. Air sparging by itself has not been adequate in maintaining mixing for slurry reactors. Most soils and sludges are either too viscous or have settling and scour velocities too great for suspension by sparging equipment. Air sparging also causes excessive volatile emissions.

Lagoon solids frequently require physical breaking and continuous dispersion for reasonable rates of biodegradation. The use of centrifugal pumps has been a significant benefit for the breaking, dispersion, and emulsification of lagoon sludges. The continuous pumping of lagoon bottom deposits for resuspension of soils within the biological active zone has been reported as the operational key to successful bioremediation of viscous slurries. Centrifugal pumps have been used on floating platforms and mounted to boom arms (Figs. 8.26 and 8.27). Pumps attached to truck boom arms are suitable for smaller lagoons. The pumping unit must be capable of moving throughout the bioremediation area. Pumping units of 1500 gal/min have been used for lagoon bottom dispersion. Once suspended, material must be distributed through the lagoon. For large lagoons the low-shear banana mixers (FLYGT) have been successfully used to maintain desired circulation patterns in lagoons. These units can move 50,000 gal/min with low-energy, 5-hp motors. They have been mounted on floating barges designed to allow adjustment for depth (Figs. 8.28 and 8.29).

Figure 8.26 Centrifugal pumps mounted on small self-propelled barges for maintaining suspension of lagoon solids during bioremediation. (*Courtesy of Clark Technology Systems, Inc., Milton, Pa.*)

Figure 8.27 Centrifugal pump mounted to truck boom for maintaining suspension of lagoon solids during bioremediation. (*Courtesy of ENSR Consulting and Engineering, Houston, Tex.*)

After lagoon sludges have been bioremediated, the contaminated subsoil can be lifted into the lagoon's active treatment liquor. Suspension of contaminated subsoil has been successfully performed with small dredges. Small portable hydraulic dredges were used to achieve treatment of contaminated subsoil at the French Limited Superfund site (Fig. 8.30). Dredging allows in-place mixing of bottom sediments and lifts contaminated soil into the treatment zone with minimum materials handling and human contact.

The hydraulic dredge should be self-propelled and capable of 360° in place turns to move around the site's obstacles. Anchors or cables are undesirable and rigging spuds can be hydraulically lowered to fix the dredge in place. A design used for the French Limited site is illustrated in Figs. 8.31 and 8.32. A swinging ladder design is used for the

Figure 8.28 "Banana mixer" (FLYGT) mounted on floating platform for maintaining slurry circulation during bioremediation in large lagoons. (*Courtesy of ENSR Consulting and Engineering, Houston, Tex.*)

Figure 8.29 Floating platform for holding adjustable "banana mixers" (FLYGT). (*Courtesy of Clark Technology Systems, Inc., Milton, Pa.*)

Figure 8.30 Dredge cutter head used to suspend contaminated soil in lagoon supernatant for bioremediation. (*Courtesy of Dredging Supply Company, Inc., Harvey, La.*)

dredge so that a hydraulically operated cutter head can be moved in an arc. The cutter head lifted excavated soil into the suction side of a 2000 gal/min diesel-driven pump. This soil was discharged out of the back of the barge into the lagoon supernatant. The design provides for removal of the swinging ladder and cutter head so a tubular boom with an attached shearing pump can be suspended. This attachment provided for emulsification of the soil and tarlike sludges.

An important consideration when designing the aeration system is to consider the need to move pumping and mixing equipment throughout the lagoon. Air sparger pipes extending into the lagoon hinder the movement of equipment. Downdraft mechanical aerators and liquid-liquid eductors have been successfully employed.

Lagoon aeration systems

Lagoon aeration systems have used air sparging, mechanical aerators, submersible aspirators, and pressurized pure oxygen eductors (Mixflo). Design considerations include rate of oxygen injection, degree of volatilization, minimization of maintenance, ability to handle high solids levels, as well as cost (Table 8.16). The advantages and disadvantages of major aeration types are provided in Table 8.17.

Factors that usually determine a system's selection are rate of oxygen delivery, volatilization potential, and flexibility. Rate of delivery and volatilization are significantly influenced by the oxygen source.

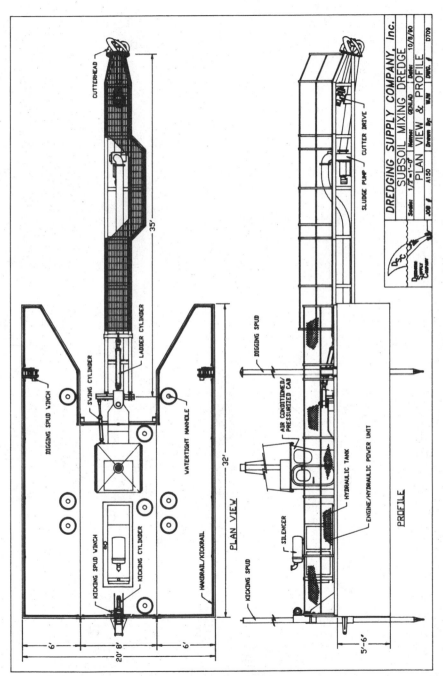

Figure 8.31 Plan and profile of soil and sludge-mixing dredge used at French Limited Superfund site. (*Courtesy of Dredging Supply Company, Inc., Harvey, La.*)

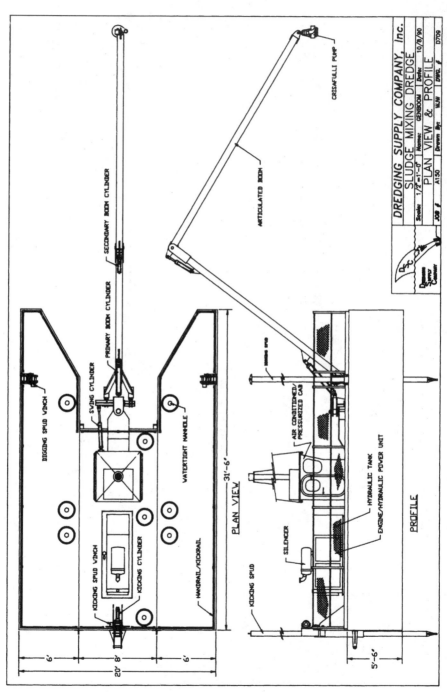

Figure 8.32 Plan and profile dredge equipped with a pump for sludge mixing. (*Courtesy of Dredging Supply Company, Inc., Harvey, La.*)

TABLE 8.16   Design Consideration for Lagoon Aeration Systems

Rate of oxygen dissolution
Percent of injected oxygen that is dissolved or retained in the lagoon
Amount of total gas released from solution and chemical volatilization potential
Ability to handle high solids level (6–15%)
Ability to mix dissolved oxygen through the lagoon
Ability to maintain solids in suspension
Flexibility of equipment to variations in liquid levels and movability
Power requirement and cost
Worker safety

TABLE 8.17   Advantages and Disadvantages of Major Aeration Delivery Mechanisms

Performance characteristics	Sparger coarse bubble	Mechanical downdraft or propeller	Diffuser fine bubble	Aspirator turbine
Percent dissolution of oxygen	Poor	Fair	Good	Good
Potential for volatiles emissions	Poor	Poor	Good	Fair
Ability to handle high solids	Poor	Good	Fair to good	Poor
Flexibility	Poor	Good	Fair	Fair
Power requirement*	110	160	100	150

*Relative horsepower difference adapted from Bergman, 1992.

Atmospheric air limits the maximum dissolved oxygen concentrations to 9 to 10 mg/L. Pure oxygen and hydrogen peroxide have significant advantages when high delivery rates are necessary. They also minimize volatile organic emissions. Since air contains 79 percent nitrogen, significantly larger volumes of air must be injected for delivery of a given amount of oxygen. This excess gas is released from the lagoon with the transport of volatile emissions. The relative volumes of oxygen delivered and resulting off-gases for various delivery systems are illustrated in Table 8.18.

Coarse bubble sparging systems have experienced major drawbacks. They result in excessive emission of volatile organic compounds. The chlorinated aliphatic hydrocarbons are not readily degraded by aerobic systems, and as much as 95 percent can be released as air emissions within relatively short aeration periods. The second major deficiency of coarse bubble sparging is the inability to maintain adequate mixing for solids suspension. Air sparging systems also limit operational flexibility for moving equipment throughout the lagoon. The need to provide air distribution requires a grid of sparger pipes. These pipes become a major obstruction for the use of sludge and soil mixing equipment. Movable pumps and dredging equipment are frequently required for continued solids dispersion,

TABLE 8.18   Oxygenation Efficiencies and Associated Volumes of Off-Gas

Delivery method	Oxygen source	Percent injected O_2 dissolved	Relative gas volume delivery rates,* surface ft^3/min	Relative off-gas volumes, surface ft^3/min
Fine bubble diffuser	Air	14	3418	3318
Coarse bubble diffuser	Air	7	6835	6735
Jet aeration	Air	12	3987	3887
Pressurized eductor	Oxygen	90	112	12
Water pump	Hydrogen peroxide	90–100	0	0

SOURCE: Adapted from Bergman, 1992.
*To deliver 100 surface ft^3/min of oxygen.

emulsification, and subsoil mixing. Sites that have used pipe distribution systems have replaced them with mechanical mixers (downdraft or propeller) to provide for the needed solids circulation equipment.

When sparging systems are appropriate, design specifications should require steel pipes and an anchoring mechanism. The use of flexible rubber connectors to couple each sparger pipe to air headers has been inadequate for maintaining location stability of the pipes under the mixing action necessary for solids suspension. Pipes of PVC do not stay submerged even with anchoring procedures, and they are frequently broken by other equipment.

Mechanical aeration systems have been used on floating modules. The downdraft aerators have provided good oxygen transfer and provide good mixing of viscous tars and oily waste. The use of floating platforms provides the mobility needed to access the lagoon bottom for resuspension of sludge. Major disadvantages have been the high emissions for volatile compounds and the need to provide intervals of shutdown to prevent foam from reaching the aerator's motors.

Fine bubble diffusers provide for greater transfer efficiency and therefore minimize emissions. Most high-transfer-efficiency diffusers are fine-pore diffusers made of ceramic plates, or flexible porous membranes. Many of these systems are difficult to maintain. They are subject to interior clogging and external fouling. For large lagoons the delivery pipes interfere with equipment movement. Application of fine-pore diffusers to treatment of lagoon slurries has not been reported. To increase the efficiency for oxygen transfer to liquids, liquid pumping systems with various mixing techniques have been developed. In liquid pump systems, a portion of the liquid to be oxygenated is pumped under pressure (2 to 7 atm) and pure oxygen introduced. This two-phase mixture is released into the bioreactor through mixing ejectors. One system is a two-stage pressurized eductor (Mixflo,

TABLE 8.19   Pressurized Eductor Design for French Limited Site

Bioreactor volume	14 million gal
Oxygen requirement	2500 lb/h
Pipeline contactor:	
Size	18-in carbon steel
Flow velocity	11.5 ft/s
Pressure	30 lb/in^2 gauge
Dissolved oxygen saturation	82.4 mg/L
Slurry pumping equipment:	
Pumping rate	60,000 gal/min
Required pumping power	1500 hp
Design pumping power	2400 hp
Number of pump units	8
Eductor:	
Number of units	24
Minimum pressure loss	25 lb/in^2
Oxygen:	
Total volume injected	280 ft^3/min
Cryogenic storage	11,000 gal

SOURCE: Bergman, 1992.

Praxair Technology, Inc.) that has been successfully used at the French Limited site in Crosby, Tex. The design criteria are provided in Table 8.19.

The pressurized eductor aeration system (Mixflo) uses a two-stage pure oxygen dissolution process (Fig. 8.33). The first stage consists of a pressurized pipeline (2 to 4 atm) in which the waste or slurry is pumped. Pure oxygen is injected into the pipeline as finely dispersed bubbles. The gas-water mixture turbulently flows through an adequate length of pipeline (providing approximately 30 s of holding time) to provide for 60 percent of the oxygen to dissolve. Because of the pressurized pipeline, the solubility of oxygen is increased by 1.5 to

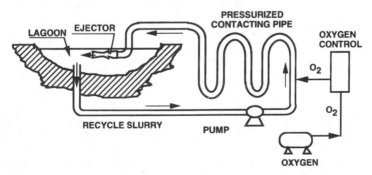

Figure 8.33 The two-staged pressurized eductor (Mixflo) for oxygenation of slurries.

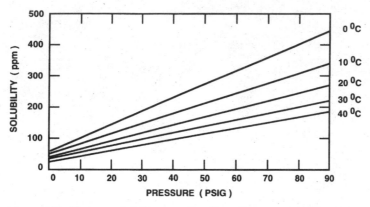

Figure 8.34 Oxygen solubility as a function of pressure and temperature.

2 times (Fig. 8.34). The pressurized system allows for a decreased contact time in the pipe since the rate of oxygen dissolution is increased. Pressurization also reduces the amount of waste slurry that must be pumped since the liquid has a higher dissolved oxygen carrying capacity.

The second stage of the pressurized eductor system is the liquid-liquid eductor (Fig. 8.35). The eductor reinjects the oxygen-carrying waste slurry into the bioreactor. The aspiration effect of the eductor pulls in unoxygenated slurry and mixes it with the recycled oxygenated slurry before dispersing the mixture into the bioreactor. Installed eductors are shown in Figs. 8.36 and 8.37. These can be adjusted in height and angle of discharge from their support pilings. A typical ratio of unoxygenated slurry to oxygenated slurry is 3:1. This provides for an overall dissolution of approximately 90 percent of the oxygen. The energy of injection and pressure release creates a fine-

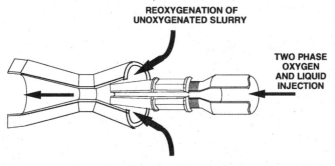

Figure 8.35 Schematic of the liquid-liquid eductor. (*Praxair, Inc.*)

Figure 8.36 Field installation of eductors at the French Limited Superfund site before lowering in lagoon. (*Courtesy of ENSR Consulting and Engineering, Houston, Tex.*)

Figure 8.37 Field operation of eductor aeration system at French Limited Superfund site. (*Courtesy of Clark Technology Systems, Inc., Milton, Pa.*)

bubble dispersion. The aeration efficiency yields approximately twice the pounds of oxygen dissolved per horsepower per hour as bubble diffusers (Storms, 1993).

The diluting of the pressurized oxygen slurry with nonoxygenated slurry has two advantages. When the oxygenated slurry is diluted, the dissolved oxygen level decreases to below saturation at atmospheric pressures. Therefore, dissolved oxygen does not come out of solution, reducing volatilization and maintaining a high oxygen transfer efficiency. Second, oxygen not dissolved in the eductor is well dispersed with nonoxygenated slurry, minimizing bubble coalescence downstream of the injection.

The use of in situ lagoon bioremediation has been successfully applied to complicated chemical mixtures. It minimizes soil and sludge handling and has been performed at costs significantly less than alternate technologies.

Liquid-Phase Bioremediation

Introduction

Liquid-phase bioremediation is the application of surface bioreactors to the treatment of water contaminated with hazardous chemicals. They support the growth and retention of desired microorganisms under optimized process conditions. Reactor design requires the integration of biological concepts with reaction kinetics, mass transfer, and the flow characteristics of the contacting unit. The design principles for liquid-phase bioremediation originate with municipal and industrial wastewater treatment. However, the application of biological processes to the treatment of hazardous chemicals frequently necessitates changes in equipment and process configuration.

It is not the intent to provide the theory and principles of bioreactor design in this text. Texts are available that address design principles for suspended and fixed-film biological reactors. However, a review is provided on the kinetics and predictive models for bioreactor design. The objective of this chapter is to emphasize differences in design when applying biological processes to hazardous chemicals in groundwater versus municipal-industrial wastewater. Available information is provided on design criteria and system performance for bioremediation of hazardous chemicals.

Bioreactor design criteria for the degradation of hazardous compounds are few and frequently site-specific. Unlike classical wastewater treatment design, where the goal is overall reduction in total organic compounds (BOD, TOC, or COD), the goal for bioremediation is the degradation of specific target compounds. Coupled with this is the sensitivity of treatment response to site-specific properties, physical-chemical interactions, and microbial interactions. The complexity of these variables usually requires treatability studies for the development of design criteria. Few full-scale facilities have been evaluated in a manner that addresses design criteria and operational variables. Most available design criteria for hazardous compounds have been developed in laboratory and pilot studies. This information should be used as a starting point.

For many systems, treatability studies are extremely important for the design and optimization of process operations. Each situation is unique because of differences in:

- Predominant microbes
- Water and soil chemistry
- Mixture of hazardous chemicals
- Predominant microbial metabolism mode
- Influence of enhancing or inhibiting chemicals

Bioreactors consist of two basic physical systems: suspended growth and fixed film. Suspended growth bioreactors can be designed as a batch, plug flow, or complete mix system. Fixed-film bioreactors are designed as fixed beds, fluidized beds, or rotating media. With suspended growth reactors, the microorganisms are suspended as microbial aggregates in the liquid. This suspension is called biomass, activated sludge, mixed liquor suspended solids (MLSS), or mixed liquor volatile suspended solids (MLVSS). For these systems, the biomass must be removed, usually by settling, and a portion recycled back to the bioreactor. With fixed-film reactors, the biomass growth occurs on or within a solid medium that acts as a support. These may be submerged in the wastewater, or wastewater may contact the support at periodic intervals.

Issues of concern in reactor design and process control are:

- Efficient degradation of the target compounds
- Meeting effluent quality in ppb versus ppm
- Buildup and retention of adequate biomass
- Biomass management and disposal
- Stripping of volatile compounds
- Potential system upsets
- Variability of influent chemical composition and concentration
- Maintaining biological transformation capacity for cometabolism

Maintaining an effective biomass level is difficult when treating hazardous chemicals. First, these chemicals are typically found at low concentrations and may have little energy for supporting biomass growth.

Second, most hazardous chemicals provide for a low growth rate. Third, influent chemical concentrations can decrease significantly during the life of the project. The ability of a bioreactor to maintain an adequate biomass or solids retention time (SRT) is one of the main challenges in design and facility operation. Solids retention time is a measure of the average residence time that organisms are in the bioreactor. This parameter is also called mean cell residence time (MCRT).

Process Design Parameters and Relationships

The design parameters for bioreactors are well known to the wastewater engineering field, but they usually require defining for most others. Three process design parameters are particularly important. The following discussion introduces these parameters and the predictive modeling approach for describing bioreactor performance. These principles are important for understanding the necessary changes in design emphasis for treating groundwater containing hazardous compounds versus the classical design approach for municipal wastewater.

Process design parameters of importance include the specific utilization rate of substrate U, the mean cell residence time θ_c, and the food to microorganism ratio F/M. The specific substrate utilization rate is expressed as

$$U = \frac{1}{X}\frac{dS}{dt} \tag{9.1}$$

where dS/dt = rate at which the substrate is utilized or degraded, mass/unit volume-time

X = concentration of microorganisms, mass/unit volume

The mean cell residence time (MCRT) is similar to a holding or detention time for biomass. It is defined as the mass of organisms in the reactor divided by the mass of organisms removed each day. For a bioreactor with a typical activated sludge process configuration (Fig. 9.1), the mean cell residence time θ_c is expressed as

$$\theta_c = \frac{VX}{Q_w X_r + Q_e X_e} \tag{9.2}$$

where V = volume of the reactors (for multiple reactors as in Fig. 9.1, V is the total volume)

X = concentration of microorganisms in the reactor, mass/unit volume

X_r = concentration of microorganisms in the return sludge, mass/unit volume

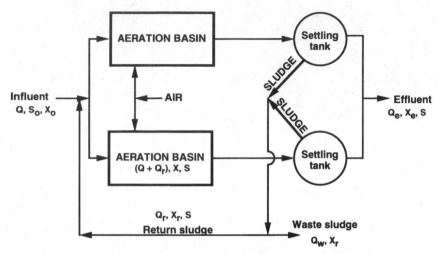

Figure 9.1   Schematic of a typical activated sludge process.

Q_w = flow rate of sludge that is wasted from the reactor with microorganisms of concentration X_r volume/unit time

Q_e = flow rate of liquid effluent from the bioreactor's solid separation process, volume/unit time

X_e = concentration of microorganisms in the liquid effluent from the solid separation process, mass/unit volume

The food to microorganism ratio is defined as the rate that substrate is provided to the reactor per unit mass of biomass in the reactor, expressed as

$$\frac{F}{M} = \frac{S_0 Q}{VX} \tag{9.3}$$

where S_0 = influent substrate concentration, mass/unit volume
    Q = influent flow rate, volume/time
    V = volume of the reactor, volume

Enzyme-substrate kinetics

To predict a bioreactor's performance for the degradation of compounds, a mathematical model based on microbial growth and substrate utilization is necessary. The basis of this model is derived from the reaction between an enzyme and substrate as depicted in Fig. 1.2. Figure 1.2 illustrates an association between enzyme and substrate. This is a reversible relationship that forms an enzyme-substrate complex. The forward rate for this association is designated k_1 and the reverse k_2. The enzyme-substrate complex can dissociate back to their separate entities or react to produce a degradation product and release the enzyme. The rate of degradation is designated k_3. This reaction is expressed by the following:

$$E + S \underset{k_2}{\overset{k_1}{\rightleftarrows}} (ES) \overset{k_3}{\rightarrow} E + \text{metabolites}$$

where E = concentration of enzyme, mass/unit volume
S = concentration of substrate, mass/unit volume
ES = concentration of enzyme complex, mass/unit volume

The above expression is the Michaelis-Menten representation of enzyme-substrate reaction. It was first proposed in 1913 and forms the basis for analyzing the substrate concentration effect on the reaction's transformation velocity.

Since the free enzyme concentration is $E - ES$, one can write the dissociation constant of ES, defined as K_m, as

$$K_m = \frac{[E - (ES)](S)}{ES} \tag{9.4}$$

solving for ES,

$$ES = \frac{(E)(S)}{K_m + S} \tag{9.5}$$

The rate of substrate degradation dS/dt is given by the rate of ES going to metabolites, which is expressed as

$$\frac{dS}{dt} = -k_3(ES) \tag{9.6}$$

where dS/dt is the velocity or rate of degradation and k_3 is the rate constant. The negative sign denotes a decreasing concentration of substrate. Substituting Eq. (9.6) into (9.5), one obtains

$$\frac{dS}{dt} = \frac{-k_3(E)(S)}{K_m + S} \tag{9.7}$$

The maximum rate of degradation V_{max} is obtained when the concentration of enzyme complex is maximal, when all available enzyme is bound by substrate and $ES = E$. Under these conditions, the maximum velocity of the reaction V_{max} is given by

$$V_{max} = -k_3(ES) = -k_3(E) \tag{9.8}$$

and the Michaelis-Menten equation is obtained:

$$\frac{dS}{dt} = \frac{V_{max}/S}{K_m + S} \tag{9.9}$$

Equation (9.9) can be written as

$$K_m = (S)(\frac{V_{max}}{dS/dt} - 1) \tag{9.10}$$

and when $V_{max}/dS/dt = 2$, the degradation rate dS/dt is one-half the value for maximum rate V_{max}, and then $K_m = S$.

Monod equation

In the engineering field, the most widely used expression for biodegradation kinetics is the Monod equation and its modifications for nonlog growth and an inhibitory substrate. The growth rate for microorganisms can be expressed as

$$\frac{dX}{dt} = \mu X \tag{9.11}$$

where X = concentration of microorganisms, mass/unit volume
     t = time
     μ = specific growth rate, per time

The Monod equation relates the dependence of μ on substrate concentration and is expressed by

$$\mu = \frac{\mu_m S}{K_s + S} \tag{9.12}$$

where μ = specific growth rate, per time
     μ_m = maximum specific growth rate (condition for which there is no limiting factor on microorganism growth), unit per time
     S = concentration of growth-limiting substrate in solution, mass per unit volume
     K_s = half-velocity constant (this correlates to the substrate concentration at one-half the maximum growth rate, mass per unit volume)

Equation (9.12) describes the relation observed in many studies of the effect of substrate concentration on degradation rate (Fig. 9.2). The Monod equation is another form of the Michaelis-Menten equation [Eq. (9.9)] as demonstrated below.

Since a portion of the substrate is converted to the microbial cell growth, and this quantity is reproducible for a given microorganism and substrate system, the microbial growth rate can be related to the substrate degradation rate by

$$\frac{dX}{dt} = -Y\frac{dS}{dt} \tag{9.13}$$

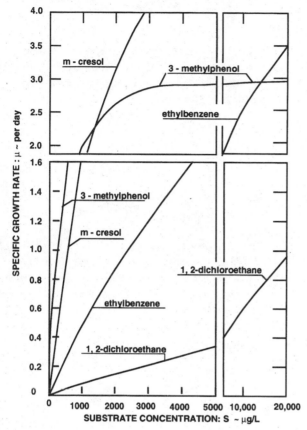

Figure 9.2 Effect of kinetic constants on the specific growth rate relative to substrate concentration.

where Y = yield coefficient, mg/mg
dX/dt = microbial growth rate
dS/dt = substrate degradation rate

Combining Eqs. (9.11), (9.12), and (9.13) yields

$$\frac{dS}{dt} = -\frac{\mu_m}{Y}\frac{SX}{K_s + S}\qquad(9.14)$$

The term μ_m/Y is often replaced by a rate constant k defined as the maximum rate of substrate utilization per unit mass of microorganisms:

$$k = \frac{\mu_m}{Y}\qquad(9.15)$$

using term (9.15) yields

$$\frac{dS}{dt} = -\frac{kSX}{K_s + S} \tag{9.16}$$

The parameters in Eqs. (9.14) or (9.16), the maximum specific growth rate μ_m, the half saturation constant K_s, the yield coefficient Y, and the maximum rate of substrate utilization k, are intrinsic. Their values are independent of the biomass concentration and bioreactor system. More important, they are only dependent on the compound being degraded and the microbial consortium performing the transformation. Their values impact on the ability to effectively degrade the compound and the design and operation of the bioreactor. Since they are independent of bioreactor configuration, their magnitude for various compounds and microbial systems can be established in treatability studies. These kinetic constants have been measured for some hazardous compounds as tabulated in Table 9.1.

Various modifications to the basic Monod kinetic expression have been used to account for inhibitory substrates, mixed substrates, substrates that don't support growth, and specialized conditions that reduce the Monod equation to a simple second-order or first-order reaction.

For a substrate that is inhibitory to its own biodegradation, Andrews (1968) formatted a modification of the Monod equation as

$$\mu = \frac{\mu_m S}{K_s + S + S^2/K_1} \tag{9.17}$$

where K_1 = inhibition coefficient

An important modification is the recognition that the microbial cells are not all in the log growth phase. Some are in endogenous metabolism. Under this condition, energy is used for cell maintenance without growth yield. Other factors such as death and predation must be accounted for. Equations (9.12) and (9.14) do not provide for this condition. These decreases in microorganism growth rate are lumped into one term and assumed to be proportional to the concentration of organisms in the system. This decay term is expressed as

$$D = k_d X \tag{9.18}$$

where k_d = endogenous decay coefficient, per time
D = rate of biomass decay, mass/unit volume-time

Adding this term to Eqs. (9.12) and (9.15) results in the net rate of growth expressed as

$$\frac{dX'}{dt} = \frac{\mu_m SX}{K_s + S} - k_d X \tag{9.19}$$

TABLE 9.1 Kinetic Constants for Microbial Transformations

Compound or waste	Microbial system	K_s or K_m	Y	μ_m	V_{max}	Reference
Domestic wastewater	UAC	25–100 mg/L	0.04–0.08 mg VSS/mg BOD	0.8–8 mg VSS/mg BOD/day		1
Acetate	Methanogenic	399 μM				5
Acetone	UAC	1.00 mg/L	0.55 mg/mg	0.22/h	0.33 mmol/L-h	2
4-Amino-3,5-dichlorobenzoate	Methanogenic	29.9 μmol*			1.49 μmol/L-h*	3
4-Amino-3,5-dichlorobenzoate	Methanogenic	29.7 μmol†			0.65 μmol/L-h†	3
Aniline	UAC	28.9 mg/L	0.43 mg/mg	0.25/h		2
Benzene	UAC	6.57 mg/L	0.27 mg/mg	0.28/h		2
Benzene	UAC	0.98–3.5 mg/L	0.39 mg/mg	0.067–0.076/h		15
bis(2-Ethyl hexyl)-phthalate	UAC	0.32 mg/L	0.56 mg/mg	0.13/h		15
Butylbenzene	UAC	10.00 mg/L	0.23 mg/mg	0.11/h		2
sec-Butylbenzene	UAC	10.00 mg/L	0.62 mg/mg	0.11/h		2
tert-Butylbenzene	UAC	10.00 mg/L	0.22 mg/mg	0.02/h		2
Butylbenzyl phthalate	UAC	0.40–1.81 mg/L	0.56–0.60 mg/mg	0.12–0.15/h		15
2-Butanone	UAC	10.00 mg/L	0.45 mg/mg	0.24/h		2
Butylate	Methanogenic	76.4 μM			0.20 mmol/L-h	5
Catechol	UAC	5.40 mg/L	0.28 mg/mg	0.495/h		2
Chloroform	*M. trichosporium*	548 μM			34 nmol/mg cells/min	4
3-Chlorobenzoate	Methanogenic	67.1 μM		0.088	23.7 μmol/L-h	3
1-Chlorobutane	Methanogenic	243 μM		0.158		4
2-Chloroethanol	Methanogenic	6 μM		0.158		4
2-Chloroethanol	*X. autotrophicus*	7 μm				5
2-Chloroethanol	*Ancylobacter aquaticus*	4 μm		0.140/h		5

TABLE 9.1 Kinetic Constants for Microbial Transformations *(Continued)*

Compound or waste	Microbial system	K_s or K_m	Y	μ_m	V_{max}	Reference
2-Chlorophenol	*P. chrysosporium*	132 mg/L			26–45 mg/L-h	11
2-Chlorophenol	UAC	16.0–17.0 mg/L	0.35–0.49 mg/mg	0.02–0.025/h		14
4-Chlorophenol	UAC	2.66–7.2 mg/L	0.31–0.41 mg/mg	0.25–0.29/h		14
Cumene	UAC	10.03 mg/L	0.23 mg/mg	0.194/h		2
m-Chlorobenzoate	UAC	2.0 mg/L	0.14 mg/mg	0.6/day		13
p-Chlorobenzoate	UAC	1.1 mg/L	0.25 mg/mg	1.2/day		13
o-Chlorobenzoate	UAC	2.4 mg/L	0.22 mg/mg	1.0/day		13
m-Cresol	UAC	6.50 mg/L	0.44 mg/mg	0.542/h		2
m-Cresol	UAC	0.93–3.50 mg/L	0.30–0.34 mg/mg	0.19–0.21/h		14
o-Cresol	UAC	1.75 mg/L	0.53 mg/mg	0.294/h		2
p-Cresol	UAC	6.00 mg/L	0.39 mg/mg	0.321/h		2
Dibutylphthalate	UAC	7.89 mg/L	0.38 mg/mg	0.180/h		2
1,2-Dichrobenzene	UAC	1.99 mg/L	0.40 mg/mg	0.054/h		15
1,3-Dichrobenzene	UAC	3.74 mg/L	0.41 mg/mg	0.066/h		15
1,4-Dichrobenzene	UAC	2.69 mg/L	0.39 mg/mg	0.053/h		15
2,5-Dichlorobenzoate	UAC	1.5 mg/L	0.16 mg/mg	0.6/day		13
3,5-Dichlorobenzoate	UAC	25.3 mg/L		0.05/day		13
3,5-Dichlorobenzoate	Methanogenic	47 μM			7.7 μmol/L-h	3
1,2-Dichloroethane	*Ancylobacter aquaticus*-25	24 μM		0.098/h		5
1,2-Dichloroethane	*Ancylobacter aquaticus*-20	222 μM		0.117/h		5
1,2-Dichloroethane	*X. autotrophicus*	260 μM		0.104/h		6
1,2-Dichloroethane	*M. trichosporium*	65 μM			77 nmol/mg cells/min	4
1,2-Dichloroethane	Methanotrophic	150 μM		0.102		4
1,1-Dichloroethylene	*M. trichosporium*	6 μM			5 nmol/mg cells/min	4
cis-1,2-Dichloroethylene	*M. trichosporium*	182 μM			30 nmol/mg cells/min	4
trans-1,2-Dichloroethylene	*M. trichosporium*	331 μM			148 nmol/mg cells/min	4

Compound	Process/Organism	Concentration	Yield	Rate	Special rate	Ref.
Dichloromethane	M. trichosporium	33 µM		0.198/h	4 nmol/mg cells/min	4
1,3-Dichloro-2-propanol	Pseudomonas sp.	12 µM				5
2,4-Dichlorophenoxyacetate	UAC	5.4 mg/L	0.14 mg/mg	2.3/day		13
Diethyl phthalate	UAC	7.49 mg/L	0.36 mg/mg	0.281/h		2
Diethyl phthalate	UAC	4.64 mg/L	0.32 mg/mg	0.10/h		15
2,4-Dimethylphenol	UAC	6.99 mg/L	0.32 mg/mg	0.303/h		2
2,4-Dimethylphenol	UAC	0.95–2.16 mg/L	0.20–0.59 mg/mg	0.063–0.19/h		14
Dimethylphthalate	UAC	7.00 mg/L	0.49 mg/mg	0.317/h		2
Dimethylphthalate	UAC	7.20 mg/L	0.43 mg/mg	0.090/h		15
Di-n-butylphthalate	UAC	0.45 mg/L	0.48 mg/mg	0.14/h		15
Dipropylphthalate	UAC	6.50 mg/L	0.23 mg/mg	0.220/h		2
Epichlorohydrin	Pseudomonas sp.	8 µM		0.146/h		5
Ethylbenzene	UAC	10.07 mg/L	0.34 mg/mg	0.216/h		2
Hydrogen	Methanogenic	8.5 µM			5.4 mmol/L-h	5
Isophorane	UAC	10.00 mg/L	0.54 mg/mg	0.102/h		2
Methane	M. trichosporium	363 µM			92 nmol/mg cells/min	4
Methanol	Methanotrophic		0.126			4
2-Methylphenol	Sulfate-reducing	0.27 mg/L	0.003 mg/mg	0.040/day		7
3-Methylphenol	Sulfate-reducing	0.40 mg/L	0.002 mg/mg	0.122/day		7
4-Methylphenol	Sulfate-reducing	1.90 mg/L	0.052 mg/mg	0.095/day		7
4-Nitrophenol	UAC	10.1–11.7 mg/L	0.37–0.38 mg/mg	0.31–0.42/h		14
Pentachlorophenol	UAC	60 µg/l	0.136 g/g	0.074/h		9
Phenol	UAC	2.41–3.39 mg/L	0.34–0.48 mg/mg	0.18–0.47/h		14
Phenol	Sulfate-reducing	2.00 mg/L	0.003 mg/mg	0.104/day		7
Phenol	Methanogenic	78 µg/mL			14 µg/mL-h	8
Phenol	UAC	3.00 mg/L	0.30 mg/mg	0.283/h		2
1-Phenylhexane	UAC	7.47 mg/L	0.53 mg/mg	0.135/h		2
Resorcinol	UAC	5.87 mg/L	0.28 mg/mg	0.392/h		2

TABLE 9.1 Kinetic Constants for Microbial Transformations *(Continued)*

Compound or waste	Microbial system	K_s or K_m	Y	μ_m	V_{max}	Reference
Tetrachloro-ethylene	Methanogenic	0.25 mg/L			22.9 mg/gVS/day	12
Toluene	UAC	7.75 mg/L	0.36 mg/mg	0.523/h		2
1,2,3-Trichloro-benzene	UAC	3.28 mg/L	0.43 mg/mg	0.063/h		15
1,1,1-Trichloro-ethane	*M. trichosporium*	24 µM			214 nmol/mg cells/min	4
1,1,1-Trichloro-ethane	*Clostridium* sp.	27–35 µM			0.28 µmol/mg/day	10
Trichloro-ethylene	*M. trichosporium*	290 µM			145 nmol/mg cells/min	4
m-Xylene	UAC	0.75 mg/L	0.26 mg/mg	0.123/h		2
p-Xylene	UAC	2.47 mg/L	0.36 mg/mg	0.140/h		2

*99% sediment.
†50% sediment.
UAC = unidentified acclimated culture from natural sources (sewage, etc.).

References:
1. Metcalf & Eddy, 1991.
2. Tabak, 1991.
3. Suflita, 1983.
4. Janssen, 1991.
5. Ahring, 1987.
6. vanderWijngaard, 1993a.
7. Colberg, 1991.
8. Dwyer, 1986.
9. Klecka, 1985.
10. Galli, 1989.
11. Lewandowski, 1990.
12. Chu, 1994.
13. Shamat, 1980.
14. Brown, 1990.
15. Grady (unpublished paper).

where dX'/dt = net rate of growth adjusted for the microbial decay rate, biomass/unit volume-time

Models for reactor design

Predictive models for bioreactor performance are based on the microbial growth and substrate utilization kinetics discussed above. These kinetic relationships are combined with a mass balance expression for the flow configuration of the bioreactor to yield predictive models. Flow configurations vary according to process hydraulics, rapid mix versus plug flow, and the presence or absence of various sludge (biomass) recycling patterns. The first attempt to unify the design of biological reactors for various design configurations was proposed by Lawrence and McCarty (1970). The fundamental approach to these predictive models is reviewed below, but those designing treatability studies for treating hazardous wastes in bioreactors should review the details of these relationships for proper process scale-up and process operation. Example sources are Gaudy (1988), Metcalf & Eddy, Inc. (1991), and Lawrence (1970).

The mass balance for the mass of microorganisms in a complete mix bioreactor without sludge recycling is shown in Fig. 9.3a and can be written as

$$\frac{dX}{dt} V = QX_0 - QX + V\frac{dX'}{dt} \qquad (9.20)$$

where dX/dt = rate of change of microorganism concentration in the bioreactor, mass/unit volume-time

Q = flow rate, volume/time

V = reactor volume

X_0 = concentration of microorganisms in the influent, mass/unit volume

X = concentration of microorganisms in the reactor, mass/unit volume

Since the concentration of microorganisms in the influent to a bioreactor is usually insignificant, X_0 can be assumed to be zero. Equation (9.20) simplifies to

$$\frac{dX}{dt} V = -QX + V\frac{dX'}{dt} \qquad (9.21)$$

substituting Eq. (9.19) into (9.21) yields

$$\frac{dX}{dt} V = -QX + V\left(\frac{\mu_m SX}{K_s + S} - k_d X\right) \qquad (9.22)$$

For steady-state conditions, $V\, dX/dt = 0$ and Eq. (9.22) becomes

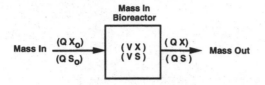

(a)   COMPLETE MIX BIOREACTOR

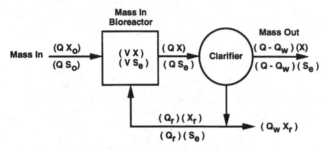

(b)   COMPLETE MIX BIOREACTOR WITH
CLARIFICATION AND SLUDGE RECYCLING

Figure 9.3 Mass balance diagram for biomass X and substrate S for two bioreactor configurations.

$$\frac{Q}{V} = \frac{\mu_m S}{K_s + S} - k_d \tag{9.23}$$

Since the hydraulic detention time θ is given by

$$\theta = \frac{V}{Q} \tag{9.24}$$

Eq. (9.23) can be written as

$$\frac{1}{\theta} = \frac{\mu_m S}{K_s + S} - k_d \tag{9.25}$$

By definition, Eq. (9.2), the mean cell residence time, can be written as

$$\theta_c = \frac{VX}{QX} = \frac{V}{Q} \tag{9.26}$$

Since there is no sludge recycling the mean cell residence time is equal to the hydraulic detention time, and

$$\frac{1}{\theta_c} = \frac{1}{\theta} = \frac{\mu S}{K_s + S} - k_d \tag{9.27}$$

Writing the mass balance for the substrate S yields

$$V \frac{dS}{dt} = QS_0 - QS - VX k \frac{S}{K_s + S} \qquad (9.28)$$

Assuming steady-state conditions $dS/dt = 0$ yields

$$S_0 - S = \frac{\theta_c (\mu_m SX)}{Y (K_s + S)} \qquad (9.29)$$

Equations (9.27) and (9.28) are combined and solved for X to yield

$$X = \frac{\mu_m(S_0 - S)}{k(1 + k_d \theta_c)} = \frac{Y(S_0 - S)}{1 + k_d \theta_c} \qquad (9.30)$$

and the effluent substrate concentration is given by

$$S = \frac{K_s(1 + \theta k_d)}{\theta_c(Yk - k_d) - 1} \qquad (9.31)$$

For any given kinetic coefficients and mean cell residence time, the effluent substrate concentration and biomass concentration in the bioreactor can be predicted. The MCRT has a significant effect on treatment efficiency (Fig. 9.4). At low substrate concentrations and low microbial growth yield, it becomes difficult to support an adequate MCRT for effective growth and low effluent concentrations in suspended reactors. The kinetic constants require MCRT values much larger than those necessary for the treatment of municipal wastewater.

A bioreactor with sludge recycling improves the potential for retaining adequate biomass in the system (Figs. 9.1 and 9.3b. The mean cell residence time for sludge recycling and wasting from the bioreactor is now a function of the recycle ratio of the return biomass. For the recycle system in Fig. 9.3b, the relationship for mean cell residence time becomes

$$\frac{1}{\theta_c} = \frac{-\mu_m SX}{K_s + S} - k_d \qquad (9.32)$$

Mean cell residence time θ_c is no longer equal to hydraulic detention time. For the recycle system illustrated in Fig. 9.1, MCRT also becomes a function of the recycle ratio of the returned sludge. This provides the operator flexibility in meeting a given level of effluent quality, since the MCRT has some degree of control. Comprehensive discussions on reactor performance relationships for mix flow, plug flow, and various recycling schemes are provided by Lawrence (1970) and Gaudy (1988).

At the introduction of this section, it was noted that bioreactor design parameters consisted of

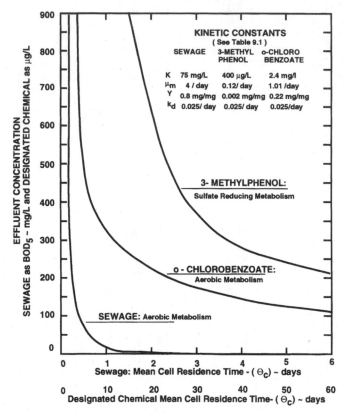

Figure 9.4 Comparison of effluent concentration to mean cell residence time for sewage, 3-methylphenol and o-chlorobenzoate.

Mean cell residence time θ_c

Specific substrate utilization rate U

Food to microorganism ratio F/M

From Eqs. (9.1), (9.14), and (9.32), U is directly related to mean cell residence time θ_c as follows

$$\frac{1}{\theta_c} = YU - k_d \qquad (9.33)$$

The food to microorganism parameter is also related to U. The relationship to U is typically expressed in terms of process efficiency E.

$$E = \frac{S_0 - S_e}{S_0} \times 100 \qquad (9.34)$$

where E = process efficiency, percent
  S_0 = influent substrate concentration, mass/unit volume
  S_e = effluent substrate concentration, mass/unit volume

The overall removal rate in a bioreactor of detention time θ is given by

$$\frac{dS}{dt} = \frac{S_0 - S_e}{\theta} \qquad (9.35)$$

Substituting the detention time θ for Q/V in Eq. (9.3) yields

$$\frac{F}{M} = \frac{S_0}{\theta X} \qquad (9.36)$$

Combining Eqs. (9.34), (9.35), and (9.36) with the definition of specific substrate utilization rate, Eq. (9.1), yields

$$U = \frac{F}{M}\frac{E}{100} \qquad (9.37)$$

Since the kinetic coefficients K_s, Y, μ_m, and k_d are only dependent on the chemical compound being transformed and the microbial system, effluent quality is only a function of mean cell residence time θ_c or the specific substrate utilization rate. As illustrated by combining Eqs. (9.1) and (9.14),

$$U = \frac{\mu_m}{Y}\frac{S_e}{K_s + S_e} \qquad (9.38)$$

or

$$S_e = \frac{K_s U}{(\mu_m/Y) - 1} \qquad (9.39)$$

There is a minimum mean cell residence time θ_{cmin}. At this value, the microbial cells are washed out or lost from the bioreactor faster than they reproduce. This θ_{cmin} is a direct function of these kinetic coefficients and the influent substrate concentration. At θ_{cmin} the influent substrate concentration S_0 equals the effluent concentration S_e and θ_{cmin} is given by

$$\frac{1}{\theta_{cmin}} = \frac{\mu_m S_0}{K_s - S_0} - k_d \qquad (9.40)$$

For the treatment of hazardous compounds this critical θ_{cmin} may be much larger than the value for most municipal wastewater. Clarifiers are not 100 percent efficient in settling biomass, resulting in loss of

TABLE 9.2   Biological Process Design Considerations That Cause Departure from Classical Municipal-Industrial Design

1. Designing for the mineralization or detoxification of specific target compounds
2. Designing for a wastewater with a low carbon and energy source
3. Designing for low specific growth rate yielding substrates
4. Designing a microbial process that potentially requires a specific type and amount of electron acceptor
5. Designing for cometabolism with primary and secondary substrates
6. Designing for sequential microbial metabolism modes
7. Designing for a specific microbial seed or consortium
8. Designing for volatilization control of hazardous chemicals
9. Designing for nutrient and pH adjustment
10. Designing enricher reactors for bioreactor start-up and bioaugmentation during treatment
11. Designing for anticipated changes in treatment performance with time (life cycle design)

biomass even with no wasting. For low substrate concentration it is difficult or even impossible to maintain a $\theta_c > \theta_{cmin}$.

Substrate concentrations in groundwater are significantly lower than most municipal and industrial wastewater. This condition combined with the typical kinetic coefficients for hazardous compounds and the requirement for ppb levels in the effluent makes biomass retention one of the more difficult problems in the use of suspended bioreactors.

Design and Operation Emphasis

The design of liquid-phase bioreactors for treating hazardous chemicals emphasizes points that are seldom, if ever, considered for typical wastewater plant design (Table 9.2). This is a result of the uniqueness of groundwater and soil remediation coupled with the goal of degrading select chemicals.

Low energy and low growth yielding substrates

Contaminated groundwater has low carbon and energy sources compared with municipal and industrial wastewater. Frequently the total organic concentration of contaminated groundwater is less than the effluent from most secondary wastewater treatment plants. In addition, all municipal and most industrial wastewater processes deal with moderate- to high-growth-rate substrates. Most hazardous compounds are low-specific-growth-rate substrates. This is specially true for those chemicals that are treated by anaerobic systems and those chemicals requiring cometabolism. As a result of the lower biomass generation, biological reactors must be designed to maintain higher mean cell residence times (MCRT) than typical for municipal systems.

The importance of electron acceptors, nutrients, and pH control is far more critical for treating hazardous compounds. The control of electron acceptors is discussed in Chaps. 3 and 4. For most contaminated groundwater, nutrient addition is important and pH control is critical. Municipal wastewater has a higher nutrient content and buffer capacity than groundwater. As a result, the ability to control pH throughout the treatment system must be considered.

Cometabolism and sequencing metabolic modes

Designing for primary and secondary substrates, cometabolism, has never been a significant part of municipal or industrial wastewater plant design. When specific target compounds must be detoxified or mineralized, cometabolism may be the only potential biological system. Bioreactors for cometabolism have significant design and operational differences from conventional biological processes. With cometabolism the primary substrate usually hinders or competes for enzymes, making the enzyme unavailable for transformation of the target compounds. However, in the absence of the primary substrate, the ability to maintain this transformation capacity is exhausted.

For some organisms, such as methanotrophic bacteria, the loss transformation capacity also results from damaged cellular material. For example, transformation products of trichloroethylene cause toxicity. To deal with these problems, process design must provide for regrowth or repair of damaged microbial cells to maintain transformation capacity. Reactor configurations that allow the use of staged microbial phases can separate these two metabolic phases.

Design and operation of a cometabolism process require the optimization of two or more treatment stages that are different in objectives and process monitoring requirements. To aid in process optimization, bioreactors may be staged so biomass alternates between the primary substrate growth phase and the target compound transformation phase. This can be accomplished with multiple reactors, or by individual phases operated in a sequencing manner, or by cyclic feeding of the primary substrate. This approach results in the microorganisms alternating between growth and starvation. Biomass activity varies over time, resulting in a cycling non-steady-state process. The food to microorganism ratio is a maximum at the primary substrate feeding followed by a minimum at the end of the target compound transformation phase (Fig. 9.5). The process is a non-steady-state operation.

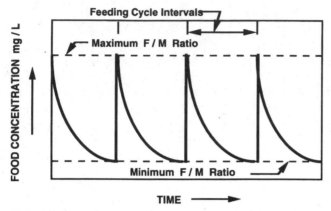

Figure 9.5 The non-steady-state feed to microorganism (F/M) ratio cycling for cometabolism.

Reactor design for cometabolism with methanotrophic bacteria must attempt to counter the loss of transformation capacity resulting from exposure to toxic intermediates from TCE degradation. One method utilizes two reactors. Methanotrophic bacteria are grown in a separate culture reactor with primary substrate and then brought in contact with the contaminated water in the main reactor. This is performed without or at reduced primary substrate levels. The culture reactor is maintained throughout the bioremediation program to regrow or reclaim loss transformation capacity and augment biomass in the main reactor.

The cycling food to microorganism ratio requires design and operational strategies that may be answered only by treatability studies. Design and operational criteria must be established for:

1. Frequency of the primary substrate feeding period

2. Duration of the primary substrate feeding period

3. Concentration level of the primary substrate feed

4. Hydraulic detention time for the secondary substrate (target compounds)

5. Oxygen or other electron acceptor concentration levels

6. Maximum and minimum biomass loading

7. Optimum range for the mean cell residence time

8. Minimum mean cell residence time

The cleanup of hazardous waste sites may require the application of more than one microbial metabolism mode. As discussed for many of the chlorinated aromatics (Chaps. 4 and 5), sequential processes of anaerobic followed by aerobic may be necessary for detoxification and mineralization of the target compounds.

Process designs for sequential anaerobic-aerobic are not typical of municipal systems, resulting in some significant design modifications to typical biological reactors and providing new life to some almost forgotten processes.

Bioreactor start-up

Municipal wastewater biological processes have always relied on the indigenous microorganisms for treatment. Although many commercial products have been produced, little measurable improvement has been reported during the treatment of municipal wastewater. The only possible exception has been for anaerobic digesters, but these have been seeded with poorly defined organisms from well-operating digesters. Degradation of some of the more resistant compounds is enhanced with the proper organism.

Bioaugmentation to both suspended and fixed-film bioreactors has demonstrated significant improvement in treatment response for hazardous compounds (Babcock, 1992). An inoculation or culture reactor should be used for developing and supplementing microbial seed during the bioremediation program. These also have been called "enricher-reactors." The amount of bioaugmentation varies with the microbial species, process chemistry, and reactor configuration. Continuous-flow suspended reactors require a larger bioaugmentation mass than batch reactors or fixed-film reactors. The effect of bioaugmentation on batch and continuous-flow reactors for degrading 1-naphthylamine is illustrated in Table 9.3.

Typical start-up procedures are to inoculate the main reactor by developing growth in a culture reactor. These are usually designated

TABLE 9.3   Effect of Bioaugmentation on the Transformation of 1-Naphthylamine

Quantity of bioaugmentation, mass of inoculation per mass of indigenous microorganisms as MLVSS/MLVSS	Increase in rate of degradation (continuous-flow suspended reactor), percent increase
1	33
2	100
5	100
10	100
20	100
50	300

as batch reactors. The batch reactor's biomass is transferred as a suspension to the main reactor. This suspension is usually in a controlled culture solution containing the necessary substrates for the desired microorganisms. The culture solution may contain some of the hazardous compounds or analog compounds. An influent flow of culture solution to the main reactor is maintained until adequate biomass has developed. Once the biomass is developed in the main reactor, the feed solution is stopped and the contaminated water introduced.

Fixed-biofilm reactors are inoculated in a similar manner. The biofilm is developed by recycling a suspension of seed microorganisms from a culture reactor with nutrients and primary substrate at concentrations 2 to 10 times greater than that typically used during actual treatment. Once the biomass has developed to a sufficient level for achieving the desired mean cell residence time, 2 to 5 weeks, the flow from the culture reactor is stopped, and the treatment cycle implemented. Even with fixed-biofilm reactors it is wise to maintain growing biomass in the culture reactor to reestablish the biomass during any unexpected biomass loss or population shifts.

Bioreactors, after the original inoculation, will undergo a continuous shift in microbial population. This is often due to the continuous inoculation of indigenous microorganisms from processing groundwater or from less than optimum environmental conditions for the inoculated microorganisms. To maintain the desired population, the bioreactor is bioaugmented from the culture reactor.

Life-cycle design

Design of bioreactors for treatment of contaminated groundwater must address the time effect, life-cycle design, on performance. The most important factors to consider in life-cycle design are the changes in contaminant concentration, nature, and loading rate on the biological treatment facility.

Treatment performance is a function of project life span as a result of variations in hydraulic loading and the distribution of contaminants in the subsurface. The phase distribution of a typical gasoline spill is illustrated in Table 9.4. Contamination varies in the subsurface because of:

- Variations in quantities spilled
- Variations in spill release rates
- Mechanisms of subsurface transport
- Subsurface homogeneity

TABLE 9.4   Subsurface Phase
Distribution of Gasoline,
Gallons

Solid phase	960
Soil gas phase	28
Groundwater phase	12
Total	1000

The design of bioreactors and auxiliary equipment for groundwater cleanup requires the ability to predict the expected changes received by the treatment facility. The ability to make these predictions requires the consideration of those parameters that influence contaminant distribution and transport in the subsurface, and the expected operational modes of the process. Aquifers and soil are not homogeneous. Pockets of contaminants exist and contaminants are transported at different rates. This causes a change in total loading rate as well as changing chemical composition of the influent to the bioreactor. Contaminant subsurface transport occurs by advection and dispersion. Advection is a process in which dissolved organic compounds are transported by the bulk liquid motion of flowing groundwater. Dispersion occurs from diffusion and mechanical mixing. The importance of these transport mechanisms is significantly influenced by:

- Solubility of the organic compound
- Density of the organic compound
- Viscosity of the organic compound
- Adsorption on aquifer solids

Excellent reviews on subsurface transport of organic contaminants are available and should be utilized to predict potential variations when pumping contaminated aquifers.

Life-cycle design evaluates the performance of a treatment system over potential flow rates, changes in the nature of the contaminants, and changes in concentration. For example, a well at the migrating front of a plume will have a steady increase in concentration as the plume is drawn toward the well followed by a decline as the site is remediated. Wells placed in the heart of a plume will experience an initial high organic loading followed by a rapid decline. The rate of organic yield is frequently dependent on the desorption of contaminants from solids within the aquifer.

Changing contaminants and mass loading rates over the life of a project can have detrimental effects on the performance of a biological process. It is inappropriate to assume that lower concentrations will result in better treatment. Often biological processes lose efficiency at lower concentrations and may have inadequate microbial growth, resulting in loss biomass. Suspended-growth reactors are more susceptible to these problems than fixed-bed reactors. A biological reactor that receives a decreasing concentration of organic load will respond by producing less biomass. To maintain treatment efficiency and the minimal amount of biomass in the reactor, it is necessary to increase mean cell residence time (MCRT). This is achieved by wasting less sludge, yielding an overall increase in sludge age. If sludge age becomes too high, bacteria lose their ability to produce polymers, flocculate, and settle. This results in the inability to separate the biomass from the treated groundwater, resulting in failure not only of the settling process but also of the biological process.

The hydraulic loading to treatment systems also varies during the remediation period. Frequently water is reused on-site for such purposes as increasing hydraulic head, hydraulic isolation, and flushing contaminants from the unsaturated zone. One solution is to provide for multiple treatment units that can be brought on- and off-line. This provides greater flexibility for changing the hydraulic detention time and dealing with upsets. It also allows operators to change detention times for degrading a solution of mixed contaminants that don't degrade uniformly. For example, simple aromatics are degraded before the more complex aromatic structures. By changing reactor volume through the number and operating sequence of units, effluent quality can be maintained.

Volatilization control

Bioreactor design also must evaluate the potential for volatilization of hazardous compounds. Typical aeration systems for open suspended reactors will volatilize over 90 to 98 percent of many EPA listed volatile organic compounds (VOC) during normal detention times (Sutton, 1988; Cookson, 1987). The use of hydrogen peroxide or pure oxygen under controlled feed in response to oxygen needs can minimize VOC cmissions. Methods for improved transfer efficiency with less volatilization have been developed to reduce the release of volatile compounds. One such system is the liquid-liquid eductor as discussed in Chap. 8. See Chap. 8 for a discussion on other aeration procedures and their transfer efficiency.

Off-gas treatment systems may be required depending on chemicals, daily discharge quantity, and agency regulations. For some systems, a closed reactor design, as typical of fixed-bed reactors, may be more cost-effective than emission control facilities. Air emissions treatment represents 40 to 50 percent of VOC air stripping costs.

Some compounds can be treated by either aerobic or anaerobic systems. The selection of an anaerobic system will eliminate the need for aeration and minimize volatile emissions. Contaminated wastewater containing mixtures of halogenated aliphatic compounds and hydrocarbons can be anaerobically dehalogenated before the aerobic system is employed for degradation of the hydrocarbons. This would prevent the air stripping of these chlorinated solvents and reduce oxygen delivery costs. However, attention must be given to the potential of generating odorous sulfur compounds during the anaerobic stage.

Suspended Bioreactors

Activated sludge process

Since its first development, the activated sludge process and its variations have become the most widely utilized biological process for wastewater treatment. The activated sludge process is a suspended-growth reactor that operates in the aerobic metabolism mode. The process contacts a waste stream with preformed biomass in an aeration tank. The biological reactor must be combined with a sedimentation process to remove the suspended biomass, frequently called mixed liquor suspended solids (MLSS), mixed liquor volatile suspended solids (MLVSS), or activated sludge from the processed liquid. A portion of this biomass is recycled to control the amount of biomass in the bioreactor. The term activated is to imply that the microorganisms are conditioned and ready for substrate sorption and degradation.

Process principles.   The complete process consists of two unit processes: an aeration tank and a sedimentation tank (Fig. 9.1). In the bioreactor, the aeration tank, the waste, and the biomass are completely mixed and aerated. Mixing action from air sparging in an activated sludge process is shown in Fig. 9.6. Under proper conditions, the activated biomass is capable of degrading most organic waste as an energy source with mineralization to H_2O, CO_2, and cell growth.

After adequate contact time in the aeration tank, the activated sludge, or biomass, flows to a sedimentation tank in which the biomass is allowed to coagulate and settle. The clarified liquid is collected by a surface weir, and the biomass (sludge) is drawn off the tank's bottom (Fig. 9.7). The supernatant in the sedimentation tank may be discharged or subjected to further treatment. A good portion of the biomass is recycled back to the aeration tank to maintain an adequate level of organisms in the reactor. Excess biomass is wasted for further treatment or disposal. The activated sludge process is normally operated in a continuous mode.

Figure 9.6 Aeration within the bioreactor of an activated sludge process.

There are many variations to the activated sludge process. Variations include the conventional process, complete mix, step aeration, contact stabilization, extended aeration, high-rate aeration, and others. Except for various flow regimes from plug flow to rapid mix, the major differences in these process modifications are the food to microorganism ratio (F/M), mean cell residence time (MRCT, θ_c), and changes in aeration patterns.

For a given flow regime, organic substrates, and microbial population, the degree of mineralization of organic contaminants is a function of the amount of mean cell residence time, as discussed earlier [Eq. (9.31)]. A decrease in biomass or mean cell residence time results in a decrease in treatment response. Since the rate of degradation for many hazardous chemicals is very slow, suspended bioreactors must be designed to maintain a higher MCRT. The MCRT can be increased

Figure 9.7 Sedimentation tank of an activated sludge treatment system: effluent weir collection of treated wastewater.

by reducing the rate at which sludge is wasted. This reduces the food to microorganism (F/M) ratio. With municipal wastewater having high-organic-energy sources, these operating parameters can be easily controlled. For suspended reactors treating groundwater contaminated with hazardous chemicals, however, it is difficult and in some cases impossible to maintain an adequate MCRT. The net energy yield from groundwater is often too low to support a suspended biomass reactor such as the activated sludge process.

The success of a suspended biological reactor also depends on the ability to separate the MLSS from the treated liquid. This separation is achieved by settling. Settling characteristics are influenced by several variables, including the microbial species and their age or growth stage. The settleability of the biomass is measured by the sludge volume index (SVI). The sludge volume index is the volume (in milliliters) occupied by the sludge containing a unit weight (in grams) of suspended biomass after settling for 30 min. Low SVI (<100) indicates good settling sludge and high SVI (>200) indicates poor settling sludge.

Applications. Because of the difficulty in maintaining biomass, the activated sludge process has only limited usefulness for treating groundwater contaminated with hazardous chemicals. The main application of this process has been existing plants that are also receiving a higher-energy-source waste. These plants are typically designed to treat an industrial wastewater and may or may not function effectively on degrading target compounds in a combined industrial waste and groundwater mixture. These systems are limited in their ability to promote cometabolism or other specialized redox reactions. The organic load of the industrial waste will serve as the primary substrate and compete with the target compound. For effective degradation of the target compounds, this primary feed concentration must be controlled. This is frequently difficult if not impossible for a plant treating an existing wastewater discharge.

Powder-activated carbon activated sludge process

The addition of activated carbon to the activated sludge process results in a suspended growth reactor with some benefits of a fixed-film system. Activated carbon provides an adsorption mechanism for the removal of hazardous compounds, and it improves biomass retention. The improved sludge retention is the major advantage of this process. This mode of operation for a bioreactor, frequently referred to as the "PACT" system, is available from Zimpro/Passavant, Inc.

Process principles.   The powder-activated carbon system is a two-stage process consisting of a suspended bioreactor followed by a settling process. Influent to the bioreactor is mixed with powder-activated carbon (PAC) and recycled biological-PAC solids (Fig. 9.8). An adequate mixing intensity is important to ensure intimate contact

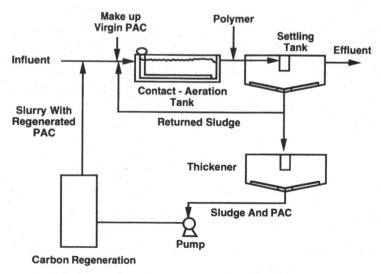

Figure 9.8   Powder-activated carbon activated sludge process schematic.

TABLE 9.5   Powder-Activated Carbon Activated Sludge Process Operating Conditions

Parameter	Purpose
Mixing intensity	Provide intimate contact between waste and biomass, in general between 15 and 30 kW/1000 m^3
Dissolved oxygen	Serve as an electron acceptor; not less than 1 mg/L
MLVSS	Provide adequate microbial population, 2000–8000 mg/L depending on mean cell residence time
MCRT	One of the more important design and operating parameters, greater than 10 days (greater than 20 is not uncommon) depending on the type of waste and microbial system
F/M	Food/MLVSS ratio is related to MCRT, and used as a design and operational parameter
SVI	Sludge volume index: indication of sludge settleability, less than 100 mL/g
PAC/biomass	0.8–1

among waste compounds, biomass, and PAC. After aeration in the bioreactor, the mixed-liquor solids are settled and separated from the treated effluent. Settling is sometimes enhanced with coagulants such as alum, ferric chloride, and polymers. Some settled biomass-carbon mixture is recycled to maintain the desired MCRT. The biological characteristics of the PACT process are similar to the conventional activated sludge process. Important operating parameters are listed in Table 9.5. Virgin activated carbon must be added daily. For large facilities that use over 2000 lb of carbon per day, on-site carbon recovery becomes practical.

Three removal mechanisms have been proposed for the PACT process: biodegradation with irreversible adsorption, biodegradation with reversible adsorption, and enhancement of biofloc settling. Aromatic structures have a greater affinity for most activated carbons than aliphatic structures. Many large aromatic compounds are irreversibly adsorbed. These compounds are concentrated by adsorption and degraded on the PAC surface.

For biodegradation with reversible adsorption, target compounds desorb back into the bulk solution as their concentration in the bulk solution decreases. Typical compounds are single-ring aromatic compounds and aliphatic compounds. The third removal mechanism consists of the activated carbon becoming incorporated with the flocculating biomass. As a result, the biofloc density increases and sludge settling is enhanced. The carbon-biomass mixture is recycled to the aeration tank. Since some biomass is attached to the activated carbon and more biomass is captured during settling, a higher MCRT is achieved over that associated with the conventional activated sludge process.

For most hazardous wastes, the process must be acclimated to the wastes under proper operating conditions. When the target compounds are present in high concentrations and can serve as the carbon and energy source, the system can be operated to remove the target compounds as the primary substrate. However, most contaminated groundwater has a low carbon and energy source. Although the powder-activated carbon helps retain biomass, another primary energy source may be required. When using a primary substrate with activated carbon, select a primary substrate that has a low adsorption affinity for carbon. The nature of the primary substrate varies with the target compounds and the microbial system. Chemicals that have been used as substrates include methanol, glucose, carbohydrates from fruit processing, phenol, toluene, and organic acids. Of these, phenol and toluene have the greatest adsorption affinity for activated carbon. The activated sludge process, with or without carbon, does not provide the flexibility required for controlled cometabolism unless multiple-stage units are utilized.

Applications.   Experience with the PACT process is mainly a result of treating industrial wastewater production streams that contain persistent chemicals or chemicals at potentially toxic concentrations. Most applications have been in the petroleum refinery and petrochemical industries. These wastewaters contain aromatic derivatives, aliphatic surfactants, fluorocarbons, and petroleum additives. Compounds that pose toxicity problems to activated sludge microorganisms may be particularly suitable for PACT treatment. Toxicity levels are reduced by the adsorption mechanism of PAC, making the dissolved chemical concentrations less. Since many of these industrial applications usually contain significantly higher organic energy sources, caution must be used when transferring design and operating conditions to contaminated groundwater. Design parameters should be based on treatability studies.

Application of PACT systems in the bioremediation field has included the treatment of groundwater for such chemicals as aniline, phenolic compounds, aromatic compounds, and herbicide-related chemicals. A PACT system was applied to the treatment of both groundwater and dilute production wastes from the manufacture of herbicides (Sutton, 1988). Groundwater contaminated with dichlorobenzidene and orthochloroaniline at concentrations of 0.4 and 6.5 mg/L, respectively, was treated to meet less than 10 µg/L of these chlorinated compounds in the effluent.

TABLE 9.6   Operating Conditions for a Two-Stage PACT
Process

Parameter	First stage	Second stage
Powder-activated carbon	70 ppm	23 ppm
MCRT	29 days	16 days
Hydraulic detention time	8 h	2.6 h
Airflow	0.67 Lpm/L	0.22 Lpm/L
Temperature	20°C	20°C

SOURCE: O'Brien, 1990.

A pilot two-stage powder-activated carbon activated sludge system successfully treated wastewater containing chlorinated compounds of 1,2-dichloroethane, chloroethane, 1,2-dichlorobenzene, and chloroform. Concentrations ranged between 2000 and 6000 ppb. Methylene chloride concentrations were 12,000 ppb (O'Brien et al., 1990). The pilot studies indicated that a mean cell residence time in excess of 20 days was desirable to maximize priority pollutant removal. However, a MCRT longer than 20 days produced foaming problems. Increasing hydraulic detention time beyond 8 h did not significantly improve effluent quality. Temperature variations at 10 to 35°C had a selective effect to some priority pollutants. For example, dichloroethane removal was severely impacted by temperature, while chloroethane was not. The operating conditions are presented in Table 9.6.

Alternative anoxic oxic process

Alternative anoxic oxic (AAO) process is a modified activated sludge process that provides an anaerobic and aerobic sequence treatment (Fig. 9.9). The process was originally developed for phosphorus removal. However, the ability to provide anaerobic dehalogenation before aerobic mineralization has raised interest in this process for hazardous chemicals. Adaptation of the AAO process to the treatment of hazardous compounds has resulted in more variations for process operation. The process can be operated as a complete oxic process, as an anoxic process, or as a combined anaerobic-aerobic system (Hong, 1989). The anoxic process can be operated as a nitrate-reducing, sulfate-reducing, methanogenic metabolism, or sequence combinations. A sequencing operation between oxic and anoxic can be done in the same basin under automatic controls using redox potential and dissolved oxygen levels for operation. A monitoring probe for automatic process cycling between oxic and anoxic is illustrated in a plant in the Netherlands (Fig. 9.10).

Process configurations that provide for the optimization of multiple metabolic modes are provided in Fig. 9.11. The anaerobic reactor is

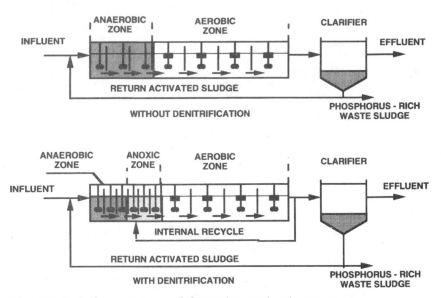

Figure 9.9 Typical arrangements of alternative anoxic oxic processes.

Figure 9.10 An in-line probe to control an alternating sequence of oxic and anoxic metabolism in a bioreactor.

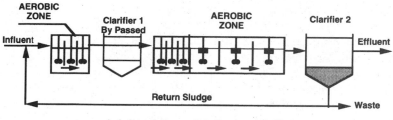

(a) Operated as an Oxic Treatment Facility

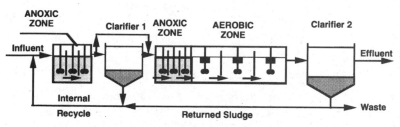

(b) Operated as an Anoxic - Oxic Sequence Treatment Facility

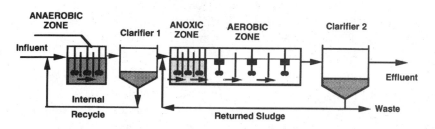

(c) Operated as an Anaerobic - Anoxic - Oxic Sequence Treatment Facility

Figure 9.11 Process configurations that provide the flexibility for selecting the most appropriate metabolism modes.

designed for dehalogenation of the more complex aromatic compounds, and the aerobic reactor is designed for mineralization of the transformation products. The first reactor also can be used in an anoxic mode for those halogenated compounds that respond to dehalogenation under sulfate-reducing or nitrate-reducing conditions.

When all zones are operated as a complete oxic process (Fig. 9.11a), the system is identical to the typical activated sludge process. The first settling tank is bypassed. This provides a flow regime that is between rapid mix and plug flow. In Fig. 9.11b the first and second biological contact units are operated in an anoxic mode. The electron acceptor is nitrate or sulfate. The first settling tank can be bypassed or used to recycle the biomass of the first contactor back to the reactor. If the first settling tank is bypassed, then the major microbial species in the biomass will be facultative microorganisms. The strict anaerobes are hindered by the presence of molecular oxygen. The use of settling tank one in the flow configuration will depend on the needed microbial consortium.

In the anaerobic selector mode (Fig. 9.11c) the first bioreactor is operated to support methanogenic metabolism with carbon dioxide as the electron acceptor. Under this mode of operation, settling tank one should not be bypassed. It is important to separate the methanogens from the other biomass, since oxygen is toxic to these microorganisms. Exposing this biomass to the aerobic unit will delay the methanogenic response when the sludge is recycled back to the anaerobic unit.

Sequencing batch reactor

The sequencing batch reactor (SBR) is a suspended-growth fill-and-draw system. This process originated in the early 1900s and has recently been applied to the successful treatment of hazardous compounds. The process was evaluated on landfill leachate that contained hazardous compounds in the early 1980s (Sutton, 1988). Since then, numerous SBR systems have been utilized for the treatment of hazardous waste. The U.S. EPA has recognized the technology under the Innovative and Alternative Technology Assessment (IATA) program.

Process principles. The SBR system consists of a single tank in which timed processes take place sequentially (Fig. 9.12). A complete cycle of unit operations is performed within a single tank by holding the influent waste stream in a batch mode and treating it through a succession of steps. Typical steps consist of fill, react, settle, draw, and for some applications, an idle or regrowth mode. The waste volume inside the SBR is increased to a maximum volume during the fill cycle. Addition of nutrients, primary substrate, and seed microorganisms may be provided during the fill period. Mixing is provided during the biodegradation period. If the desired microbial system is aerobic, aeration is usually the mixing mechanism. However, mechanical mixing is used for anoxic or anaerobic systems.

During the react period, flow to the tank is discontinued. This mixture is held for the desired biodegradation period. The react period starts with a maximum food to microorganism (F/M) ratio. At the end of the react cycle, the F/M ratio reaches a minimum. The cycling F/M ratio is typical of that represented by Fig. 9.5 illustrating the feeding for cometabolism. Settling characteristics are a function of mixed liquor suspended solids and the F/M ratio. For anaerobic systems, the biomass frequently settles better at low F/M ratios. After the biodegradation is complete, the mixing is stopped and the liquid is allowed

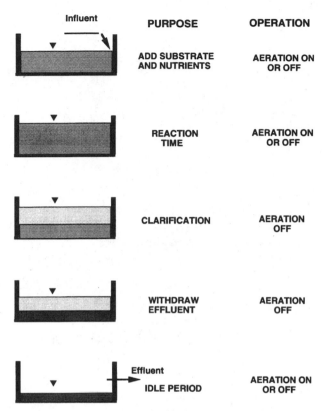

Figure 9.12 Typical sequencing batch reactor operating modes.

to settle under quiescent conditions. If necessary, coagulants can be added during the final mixing stages to aid clarification.

Several decant mechanisms have been used. The most popular is a floating or adjustable weir (Fig. 9.13) (EPA/625/8-86/001, 1986). Clarification time can range from 10 min to several hours. Times are usually 1 h or less for aerobic sludges to prevent anaerobic conditions and gas production. The biomass should settle to a predetermined elevation during each settling cycle before decanting. After settling, the clarified supernatant is removed from the tank and excess biomass may be removed. The tank can be refilled to start another cycle or placed in an idle period. This idle stage can be used for evaluation of the sludge (biomass) quantity and establishing sludge wasting rates or used to recover transformation capacity.

The SBR process is more flexible in operation and performance control than the other suspended-growth processes. The system can be operated to bring about almost any microbial metabolism mode, and

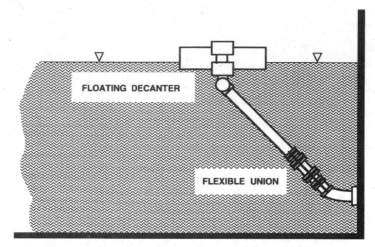

Figure 9.13 Floating weir for decanting a batch reactor. (*EPA/625/8-86/C01, 1986.*)

these can be easily changed to suit the demands of site cleanup. The process can be operated as an aerobic contactor, anoxic contactor, anaerobic contactor, and anaerobic-aerobic sequencing bioreactor. Metabolism modes of nitrate reducing, sulfate reducing, fermentation, or methane fermentation can be easily employed as needed.

Even more important than metabolism mode flexibility is the control on process holding time and MCRT. Since the discharge of effluent is periodic, the reaction period can be extended until the effluent meets the specified requirements. As discussed above, it is wise to design smaller multiple bioreactors. This is particularly important for the SBR, since the holding time is varied based on performance relative to that aliquot of contaminated groundwater. Multiple tanks allow continuous pumping of groundwater.

The SBR process has several major advantages over other suspended reactors. The MCRT can be increased as needed since biomass is held in the tank as long as necessary. The biomass cannot be washed out by hydraulic flows, since short circuiting is minimized and near perfect settling occurs. The batch operation allows the sludge to separate under near quiescent conditions. Poor floc formation and settling can be recognized and corrected by addition of a coagulant before the decant cycle. The SBR process provides the control needed for achieving the MCRT necessary for persistent and hazardous chemical mineralization.

The SBR process also can be used with activated carbon. The addition of powder-activated carbon to SBR systems (PAC-SBR) usually improves the treatment performance on hazardous compounds. Many of the same advantages found with PACT systems also apply to PAC-SBR systems. This combination has been applied to the treatment of hazardous waste streams with relatively mobile systems (Sutton, 1988).

The operational protocols for SBRs are determined by the actual field response and performance requirements. For example, the volume of biomass wasted after the decant cycle is determined by a desired loading of the expected fill cycle, the sludge retention time, and the hydraulic holding time. Process monitoring also can indicate the need for activated carbon addition for toxic or poorly mineralized compounds, eliminating the need for continual carbon addition.

As a suspended-growth reactor, the SBR process provides the greatest control the operator can maintain over microorganism selection and performance. The process operation can range from anaerobic through anoxic to high dissolved oxygen conditions. Substrate availability can range from low to high food to microorganism ratios, and contact times can be changed to allow for preferential growth of desirable microorganisms.

Equipment specifications must consider the potential need for operational flexibility and provide for pumping variable flow rates. Variable-speed pumps should be selected for primary substrates, sludge wasting, nutrient and pH control, carbon addition, and electron acceptor control. Aeration facilities should be designed to match the process flexibility and potential oxygen requirements.

Applications.   An aerobic sequencing batch reactor has been used to reduce concentrations of PAHs, phenols, and chlorophenols in wastes from wood preservation facilities. Influent to the SBR system consisted of creosote and pentachlorophenol sludges. Soda ash was used to control pH at 7.2, and inorganic nitrogen and phosphorus were added for nutrient supplementation. A surfactant was used to enhance solubility of the PAHs for improved rates of degradation. A 99+ percent removal occurred in the pilot plant during a 17-week reactor detention time (Bleam, 1988). The full-scale SBR was operated with a 28-day contact time to yield effective treatment of most PAHs.

An aerobic SBR demonstration project was used to treat hazardous waste from landfill leachate, groundwater remediation, and industrial wastewater at a disposal complex in Niagara Falls, N.Y. (Sutton, 1988). The SBR reactor was designed to hold 550,000 gal and processed 60,000 gal/day. The SBR system experienced occasional upsets and was converted to a PAC-SBR system for improved stability. The SBR was designed as a pretreatment step to reduce the cost of an activated carbon adsorption facility. The SBR achieved TOC removals of 55 to 81 percent and phenol removals of 97 to 99 percent. The only exception was for two heavy metal toxicity loadings (Norcross, 1985). Design and operational criteria for this facility are:

Hydraulic retention times	2.0–10 days
MLVSS level	1600 mg/L
SVI	47 mL/g

Phenol wastewater has been treated by an aerobic SBR, providing a phenol reduction of 60 to 0.5 mg/L. The process was converted to an anaerobic-aerobic system by providing a 6-h anoxic metabolic mode. Anaerobic-aerobic sequence resulted in a 30 percent overall savings in energy cost with no loss in effluent quality.

An anaerobic sequencing batch reactor (ASBR) operating under methanogenic metabolism has been patented and reported to achieve equivalent degrees of treatment independent of temperature (Dague, 1992). Methanogenic activity rates at 25°C were equivalent to those at 35°C. It is reported to improve biomass retention within the reactor over other anaerobic processes, since all treatment steps occur in a closed reactor. This maintains a constant partial pressure of the produced methane, carbon dioxide, and hydrogen during the settling step. Some anaerobic processes require degasifying before settling owing to the release of these gases when the liquid is discharged to open settling basins. The anaerobic biomass settles rapidly, allowing shorter hydraulic detention times and longer MCRT. The anaerobic SBR has been operated with the following time cycles:

Feed phase	0.5 h
React phase	21 h
Settling phase	2.0 h
Decant phase	0.5 h

The operational characteristics of SBRs make them very suitable for the application of cometabolism. The cycling F/M ratio (Fig. 9.5) for primary substrate is the standard mode of operation for SBRs.

Stensel (1992) proposed an enclosed SBR design for treating halogenated aliphatic hydrocarbons with methanotrophic bacteria. The SBR design and operation provide for the regrowth and repair of cellular material destroyed by the intermediate toxicity of trichloroethylene degradation. The loss of TCE transformation capacity, enzyme inhibition, and inaction of long-term exposure to TCE can be countered by the sequencing batch operation. The operating steps are as follows:

1. Fill the reactor with contaminated water.

2. Charge the closed reactor with methane and aerate with gas recycling.

3. React with methane addition and oxygen until TCE degradation is complete.

4. Stop gas recycling and allow the biomass to settle.

5. Decant off the treated water.

6. Biomass augmentation or growth without TCE can be conducted as needed before repeating step 1.

A potential caution with batch reactors using methanotrophs for treating trichloroethylene (TCE) is the toxicity of TCE and its epoxide intermediate to the microbial cell. Methanotrophic bacteria have a finite degradative capacity for halogenated hydrocarbons as a result of a toxic intermediate produced during the dehalogenation reaction. This degradative capacity is termed transformation capacity (Tc) with units of halogenated hydrocarbon mass removed as grams per unit of biomass, such as mixed liquor volatile suspended solids (MLVSS).

Higher Tc values are found for flow-through reactors than for batch systems (Stensel, 1992). The mechanism of the benefit associated with flow reactors is unknown. It may be the replenishment of an essential inorganic nutrient, other than nitrogen or phosphorus, or the dilution of toxic intermediates.

Transformation capacity is a function of the halogenated hydrocarbon concentration. When using methane as the primary substrate, concentrations of trichloroethylene of 8 mg/L completely inhibit methane oxidation (Stensel, 1992). Biomass concentration is related to the methane consumption rate. An increase in TCE concentration results in the inhibition of biomass growth. TCE concentration can result in the biomass wasting rate exceeding the production rate, causing reactor failure. Transformation capacity also is influenced by the presence of methane and the exposure time to trichloroethylene. Average Tc values with methane are about twice as high as for degradation without methane (Table 9.7). Methanotrophic bacteria that have had long-time exposure to TCE (several months) may undergo a population change in that the methanotrophic organisms lose their ability to degrade TCE.

Fixed Biofilm Reactors

Process design for immobilized biomass is significantly different from that for suspended reactors. The rate of removal is limited by mass transfer and diffusion within the biomass (Fig. 9.14). For fixed-film reactors, removal of contaminants is the result of sorption to the biofilm. The absorbed organic compounds are then degraded. The fixed-biofilm process can be thought of as an adsorption system that has a continuous growing and self-generating adsorbent of microbial cells. The organic substrate removal rate is controlled by diffusion through a liquid-film attachment, followed by diffusion into the cellular bio-

TABLE 9.7 Trichloroethylene Transformation Capacity Values of Methanotrophic Bacteria

TCE concentration, mg/L	Methane concentration, % gas flow	Reactor flow	TCE g/g VSS
2–5	5	Flow	0.063–0.16
2–5	0	Batch	0.019–0.043
2–5	5	Batch	0.064
5–10	0	Batch	0.028
5–10	0	Batch	0.017
10–15	0	Batch	0.014–0.036
10–15	5	Batch	0.051
15–20	5	Batch	0.038
25–30	0	Batch	0.08*
25–30	0	Batch	0.025–0.027
30–40	0	Batch	0.036
30–40	5	Batch	0.051

SOURCE: Stensel, 1992; Alvarez-Cohen, 1991.
*With 1560 mg/L of formate.

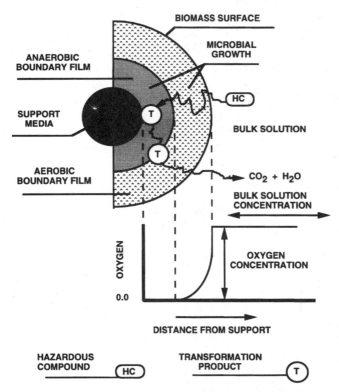

Figure 9.14 Schematic of transfer within the biomass film and oxygen concentration for aerobic systems.

mass. Transport through the biomass can be limited as a result of metabolism or diffusion.

Fixed-biofilm reactors can be designed as oxic, anoxic, or anaerobic processes. Aerobic systems treating high organic concentrations are frequently limited in their degradation rate because of oxygen transfer. Such systems support the formation of heavy biomass buildup that creates a significant oxygen demand. Oxygen transfer can be limited by both a gas phase to liquid phase and a liquid phase to biomass phase. The gas-to-liquid transfer is influenced by bubble size and dissolution pressure. The liquid-to-biomass is influenced by mixing intensity and the biomass structure.

Fixed-biofilm treatment systems have a long history of application to wastewater and include the trickling filters that were widely employed for municipal sewage. Fixed-biofilm reactors consist of many design configurations, flow regimes, and metabolism modes. The physical contacting variations for immobilized biomass reactors include:

Packed-bed reactors

Trickle-bed reactors

Fluidized-bed reactors

Air-sparged reactors

The metabolism modes include:

Aerobic

Anoxic

Anaerobic

Owing to the numerous advantages of immobilized microorganisms, the fixed-film reactors should find increasing applications and improvement in design configurations. A number of design configurations have been developed for application in the biochemical engineering of industrial products. Fixed-bed bioreactors can be skid-mounted for transportation ease to several treatment locations, or they can be constructed as a more permanent facility (Figs. 9.15 and 9.16). Skid-mounted units have been seeded with microorganisms in the laboratory and then transported to the site for start-up. Some bioreactors employ internal plates or mechanically moved internals for energy input and clogging control. Other designs use an internal draft tube to provide a defined circulation pattern for loop flow. Liquid circulation has also been provided by internal pumps (Bailey, 1986).

Figure 9.15 A fluidized bed bioreactor system mounted on skids for mobility. (*Courtesy of Envirex, Ltd., Waukesha, Wis.*)

Properties of immobilized biomass

Most microorganisms attach to almost any surface exposed to an aqueous environment. The immobilized cells grow, divide, and produce extra cellular polymers, forming a gel like material that is called a biofilm. The polymeric material is primarily polysaccharides that extend from the microbial cell (Jimeno, 1990). Once microbial cells have colonized a surface, forming an immobile biofilm, their activity depends on local surface conditions. Of primary importance is the transport of nutrients, substrate, electron acceptors, and donors to the immobilized cells.

The gel matrix traps or absorbs the components necessary for metabolism. In this manner, hazardous chemicals are held in the reac-

Figure 9.16 Fluidized fixed-bed bioreactor with activated carbon support media. (*Courtesy of Envirex, Ltd., Waukesha, Wis.*)

tor for long periods. The effective detention time may be several days or weeks, rather than hours of residence time for suspended growth reactors. This increased residence time provides for microbial acclamation to the more persistent hazardous compounds. This gel matrix also collects other microorganisms, causing a changing biofilm consortium as supported by the substrate, nutrients, and environmental conditions. Thus the microbial diversity of the biofilm consortium is ever changing with time and within the thickness of the biofilm.

The depth and composition of the biofilm are influenced by many factors (Table 9.8). One of the more important advantages of fixed-biofilm reactors is the microbial diversity they offer. The microbial diversity within the biofilm thickness changes because of concentration gradients of electron donors, acceptors, and other environmental parameters, including biological transformation products. Both aerobic and anaerobic environments can exist within the same biofilm even for an aerated fixed-biofilm reactor. The biomass adsorbs organic compounds from the bulk solution, yielding an enriched energy source at the biomass film surface. Oxygen transport through this biomass film is diffusion limited and the concentration gradient for

TABLE 9.8   Factors Influencing the Depth and Composition of Biofilm

Nature of the organic substrate

The concentration of the organic substrate

The nature of the microorganisms

The superficial velocity

Water characteristics:

- pH
- Electrolyte concentrations

The nature of the support media:

- Surface area
- Surface roughness
- Pore volume
- Pore size
- Surface charge
- Adsorption properties
- Density
- Specific gravity

TABLE 9.9   Advantages of Immobilized Biomass Reactors

1. Prevents washout of biomass
2. Operational ease relative to the separation of biomass from liquid
3. Provides for significantly more biomass per unit volume of reactor
4. Provides higher mean cell residence times (MCRT)
5. Provides more resistance to toxic loading
6. Improves microbial growth for organisms that grow poorly in free suspensions
7. Provides higher rates of biodegradation
8. Provides for a larger diversity in the microbial consortium
9. Provides for a broad spectrum of catalytic functions by physical separation of specific biomass communities
10. Allows continuous removal of reaction inhibitors for a specific biomass
11. May provide a more stable gene pool and enhanced rates of genetic transfer
12. Most immobilized reactors are closed systems that provide for minimized emissions of VOCs
13. Some studies report that trichloroethylene toxicity to methanotrophs is reduced in fixed-film bioreactors over suspended bioreactors

diffusion is significantly reduced by the biofilm's respiration activity. This results in anaerobic conditions within the biofilm (Fig. 9.14).

The ability to maintain a large microbial diversity in a relatively stable form provides significant advantages for the treatment of hazardous contaminants in groundwater or other waste streams (Table 9.9). Important advantages are high mean cell residence times, lack of sludge handling to maintain the biomass, and better response to toxicity. Since biological activity takes place on the surface rather

TABLE 9.10   Comparison of Degradation Rates for
Immobilized Biomass versus Suspended Biomass

Reactor type	Degradation rate of phenol, mmol/L-h
Fluidized-bed fixed biofilm	22
Fixed-bed biofilm	12.5
Suspended bioreactor	2.6

SOURCE: Galli, 1987.

than in bulk solution, most of the microorganisms are buffered against toxic materials or shock loads. Another advantage results from the potential combined anaerobic-aerobic metabolism. Compounds not readily degraded by aerobic metabolism will continue to diffuse into the biofilm. As they encounter the anaerobic zone, transformation products can develop which will diffuse outward to the aerobic zone where they are mineralized by the aerobic activity.

The growth for some microbial species is very poor in suspended reactors. This allows fixed-film reactors to develop specialized cultures for hazardous compound degradation. Systems that immobilize cells are particularly suited for *Phanerochaete chrysosporium*. This white rot fungus has responded poorly in suspended bioreactors. The rot fungus has been investigated for the degradation of a number of persistent chemicals, including DDT, polychlorinated biphenyls, polychlorinated dibenzo[p]dioxins, lindane, chlorinated alkanes, and PAHs (Lewandowski, 1990).

A final advantage is the finding that immobilized biomass is capable of higher degradation rates than suspended cells. The kinetic coefficients of the enzyme reactions yield higher rates. One comparative study found the rate of phenol degradation to be almost 10 times higher for a fluidized bed reactor over a suspended biomass reactor (Table 9.10) (Galli, 1987). Significantly higher rates of 2-chlorophenol biodegradation by *Phanerochaete chrysosporium* have been measured for immobilized cells over suspended reactor systems. Lewandowski (1990) established the first-order kinetic rate constants for different reactor configurations (Table 9.11). A packed-bed reactor yielded a degradation rate for 2-chlorophenol 10 times greater than suspended reactors. These advantages make fixed-biofilm reactors a prime consideration for bioremediation. Immobilized microorganisms have been applied to the treatment of a number of hazardous chemicals (Table 9.12).

The support media

For most fixed-biofilm systems, the biomass responsible for the major removal is on the surface or within the pore structure of the packing

TABLE 9.11   Comparison of Suspended and Immobilized Biomass on Degradation Rates of *Phanerochaete chrysosporium*

Reactor type	First-order degradation rate† per hour
Suspended biomass:	
Batch mobilized biomass	0.014–0.016
Immobilized biomass:	
Packed bed, berl saddles	0.086
Packed bed, silica beads*	0.015
Packed bed, balsa wood	0.019
Mixed reactor, alginate beads	0.51

SOURCE: Lewandowski, 1990.
*Manville celite catalyst carrier.
†2-chlorophenol.

TABLE 9.12   Hazardous Compounds Treated by Immobilized Biomass

Acenaphthylene
Aniline
Anthracene
Chrysene
Toluene diamine
s-Triazine herbicides
Dinitrotoluene
Dichloromethane
Di-n-butylphthalate
Di-n-octylphthalate
bis(2-Ethylhexyl)phthalate
Ethylene dichloride
Fluorene
Fluoranthene
Naphthalene
Phenanthrene
Pyrene
Trichloroethylene
Tetrachloroethylene
trans-1,2-Dichloroethylene
Vinyl chloride

SOURCE: Nelson, 1988; Galli, 1987; Melcer, 1988; Miller, 1990; Mashayekhi, 1993; Huang, 1988.

Figure 9.17 Example of synthetic media for supporting biomass. (*Courtesy of American Surfpac Corp., Downington, Pa.*)

medium. However, some systems, such as the air-sparged reactor, have predominant growth in the voids of the packing medium. At one time rock was the medium used in fixed-film bioreactors. These systems were operated as trickling filters, providing a dual-phase system of air and liquid. Synthetic medium is the most popular medium. It has been designed to enhance biomass attachment, increase surface area, and reduce clogging potential and now represents the major support for fixed-film bioreactors. The variety of configurations is illustrated by Fig. 9.17.

The physical and chemical nature of the media affects the biomass loading per unit volume of bed (see Table 9.8). Available surface area for microorganism attachment and growth is an important variable. Available area is determined by pore diameter large enough to allow microbial colonization. Solid supports with diameters of 150 to 300 μm provide the best adhesion for microorganisms (Kindzierski, 1992). Small pores, as typical of most activated carbons, are not as accessible to microorganisms.

Surface charge also influences microorganism attachment to media. Most bacteria bear a net negative charge. Cation-exchange resin is a

poor support for methanogenic phenol-degrading organisms
(Kindzierski, 1992), whereas anion-exchange resins are reported superior to activated carbon. Media utilized to immobilize biomass include:

Rock

Sand

Cellulose triacetate

Ceramic saddles

Quartzite sand

Charcoal

Kaolin particles

Anthracite

Diatomaceous earth

Cylindrical porous beads (Manville celite catalyst carrier)

Spherical diatomaceous earth beads (Manville R630 carrier)

Polystyrene

Polyurethane foam coated with activated carbon (Allied Signal)

Polyvinyl chloride

Modified cellulose

Wood

Ion-exchange resin

Activated carbon

Gelatin

For hazardous waste treatment, several support media appear to
provide superior performance in most applications. These include
activated carbon, cylindrical porous beads, and several synthetic
resins and plastics.

For hazardous waste treatment, activated carbon provides benefits
over most other support media. The large surface area of activated
carbon provides a means to increase MCRT. The adsorption properties of activated carbon provide rapid buildup of biomass under conditions of very low contaminant concentrations. It improves the reactor
response to toxic influent and microbial inhibition.

Rapid biomass buildup on activated carbon over other support
media results from enhanced adsorption of organic compounds and a
better sticking foundation for microbial growth. Adsorption on the
carbon surface not only results in removal of the contaminants from

TABLE 9.13   Advantages of Granular Activated Carbon over Sand for Biomass
Support

- Provides an optimum surface for attachment
- Provides for higher biomass buildup under low concentration conditions
- Provides for shorter start-up periods
- Improves control of biofilm thickness and uniformity
- Allows lower recirculation rates for bed expansion, resulting in lower shear forces
- Results in lower energy consumption

TABLE 9.14   Biomass Growth Performance on Different Support Media

	Activated carbon	Anthracite	Sand
Properties:			
Density, g/cm^3	1.4	1.4	2.65
Settling velocity, cm/s	4.0	5.4	11
Biomass:			
Relative amount	10	2.67	1
Shear loss coefficient	1	6	20

SOURCE: Fox, 1988.

the influent, but this accumulation of organic carbon provides a concentrated energy source for rapid microbial growth within the reactor.
Start-up times are significantly reduced. After microbial growth, the
adsorption capacity of carbon becomes insignificant relative to the
biological treatment mode. Adsorption on the biomass followed by
microbial metabolism becomes the main treatment mechanism.

The higher biomass yield with activated carbon is a benefit for
anaerobic systems (Table 9.13). Anaerobic microbial growth yields are
significantly lower than aerobic microbial systems. Biomass response
has been compared on three support media in fluidized bed reactors
operating under anaerobic conditions (Fox, 1988). Support media consisted of low-density anthracite, granular activated carbon, and sand.
Density, surface area, and surface roughness influence biomass
response and reactor performance. Under an anaerobic metabolism
mode, activated carbon support provides superior performance. The
retained biomass is significantly greater, resulting in improved
degradation efficiency (Table 9.14).

Granular activated carbon not only has the highest surface area,
but it has the roughest surface with numerous pores that support biomass growth. The growth is within and on the carbon's rough surface,
yielding a stronger biofilm-support bond. Of the evaluated media,
only activated carbon, anthracite, and anion-exchange resins could
support the low growth rate of anaerobic bacteria. With a mixed pop-

ulation, the sand support favors the faster-growth organisms over methanogenic organisms. This species distribution is to some degree influenced by the hydraulic shear force in fluidized beds. The lower settling velocity of activated carbon requires a lower bed recirculation rate for bed expansion. This reduces shear forces and maintains more attached biomass, resulting in a significantly longer MCRT. Sand has a greater density than activated carbon and the least surface roughness of the three media. These properties result in shear loss coefficients 20 times greater than that for activated carbon (Table 9.14). Energy consumption for pumping is also reduced for carbon versus sand per unit of treated influent.

The local hydrodynamic conditions at the biofilm solution interface affect the stratification of metabolic activity in the biofilm. The hydraulics can increase the growth or decrease it, depending on the level of carbon energy in the groundwater. Increasing the superficial velocity (velocity within a bed based on the open cross-sectional area) for groundwater having low organic content, less than 4 mg/L as TOC, will stimulate growth in fixed-bed reactors containing activated carbon media. The higher growth results from the continual exposure of the biofilm to substrate. The opposite has been found for groundwater containing high levels of carbon energy. Since this supports greater biofilm growth, the thicker biofilm is sheared off by increasing the superficial velocity. The optimum flow rates cannot be adequately predicted, and treatability studies are highly recommended.

For low substrate levels, sand media yield inconsistent effluent quality with anaerobic systems. This lower treatment is due to the lack of attached biomass. Contaminated groundwater usually contains very low organic loadings compared with industrial wastewater. Thus the performance of sand fluidized beds with contaminated groundwater differs from its performance with industrial wastes. Sand is frequently superior to activated carbon when treating high-strength organic wastewaters. The higher growth of biomass on activated carbon support over sand is detrimental to many aerobic fixed-bed systems because of clogging. To reduce the biomass clogging potential and retain the advantages of activated carbon, a support medium made of polyurethane foam coated with powder-activated carbon has been developed by Allied Signal. With high-energy-source influent, this medium can be mixed with polypropylene HiFlow pall rings to reduce clogging potential. The system can be used with a packed-bed design to minimize backmixing when low effluent concentrations must be achieved.

An anthracite medium provides large crevices for biomass development, but either the majority of the anthracite surface is too smooth or the hydraulic shear forces too great. Scanning electron microscopy reveals significant areas of anthracite surface without biofilm, whereas the activated carbon developed extensive growth within the carbon's crevices (Fox, 1988).

Packed-bed reactors

Process principles. The term packed bed applies to the physical contact and structural support of the media particles against other media particles. This system tends to approach a plug flow reactor. Both upflow and downflow configurations are used. Most packed beds use an upflow system to reduce the clogging potential that downflow beds are subject to.

Applications. Packed-bed reactors have been mainly applied to anoxic and anaerobic metabolic systems for two reasons. First, it is difficult to maintain adequate oxygen transfer in high-organic-strength industrial wastewater because of the high biomass production rate. Second, the aerobic metabolism mode yields more biomass per pound of contaminant, resulting in biomass clogging problems. However, the packed bed is finding a potential application in the low-organic-strength groundwater typical of hazardous waste contamination. The use of pure oxygen and hydrogen peroxide has also aided in maintaining aerobic conditions. Proper selection of the packing material can reduce biomass accumulation and clogging.

Packed-bed reactors have been used for the treatment of:

Phenolic compounds

Halogenated aliphatic compounds

Benzene

Toluene

Toluene diamine

Aniline

Dinitrotoluene

Acetone

Methylethylketone

Cyclohexanone

Methanol

1-Butanol

Isopropyl

(Nelson, 1988; Yang, 1988; Miller, 1990; Sutton, 1988.)

Chlorinated hydrocarbons. Downflow fixed-bed reactors have been used for the treatment of groundwater contaminated with halogenated aliphatic compounds of ethylene dichloride, trichloroethylene, and tetrachloroethylene (Miller, 1990). Spherical diatomaceous earth beads (Manville R630 carrier) were used as the biomass support medium. Bioreactor design and operational parameters were:

Metabolism mode	Aerobic
Bed retention time	20.5 h
Recycle flow to feed flow	22:1
Operating pH	6.5–7.5
Operating temperature	86–102°F

The process configuration provided for the groundwater feed to enter the recycle line on the suction side of the recycle pump (Fig. 9.18) (Miller, 1990). This stream was aerated in a comingling jet configuration. The aerated stream was pumped to a heated reactor maintained at 30°C and then to the packed bed. Flow rates were such that a rapid-mix flow regime with high dissolved oxygen was maintained throughout the bed.

The fixed-bed bioreactor was operated in the aerobic mode and seeded with *Xanthobacter autotrophicus. X. autotrophicus* is capable of using halogenated aliphatic compounds as a substrate (Miller, 1990). The reactor was operated in the batch mode to develop the bac-

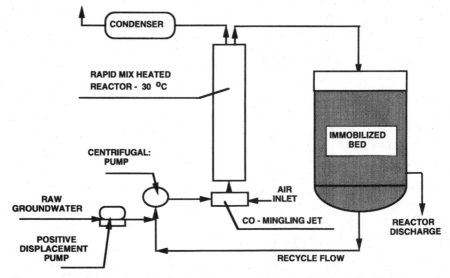

Figure 9.18 Packed-bed reactor for treating groundwater contaminated with halogenated aliphatic hydrocarbons.

terial biofilm density. Biofilm development required 42 days. After biofilm development, the bed was operated in a continuous-flow mode with a recycle ratio of 22 to 1 for recycle versus the raw groundwater flow. The feed of contaminated groundwater resulted in the development of an indigenous bacterial species that existed in conjunction with *Xanthobacter autotrophicus.*

The downflow fixed-bed reactor provided a stable and effective biomass even under varying groundwater concentrations of 0.8 to 428 mg/L. Average bioreactor efficiencies were 90.2 percent removal for ethylene dichloride and 73.6 percent removal for trichloroethylene. Losses from air stripping were calculated at 9 to 20 percent. The average feed concentration of tetrachloroethylene was only 2.5 mg/L.

Substituted aromatic compounds. Wastewater streams containing aniline, toluene diamine, dinitrotoluene, and methanol also have been treated by packed-bed reactors (Nelson, 1988). A dual-column, plug flow reactor was used with diatomaceous earth as the biomass support medium. The reactor was operated under an aerobic metabolism mode and seeded with organisms developed for detoxification of *s*-triazine herbicides. The microbial consortium was primarily *Methylobacter* sp. The microbes were continuously introduced and recycled through the reactor to aid cell attachment.

The reactor size and operational characteristics were:

Configuration	Dual columns in series
Column volume	22 ft^3
Flow rate	160–400 gal/day
Aeration rate	40–50 ft^3/h

The degradation primarily occurred in the first column with an overall process efficiency between 91 and 96 percent mineralization for the targeted compounds. The microbial system effectively removed dinitrotoluene at 90 percent plus.

Treatability studies using a packed-bed reactor with activated carbon as the biomass support were evaluated for the treatment of phenolic wastewater (Yang, 1988). The support medium was cellulose triacetate. The aerobic microbial system consisted of a consortium that contained *Pseudomonas* sp. and *Candida* sp. For these studies, the packed-bed reactor was superior to the activated sludge process for buildup of microbial mass and overall performance under loading variations.

Polynuclear aromatic hydrocarbons. A fixed plug flow bioreactor has been used to treat wastewater containing PAHs and BTEX com-

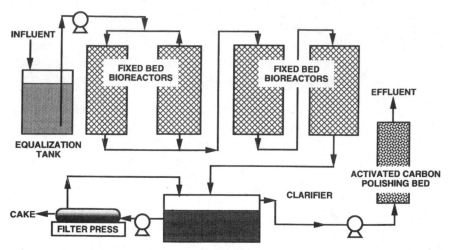

Figure 9.19 Fixed-bed bioreactor process flow diagram for the treatment of polynuclear aromatic hydrocarbons.

pounds (Mashayekhi, 1993). Four bioreactors, each of 40,000-gal bed capacity, were operated in a series-parallel configuration (Fig. 9.19).

The support medium was dual media consisting of polyurethane foam coated with activated carbon (Allied Signal) and HiFlow packing to reduce clogging potential. These materials were mixed uniformly throughout the bed in a 1:1 ratio. The bed was operated in the aerobic metabolism mode with flow rates of 70,000 to 90,000 gal/day. Performance data are illustrated in Table 9.15.

Cometabolism-chlorinated hydrocarbons. Cometabolism requires control of both a primary and secondary substrate. As discussed above, bioreactors may be staged to alternate between the growth stage and the hazardous compound removal stage. This sequencing operation has been evaluated in laboratory systems for the cometabolism of trichloroethylene (TCE) (Segar, 1992). Fixed-film bioreactors were operated as an upflow process. A phenol-degrading biomass that generated monooxygenase was first developed in a batch reactor and recycled through the packed-bed reactor with nutrients and phenol concentration of 40 mg/L. Feedwater was continuously aerated to 7 to 8 mg/L of oxygen. After the biofilm mass developed, the recycling of seed material was taken off-line and the flow scheme for TCE degradation implemented. The treatment scheme used an automatically timed sequence of pumping TCE-contaminated water for treatment followed by a period of phenol feed for biomass and regeneration of transformation capability.

Trichloroethylene mineralization of 70 to 80 percent was achieved within the acclimated biofilm reactors. Acclimation required about

TABLE 9.15   Performance of a Fixed-Bed Bioreactor on PAHs

Parameter	Influent	Effluent
BOD$_5$	2,257 mg/L	6 mg/L
TOC	1,268 mg/L	20 mg/L
TSS	37.2 mg/L	34.7 mg/L
Phenolics	699 mg/L	0.98 mg/L
Cyanide	1.9 mg/L	0.26 mg/L
Phenol	543,954 µg/L	ND
2,4-DMP	6,212 µg/L	ND
Naphthalene	23,882 µg/L	ND
Acenaphthylene	1,958 µg/L	ND
Acenaphthene	2,967 µg/L	ND
Fluorene	1,685 µg/L	ND
Phenanthrene	1,915 µg/L	ND
Anthracene	1,175 µg/L	ND
Fluoranthene	1,780 µg/L	ND
Pyrene	1,020 µg/L	ND
bis(2-Ethyl-hexyl) phthalate	24 µg/L	ND
Chrysene	310 µg/L	ND
Benzo(a)anthracene	252 µg/L	ND
Benzo(b)fluroanthene	128 µg/L	ND
Benzo(k)fluroanthene	15 µg/L	ND
Benzo(a)pyrene	72 µg/L	ND
Indo(1,2,3-cd)pyrene	25 µg/L	ND
Methylene chloride	27 µg/L	ND
Benzene	2,916 µg/L	ND
Toluene	1,700 µg/L	ND
Ethylbenzene	330 µg/L	ND

SOURCE: Mashayekhi, 1993.
ND = not detected.

2 to 4 cycles of TCE loading and transformation capacity regeneration with phenol. An important operational variable is the feeding strategy to maintain the cometabolism transformation capacity of the biomass. It is important to optimize the primary substrate feed frequency. A minimum of 2 to a maximum of 3 h of phenol feeding per day sustained the most consistent TCE removal. Also, a feed concentration of 25 mg/L resulted in higher average TCE removals than 5 or 100 mg/L. Feeding twice a day caused excessive biomass growth, resulting in channeling and decrease in TCE removal. Feeding at intervals longer than a day resulted in excessive biomass decay.

Sequencing biofilm bioreactors for the cometabolism of trichloroethylene have been evaluated and modeled with a pseudo-first-order expression. Monod kinetics reduces to a pseudo-first-order reaction when substrate concentrations are low and the reactor biomass is essentially constant. At low S values, $S \ll K_s$, Eq. (9.16) reduces to

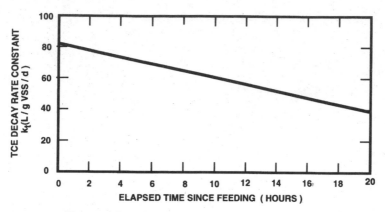

Figure 9.20 Decay of degradation rate constant during starvation phase of cometabolism.

$$\frac{dS}{dt} = k_1 X_0 S \qquad (9.41)$$

where S is the halogenated substrate concentration (mg/L); X_0 is the biomass concentration, mg VSS/mL; and k_1 is the rate constant, L/g VSS/day. For a cometabolism reactor fed with primary substrate at periodic intervals, the rate constant k_1 will decrease over time until feeding restores the biomass reducing power (see Fig. 9.5). This decay of the rate constant k_t can be expressed as exponential with time by the equation

$$k_t = k_1 \exp(-k_c t) \qquad (9.42)$$

where k_1 is the value at $t = 0$ days, t_t is the value at t days, and k_c is the decay constant with units of per day. For TCE removal and a feeding cycle of 2 h out of every 24 h k_1 and k_c values were reported as 81 L/g VSS/day and 0.83 L/day, respectively (Segar, 1992). The falloff in rate constant and percent of TCE removal as a function of time between feeding is illustrated in Fig. 9.20.

A series of packed-bed reactors using anaerobic followed by aerobic metabolism were selected for treatability studies on a complex mixture of lagoon wastewater. The waste mixture contained aromatics, chlorinated aliphatic hydrocarbons, phenolic compounds and phthalates. Anaerobic metabolism was selected as the first stage to minimize stripping and to dehalogenate the chlorinated compounds. The second-stage reactor was designed for aerobic metabolism to oxidized transition compounds plus benzene, toluene, and the phthalates. Steady-state conditions were established and complete removal was recorded for most compounds (Table 9.16). The one exception was bis (2-chloroethyl) ether (Venkataramani, 1991).

TABLE 9.16 Performance of Anaerobic and Aerobic Packed-Bed Reactors
in Series

| Compound | Influent concentration, µg/L | Effluent ~ µg/L | |
		Anaerobic reactor	Aerobic reactor
Benzene	2,660	15	ND
Toluene	38,600	109	ND
1,2-Dichloroethane	58,600	ND	ND
Ethylbenzene	1,420	ND	ND
Methylene chloride	22,000	ND	ND
bis (2-Chloroethyl) ether	26,900	8,210	3,800
Phenol	12,500	ND	ND

SOURCE: Venkataramani, 1991.
ND = not detected at limit of 10 µg/L except for phenol, which was 1.5 µg/L.

Potential use on PCBs. Packed-bed bioreactors have been evaluated
for the transformation of polychlorinated biphenyls, PCBs. A fixed-
biofilm provides the advantage of microbial diversity with the potential of supporting predominant-seeded species. This is important,
since no single naturally occurring microorganism is reported to mineralize PCBs higher than monochlorobiphenyls. The bioreactor was
bioaugmented with cultures of *Acinebacter* spp. and *Acinebacter* sp.
This consortium can use benzoate, biphenyl, and 4-chlorobiphenyl as
solid carbon and energy sources. It can dehalogenate 3,4-dichloroben-
zoate to 4-carboxyl-1,2, 2-benzoquinone by cometabolism (Adriaens,
1990).

The bioreactor support medium was polyurethane ether foam. The
system was operated in a continuous flow with a primary substrate
feed of benzoate at 500 mg/L. The contaminated water contained soluble PCB congeners including tetrachlorobiphenyl. No PCBs were
detectable in the reactor effluent, indicating that they were at least
trapped by the polyurethane biofilm if not degraded.

Although some mineralization occurred, the majority of the PCBs
were transformed to an array of intermediates that still contained
chlorinated rings.

Trickling filters

Process principles. Trickle-bed reactors contain a packed bed that is
not totally submerged. A liquid phase trickles down through the packing, contacting the biomass surface. A gas phase, either forced or natural convection, flows over the liquid-solid contacting surface. The
resulting biomass operates as an unsubmerged biofilm. The period of
biofilm development and biomass accumulation appears to be longer

for unsubmerged biofilms versus that of submerged biofilms (Jimeno, 1990). A trickle-bed reactor, the trickling filter, has been used in the United States since 1908. The reactor is packed with rocks or plastic media. Influent is applied from the top and trickles down through the surface of packing media where biofilm develops. Occasionally biofilm buildup is sheared off and flows with the water. The sheared biomass is settled in a sedimentation tank following the trickling filter. Oxygen is diffused through the liquid film from the atmosphere. In some cases, blowers are used to supply oxygen from the bottom of the reactor, but it has been typical to use only natural convection.

Although rock has been used as the support medium, it has been replaced with plastic media. Plastic media come in a number of configurations including vertical-flow media for deep columns or tower filters, cross-flow media, and several random packing materials (Fig. 9.17). These media are noted for their high hydraulic capacity and resistance to biomass plugging. The plastic medium offers surface areas of 25 to 120 ft^2/ft^3 of bed volume.

Applications.   Field application of trickling filters for treatment of groundwater contaminated with hazardous chemicals is limited. One system was used to treat trichloroethylene-contaminated ground (Wickramanayake, 1991). Trickling-filter reactors were operated in series with aerobic metabolism. Each reactor consisted of two 6-ft columns 1 ft in diameter. Support medium was $1/4$-in-diameter ceramic Manville pellets and $3/4$-in-diameter ceramic Raschig rings in reactors 1 and 2, respectively.

Each reactor was sealed and a mixture of air and methane introduced to the head space. The seed consisted of a consortium that degraded TCE and used methane as a substrate. The units were first inoculated and evaluated in the laboratory for 4 months and then moved to the site. During this move, the microbial biomass was without primary substrate and oxygen for 5 days. The reactors responded well even after this 5-day down period. After field start-up, the bioreactors were found resilient to day-to-day failures in pumps and interrupted flows of water and gas for periods of 24 h. Major operational problems were chemical rather than biological. Chemical precipitation occurred in pipelines, pumps, and metering devices. To alleviate the scaling, a water softener was installed to pretreat the groundwater.

Nutrient supplementation with phosphate, nitrate, sulfate, and trace transition metals was found beneficial. TCE degradation rates were 2 to 2.4 mg/min with oxygen and methane consumption rates of 0.1 and 50 mg/min, respectively. During the field operation, TCE removal averaged 60 percent. Volatilization loss averaged 10 percent

TABLE 9.17 Process Operational Parameter for TCE
Removal in Aerobic Trickling-Bed Reactors

Parameter	Reactor 1	Reactor 2
Flow rate, L/min	2–3	2–3
Hydraulic loading, gal/ft^2/min	2–3	2–3
Nutrient solution, mg/L	226	176
Methane, %	0.67	1.2
Oxygen, % gas	19	19
Dissolved oxygen, mg/L	5.0	4.9
Temperature, °C	33	34
pH	6.9	6.9
TCE, mg/L	0.5–1.4	0.5–1.4

SOURCE: Wickramanayake, 1991.

of the mass loading. Process operational parameters are provided in
Table 9.17.

Fluidized-bed reactors

Process principles. The fluidized bed is operated in an upflow condition to expand the support media holding the biofilm. Drag forces
caused by the fluid flow against the support media provide bed expansion. As the biomass increases in thickness on the fluidized-bed
media, significant differences in the effective particle diameter and
settling velocity can occur. Reactor design must distribute and control
the influent flow so these density changes in bed media can be
accounted for. By careful control on the flow velocity and/or the use of
expanded cross-sectional areas at the top of the bed, the biomass is
retained in the reactor. Bed expansions can be monitored optically to
evaluate expansion and buildup of biomass. For substrates that yield
high biomass growth, expansion of the bed can be controlled by
removal of the media and cleaning of biomass growth. The bioreactor
shown in Fig. 9.15 has a mechanical shearing device to reduce biomass buildup. The clean carbon is retained in the reactor while the
low-density sheared biomass flows out of the reactor. If necessary, it
is removed by a solids separation unit. The process flow for this skid-
mounted unit is typical of most fluidized-bed reactors as illustrated in
Fig. 9.21.

Popularity of the fluidized-bed reactors results from less clogging
problems than packed-bed systems. Clogging problems are frequently
chemical rather than biological. For many wastewaters, aerobic conditions are easier to maintain in a fluidized bed. The main disadvantage
is the greater vertical mixing in a fluidized bed over the packed-bed
flow regime. If physical separation of specific biomass communities is

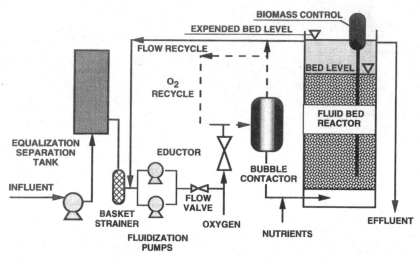

Figure 9.21 Skid-mounted fluidized-bed process flow diagram.

desired, multiple reactors must be used since physical separation cannot be controlled in a single reactor.

Applications

Chlorinated hydrocarbons. Fluidized biofilm reactors have been evaluated for the treatment of methylene chloride–contaminated water. A
12-ft fluidized bed with activated carbon support media was inoculated with *Hyphomicrobium* sp. at the General Electric Corp. Mt.
Vernon, Ind., facility. Design parameters were as follows:

Reactor size	20 in by 12 ft
Flow velocity	12 gal/min/ft^2
Flow rate (total)	27 gal/min
Flow rate (waste stream)	2.7 gal/min
Hydraulic detention time	80 min
Support media load	500 lb
Metabolism mode	Aerobic

The carbon support was fluidized by recycling flow from the bed
with the influent waste stream at a ratio of 10:1 (Fig. 9.22). Oxygen
was supplied to the recycle flow by an oxygen generator. Undissolved
gases were separated from the flow in a bubble trap and vented
before the recycle was blended with the waste stream. Data were not
provided on the methylene chloride loss by volatilization. However,

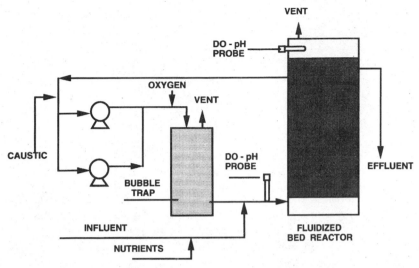

Figure 9.22 Fluidized-bed bioreactor for methylene chloride removal. (*Paone, 1992.*)

pH depression and dissolved oxygen depletion indicate significant biological activity (Paone, 1992).

The fluidized bed was inoculated by cultivating *Hyphomicrobium* sp. on 1 percent methanol in a mineral salts solution containing nutrients. The activated carbon support medium was equilibrated by recirculation of the salt-methanol solution through the bed. The operation was monitored and controlled by in-line pH and dissolved oxygen probes at the influent and effluent. The pH goal was 7.5, but rarely was a pH level above 6.8 achieved. Process control was hampered by inaccurate DO probe measurements, causing zero dissolved oxygen in the effluent.

The performance goal was to reduce the methylene chloride in the waste stream from 2000 to 5 ppm. During a 60-day performance evaluation, methylene chloride influent concentrations routinely spiked above 2000 ppm and once exceeded 4000 ppm. Given these operational problems, the effluent was maintained below 5 ppm except for four upsets. Although 5 ppm was the required effluent goal, effluent methylene chloride was between 10 and 150 ppb when analyzed with lower detection limits.

Degradation of the volatile organic dichloromethane has been reported in an aerobic fluidized-bed reactor with 99.99 percent removal (Galli, 1987). The microbial system consisted of a consortium that included the genera *Pseudomonas* and *Methylobacterium*. This study evaluated both activated carbon and sand as support media.

Activated carbon allowed a greater biomass buildup than the sand support media. Although this is usually a desired situation, for this system the biofilm thickness prevented complete oxygen transfer, resulting in anaerobic zones. The sand media gave less biomass and more consistent aerobic metabolism.

Other investigators have reported significant benefits of activated carbon in fluidized-bed reactors. The use of activated carbon as the biomass support medium yielded superior results over other media for degradation of chlorinated aliphatic hydrocarbons. Pilot studies yielded removals of 70 to 90 percent for methylene chloride (Sutton, 1988).

Aromatic hydrocarbons and ketones. Volatile organic compounds (VOCs) from automotive paint spray booths have been scrubbed into water for biological degradation in aerobic fluidized beds (Sutton, 1988). The VOC-contaminated water contained compounds of benzene, acetone, methyl ethyl ketone, cyclohexanone, and the alcohols of 1-butanol and isopropyl.

The aerobic biological reactor used quartzite sand as the biomass support medium. Removals varied from 86 to 98 percent, with the poorest removal being for benzene. Reactor design and operational parameters were:

Reactor configuration	Rectangular and fluidized
Reactor volume	3168 ft^3
Reactor depth	16.5 ft
Organic carbon loading	480 lb/day
Hydraulic retention time	4 h

BTEX compounds. A fluidized-bed reactor, operating in the aerobic metabolism mode and containing activated carbon, successfully removed BTEX from groundwater (Hickey, 1991). Oxygen was dissolved in the groundwater prior to entering the fluidized bed. Nutrients were added at a stoichiometric ratio of COD:N:P of 100:5:1. Influent temperature varied between 14 and 18°C. A hydraulic retention time of approximately 5 min and organic loading rate of BTEX compounds of 9.0 kg COD/m^3/day yielded greater than 99 percent removal with effluent concentrations below 10 ppb (Table 9.18). The removal of individual components followed the pattern of benzene > toluene > xylenes = ethylbenzene.

Suspended solids production averaged 0.7 mg/L. The mean cell residence time was estimated at 9 days and bed solids at 1.8 g/L. The system responded well to a shock loading of 10,000 ppb of BTEX. The BTEX was initially adsorbed and degraded as confirmed by oxygen consumption balance.

TABLE 9.18   Fluidized-Bed Reactor Performance on BTEX-Contaminated Groundwater*

BTEX loading rate, kg COD m³-day	Flow velocity, gal/min/ft²	HRT, min	Total BTEX concentration	
			Influent, ppb	Effluent, ppb
5	12.1	5	5420	64
3	12.1	5	3301	7
1	12.1	5	1000	<5

SOURCE: Hickey, 1991.
*Food to microorganism (F/M) ratio was estimated at 9 mg/mg-day.

Reactor for *P. chrysosporium*.   The degradation of 2-chlorophenol by the white rot fungus *Phanerochaete chrysosporium* has been used to evaluate reactor design using several immobilization techniques. Support media included ¼-in ceramic saddles, Manville celite catalyst carrier, and balsa wood chips (Lewandowski, 1990). The catalyst carrier is a cylindrical-shaped bead approximately ⅛ in in diameter and ¼ in long. The bead cylinder has an interior pore 20 μm in size. The balsa wood chips were 5 by 5 by 5 mm in size.

Reactors consisted of 2- and 4-in-diameter columns. The fungus was inoculated and allowed to develop for 5 days by operating the reactors in a recirculation mode. After 5 days for biomass buildup, an influent stream of 2-chlorophenol was continuously fed in an upflow system. Temperature was optimized at 39°C and pH at 4.0 to 4.5. Air was pumped into the bottom of each fixed-bed reactor to provide for 5 mg/L of dissolved oxygen. A glucose substrate was added to provide a primary carbon and energy source.

The microbial performance was evaluated in terms of Michaelis-Menten kinetics assuming a mass balance based on plug flow. Separate studies utilizing two bed sizes under various hydraulic loading rates yielded kinetic constants for V_{max} (maximum enzyme transformation velocity) of 45 ppm/h and 26 ppm/h and for K_m (the substrate concentration for which the reaction rate is half of V_{max}) of 132 and 133 ppm (Lewandowski, 1990). These data were collected for the catalyst carrier packing. Operational parameters are provided in Table 9.19. These Michaelis-Menten kinetic constants are the first reported for *Phanerochaete chrysosporium* in packed beds.

Biogrowth on ceramic saddles was uneven, causing fungus clumps at the top of the beds. Significantly better growth distribution was provided by the Manville celite catalyst carrier. The catalyst pores allowed growth into the porous interior of the carrier. Growth was not as uniform on wood chips as on the celite catalyst carrier.

TABLE 9.19   Operational Parameters for White Rot Fungus
Immobilized in Packed-Bed Reactors

Parameter	Reactor support media		
	Porous beads	Wood chips	Alginate beads
Flow velocity, gal/day-ft^2	3.5–15.8	3.5–15.8	1.8–18
Hydraulic retention time, h	7.4–33	4.8–21.5	6.7–67
Feed concentration, 2-chlorophenol, ppm	460	300	520

SOURCE: Lewandowski, 1990.

Chlorinated hydrocarbons—cometabolism. Design and operational information for cometabolism in fluidized-bed reactors is presently very limited. Operational information for the degradation of trichloroethylene
has been obtained from experimental facilities. Although the application of experimental data for field design must be used with caution, it
provides insights to the management of a cometabolism system.

Fluidized-bed reactors were seeded with a consortium capable of
degrading TCE with methane and/or propane as the primary substrate (Phelps, 1990). Since the primary substrate affects the degradation of the target compounds, several feeding strategies were evaluated. The degradation of trichloroethylene was performed under three
feeding regimens: continuous methane plus propane feeding, pulse
feeding, and a lengthy no-feed, or starvation, period. Head space feed
gas concentrations (volume per volume) were 5 and 3 percent for
methane and propane, respectively.

A pulse feeding mode followed by a starvation period required 50
percent as much primary substrate as the continuous-feed regime
with equal removals of TCE of 80 to 90 percent. The ratio of substrate
to TCE removed was 55 µmol/µmol and 100 µmol/µmol for pulse feeding and continuous feed, respectively. At maximum utilization efficiency the pulse feeding regimen was capable of degrading 80 mg of
TCE per day per gram of biomass. This TCE degradation was sustained for over 6 months with no detected intermediates of 1,2-di-
chloroethylene and vinyl chloride. Starvation periods of over 5 days
result in little additional TCE removal.

The addition of propane with methane was superior to methane
alone. Trichloroethylene degradation decreased by 60 percent when
methane was the only substrate. The propane utilizing organisms
were an effective participant in the degradation. The mixed microbial
consortia also provided greater process stability than those experienced with pure methanotrophic cultures.

A limiting threshold for TCE degradation has been experienced in bioreactors. For the above study this threshold was experienced at a concentration of 0.5 mg/L for both the pulse and continuous feeding regimes. This limit may be a property of the reactor, although similar thresholds have been reported for TCE degradation in trickling-filter reactors.

The microbial consortia can be very sensitive to pH changes. Efficient TCE degradation occurred at a pH of 7.2, but a pH rise to 7.5 results in a TCE degradation drop to 15 percent efficiency compared with 80 to 90 percent. A return to pH of 7.2 restored degradation rates.

A pilot fluidized bed utilizing activated carbon support media has been evaluated in an anaerobic mode for mineralization of trichloroethylene (Huang, 1988). The fluidized-bed reactor was operated as a two-stage system, two beds in series, with a portion of the second stage cycled back to the first.

The microbial consortium was seeded from an anaerobic digester and operated under methanogenic metabolism mode. The design and operational parameters for each bed reactor were:

Raw influent flow rate	1.5 gal/day
Recycle flow rate	308 gal/day
Raw to recycle ratio	1:202
Superficial velocity	9.8 gal/min-ft^2
Bed depth expanded	1.4 ft
Percent expansion	28 percent
Carbon loading	23.5 lb/ft^3
Hydraulic detention time	1 min

A high recycling flow was used to maintain a fluidized bed. The trichloroethylene (TCE) feed concentrations varied from 0.3 to 480 mg/L. Glucose was fed as a secondary substrate at a concentration of 285 mg/L. The microbial activity resulted in a pH drop through the beds of 8.35 to 7.1. Although a hydraulic retention time of 1 min provided significant dehalogenation, the pilot study illustrated the need to increase detention time.

The two-stage fluidized-bed reactor successfully mineralized 1000 µg/L of influent trichloroethylene to undetectable levels. The TCE removal in the first stage ranged from 80 to 100 percent under a variable feed concentration of 0.3 to 480 mg/L. However, the intermediate transformation products of trans-1,2-dichloroethylene and vinyl chloride were in the effluent of the first stage. These transformation products were not detected in the second-stage effluent.

Sulfate-reducing and methanogenic reactor system. Two fluidized fixed-film bioreactors were evaluated for treating hazardous compounds in land-

TABLE 9.20 Performance of Sulfate-Reducing Metabolism versus
Methanogenic Metabolism in Fluidized Fixed-Film Bioreactors

| | | Effluent | |
| | | Sulfate-reducing reactor, | Methanogenic reactor, |
Compound	Influent, μg/L	reactor, μg/L	reactor, μg/L
Acetone	10,000	189	410
Methylethyl ketone	5,000	70	220
Methyl isobutyl ketone	1,000	35	58
Trichloroethylene	400	8	5
Methylene chloride	1,200	65	46
1,1-dichloroethane	100	20	14
Chlorobenzene	1,100	67	165
Ethylbenzene	600	34	85
Toluene	8,000	436	1,102
Phenol	2,600	22	93
Nitrobenzene	500	6	8
1,2,4-trichlorobenzene	200	10	14
Dibutyl phthalate	200	26	3

SOURCE: Suidan, 1991.

fill leachate. Each bed contained activated carbon as the support medium. One operated under sulfate-reducing metabolism and one under methanogenic metabolism. The sulfate-reducing system received leachate containing 89 mg/L as SO_4. Each bioreactor operated with a 6-h hydraulic retention time (empty expanded bed). Leachate was not identical for the two systems, and direct comparison of the metabolic systems cannot be made. However, each feed was spiked to receive similar hazardous chemical loadings.

Both bioreactors performed effectively, with the sulfate-reducing metabolism mode yielding the higher rate of degradation (Table 9.20). Chloroform addition to the feed at 3.5 mg/L significantly retarded methanogenic activity but had no measurable effect on the sulfate-reducing system.

Anoxic and oxic reactor system. Two-stage fluidized beds also have been applied to process condensate from coal liquefaction by Environment Canada's Wastewater Technology Center, Burlington, Ontario (Melcer, 1988). This system utilized a quartzite sand support medium and an anoxic fluidized bed in series with an oxygenic fluidized bed (Fig. 9.23). The first stage functioned as a denitrification bed and the second stage as a nitrification bed. Typical hydraulic retention times varied from 10 to 80 h and sludge retention times ranged from 6 to 240 days. The pH of the nitrification reactor was maintained at 7.0 by addition of sodium bicarbonate.

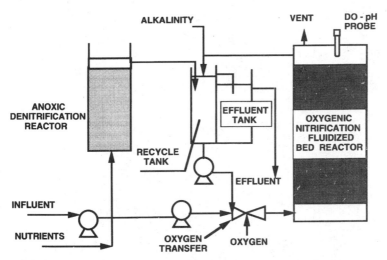

Figure 9.23 Fluidized-bed process using an anoxic reactor and oxic reactor in series.

Although periods of unstable performance were experienced in the second stage, nitrification, a consistent high degree of treatment was maintained. Removal of phthalates ranged from 84 to 97 percent. Phenolic compounds including cresol were reduced by 99+ percent. Removal of PAHs, however, yielded mixed results with little, if any, removal for some compounds.

Rotating biological contactors

Process principles.   The rotating biological contactor (RBC) consists of closely spaced circular disks that are partially submerged in the wastewater (Fig. 9.24). Disks are mounted on a shaft usually running horizontal with the tank's flow direction. These disks, usually made of polystyrene or polyvinyl chloride, provide the support media for the biomass (Fig. 9.25). Disks rotate slowly (1 to 2 r/min), alternately contacting biomass with wastewater and then with the gas phase. The rotation also creates shearing forces for removing excess solids from the disks. The sloughed solids are maintained in suspension so they can be carried from the unit to a clarifier. RBC units can be compartmentalized with baffles or separate tanks to provide for separate microbial communities (Fig. 9.26). Rotating biological reactors can provide high treatment efficiencies with lower energy inputs than suspended biomass systems, such as activated sludge reactors. Since aeration does not occur by air bubbling, foaming and floating sludge problems are essentially eliminated.

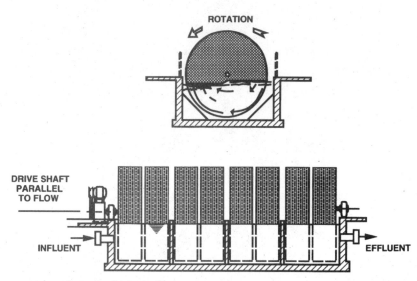

Figure 9.24   Rotating biological contactor.

Figure 9.25   Support media of a rotating biological contactor.

For aerobic metabolism the gas phase is usually the atmosphere. However, closed systems are used for developing specific microorganisms. Methanotrophic bacteria are controlled with a gas phase of oxygen and methane, and *Phanerochaete chrysosporium* with high oxygen tensions.

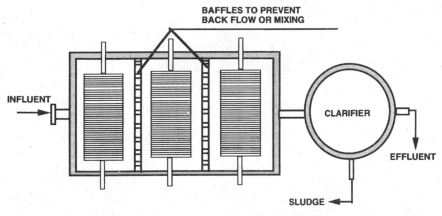

Figure 9.26 Rotating biological contactor for separate microbial communities.

Applications. Bench-scale rotating biological reactors have performed successfully on the treatment of phenolic compounds (Tokuz, 1988). Treatability studies evaluated RBC performance on a synthetic waste containing 2-chlorophenol (2-CP); 2,4-dichlorophenol (2,4-DCP); 2,4,6-trichlorophenol (2,4,6-TCP); and pentachlorophenol (PCP) at concentrations between 3 and 10 mg/L. Glucose was used as a primary substrate at a COD concentration of 480 mg/L. The hydraulic loading was 1.5 gal/ft²/day and organic loading was 6.0 lb COD/1000 ft²/day. The disks were rotated at a speed of 4 r/min. The RBC treatment resulted in a steady-state treatment efficiency of 87 percent for COD, 36 percent for 2-CP, 68 percent for 2,4-DCP, 62 percent for 2,4,6-TCP, and 53 percent for PCP. Rotating biological contactors have treated PAH-contaminated groundwater, yielding 95 percent removals as measured by COD. RBC acclimation periods of several weeks were required (Berg, 1991).

A rotating biological contactor has been modified and patented to promote the growth of *Phanerochaete chrysosporium,* the white rot fungi. The reactor modification provides for improved attachment of the fungal mycelial mass to roughened rotating disks. Fungus do not have the same means to attach to surfaces as do bacteria. The RBC is constructed as an airtight unit so the oxygen level can be maintained above atmosphere. For *P. chrysosporium,* the oxygen head space concentration should be between 20 and 100 percent with a level above 50 percent preferred (Chang, 1985). The operational parameters used for optimized degradation of mixed chlorinated aromatic compounds from bleaching of pulp and paper, for *P. chrysosporium*–seeded RBC, are provided in Table 9.21.

The RBC is inoculated with *P. chrysosporium* from a batch growth reactor. The seed material is grown under semisterile conditions in a

TABLE 9.21   Rotating Biological Contactor Operational Parameters for *Phanerochaete chrysosporium*

Disks percentage submerged	40%
Disk speed*	2 ft/min
pH level	4.0–5.0
Oxygen gas phase content	>50%
Operational temperature	40°C
Carbon and energy source:	
Glucose	
Corn syrup	
Organic sludge	
Additives:	
Nutrients, nitrogen as NH_4Cl	24–36 mg/L†
Detergent: Tween 80	0.3%

SOURCE: Chang, 1985.
*Speed at the disk periphery.
†By volume.

TABLE 9.22   Growth Medium for *Phanerochaete chrysosporium* Seed

Nutrient	Amount (mg/L)
KH_2PO_4	2,000
$MgSO_4 \cdot H_2O$	500
$CACl_2$	100
NH_4Cl	120
Thiamine	1.0
Glucose	10,000
pH	4.5

SOURCE: Chang, 1985.

stationary liquid medium containing nitrogen and trace nutrients. A proposed growth medium for *P. chrysosporium* is given in Table 9.22. The RBC is filled with growth medium and inoculated with the seed spores. A batch reactor mode is used to grow the mycelial biofilm on the disks. After two days of growth, the medium is drained and refilled with renewed growth medium. The oxygen atmosphere is maintained during all stages, and the discs rotated. However, too high a rotation speed will hinder attachment. Since the desired *P. chrysosporium* enzyme activity is expressed during a nitrogen-deficient environment, after several days of growth the fungus is allowed to deplete the nitrogen level and enter its nitrogen-deficiency stage. In this stage the wastewater is introduced to the RBC system. The rates of degradation for chlorinated aromatic compounds were increased fourfold by the addition of a detergent with nutrient minerals. However, it is likely that this response characteristic is waste-

TABLE 9.23   Organic
Compounds Degraded by *P.
chrysosporium* on Rotating
Biological Contactor

Dichlorobenzoic acid
2,4,6-Trichlorophenol
4,5-Dichloroquaiacol
6-Chlorovanillin
Trichloroquaiacol
5-Chlorovanillin
4,5,6-Trichloroquaiacol
Tetrachloroquaiacol
4,5-Dichloroveratrole
4,5,6-Trichloroveratrole
3,4-Dimethoxyacetophenone
5-Chloroveratryl alcohol
6-Chloroveratryl alcohol
Pentachlorophenol
2,4,6-Trinitrotoluene
2,4-Dinitrotoluene

SOURCE: Chang, 1985; Glaser, 1988.

specific. Compounds that have been degraded by the white rot–RBC system are listed in Table 9.23.

A pilot airtight rotating biological contactor (RBC) containing a methanotrophic fixed-film biomass was evaluated for treating chlorinated aliphatic compounds at the Kelly Air Force Base in San Antonio, Tex. (Vira, 1992). The methanotrophic biomass was developed over 3 weeks by maintaining a 20 percent methane and 10 percent oxygen concentration in the reactor head space, respectively. After biomass development, gas concentrations were reduced to 2 percent methane and 10 percent oxygen for treatment. In addition to the methanotrophic organisms a heterotrophic biomass was supported by the addition of toluene to the influent as an energy source for indigenous microorganisms. The RBC rotation speed was 1 r/min.

The pilot evaluation achieved trichloroethylene degradation to less than 1 ppb. Significant biodegradation of 1,1-dichloroethane and 1,1,1-trichloroethane also occurred. Vinyl chloride was degraded by 99 percent within 1 h of contact time. However, no perchloroethylene was biodegraded. This compound has not been demonstrated to be degraded by aerobic organisms.

Air-sparged reactors

Process principles.   Air-sparged fixed-film reactors can be divided into three types: sparged upflow submerged packed-bed reactors, downflow packed bed with in-bed sparging, and fluidized reactors with in-bed sparging. The air-sparged reactor has a large height to diameter ratio and uses a forced gas, usually air, to provide mixing. The reactor can be used in a batch mode or operated continuously with concurrent or countercurrent flow of liquid relative to the rising gas. Air-sparged fixed-film reactors are normally packed with either random or modular media. Commercial applications tend to use downflow operations for random-packing reactors.

Applications.   A rectangular submerged fixed-bed reactor with upflow cells has been developed by BioTrol, Inc. (Fig. 9.27). A full-scale system has been used for treating phenolic compounds and PAHs from a wood treating site (Chresand, 1990). The submerged reactor was packed with polyvinyl chloride tower packing to immobilize the biomass. Aeration is accomplished with a porous diffuser pipe mounted beneath the packing support grid. Hydraulic retention times varied between 4.5 and 9.0 h. Contaminated water is first adjusted in pH to 6.0 to 9.0, supplemented with nutrients, and heated, if necessary, to 80 to 90°F. Influent total phenol concentrations varied from 11 to 327 ppm and effluent concentrations varied from 0.06 to 1.11 ppm.

The BioTrol air-sparged reactor has been evaluated in the EPA Superfund Innovative Technology Evaluation (SITE) program for the degradation of pentachlorophenol (PCP) in groundwater. Using residence times of 1.8, 3, and 9 h the reactor removed 97.6, 98.7, and 99.8 percent of the influent PCP. Influent concentration varied from 27.5 to 42 mg/L (Stentinson, 1991). The bioreactor was seeded with a specific PCP-degrading bacteria, *Flavobacterium,* and indigenous micro-

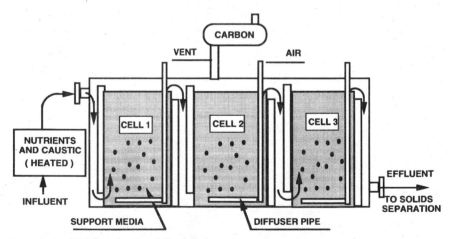

Figure 9.27 Mobile submerged fixed-bed reactor.

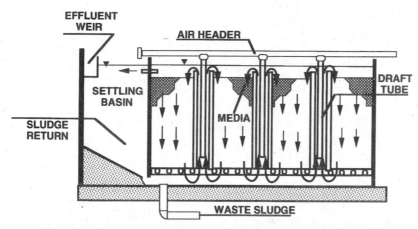

Figure 9.28   Air-sparged fixed-bed reactor using draft tubes for oxygen transfer.

organisms. After seeding, the wastewater was recycled through the system for 1 week to acclimate the biomass. Flow rates were 1, 3, and 5 gal/min. Influent was adjusted to pH 7.0 and heated to 70°F.

A submerged air-sparged fixed-bed pilot plant was evaluated for the treatment of brine groundwater contaminated with phenol at a Gulf Coast hazardous waste site (Fig. 9.28). Groundwater concentrations averaged 15,000 mg/L for dissolved solids, 1300 mg/L for total organic carbon, and 400 g/L for phenol. The reactor operated at 5000 gal/day. Aeration and mixing were done by using air-lift pumps and draft tubes vertically installed at several locations in the reactor. Acclimation required 70 days for the biomass to adapt to the high salt concentration. The system was operated at a dissolved oxygen concentration of 1 mg/L or greater and MLSS of 6000 to 7000 mg/L. A 70 percent removal of TOC was consistently obtained.

A laboratory sparged fixed-bed reactor has been evaluated for the cometabolism of chlorinated aliphatic hydrocarbons. These studies established the hydraulic retention time (HRT) for aerobic biodegradation of two chlorinated aliphatic compounds, trichloroethylene and trichloroethane (Table 9.24) (Strand, 1991). A 99 percent removal of TCE required a 17.3-h HRT (Strand, 1991). The treatment system was seeded with methanotrophs. To support the methanotrophic growth, a gas mixture of air and methane was sparged through the bed. Methane at a partial pressure of 2 percent in the gas phase yielded successful results, preventing the competitive inhibitory effects of methane. The secondary substrate, chlorinated aliphatic compounds, was monitored as COD and supplied with the primary substrate at a ratio of 4 g COD per g CH_4. Nitrogen and phosphorus were supplied at a ratio of COD:N:P of 100:5:1.

TABLE 9.24   Effect of Hydraulic
Retention Times on TCE and TCA
Removal for a Fixed-Film Reactor
Operating with Methanotrophs

Percent removal*	For TCE	For TCA
60	2.3	1.0
80	4.0	1.3
90	5.8	1.5
95	7.5	1.7
99	11.5	2.3
99.9	17.3	3.2

SOURCE: Strand, 1991.
*Based on influent concentration
of 500 µg/L, and plug flow reactor.

The application of the sparged-bed reactor was evaluated in a pilot study using groundwater spiked with trichloroethane (U.S. EPA, 1993). The reactor was seeded with *Methylosinus trichosporium,* and influent concentrations were approximately 550 µg/L of trichloroethane. Although 88 percent removal of trichloroethane was achieved under optimal conditions, the day-to-day effluent varied considerably. Only a 5- to 6-day optimal performance occurred before significant decline. This problem may be related to available enzyme and loss transformation capacity from metabolic by-products. It also can be related to a shift in the predominant microbial population in the bed. A diverse microbial group including ciliates developed.

Treatability Studies

Introduction

For many bioremediation projects, treatability studies become a required prerequisite. Few, if any, contaminated sites are identical, and identical responses are unlikely. Experience significantly aids in the design of a bioremediation project, but experience can only be applied within limits. Many variables are a function not only of the contaminated media and the contaminants, but also of the genetic variability of microbial species. Microorganisms acquire genetic material or undergo changes that may be directed by specific environmental conditions. These effects are not subtle and can determine the success or failure of a bioremediation project. They are both site-specific and microbial-specific. The same contaminant can respond differently under what appears to be identical bioremediation techniques and microorganisms.

Purpose of Treatability Studies

Treatability studies are undertaken at all levels of process sophistication. They may be simple beaker tests of a week or less to gain information on a chemical's potential for degradation. At the other end of the spectrum are pilot studies operated to develop full-scale design criteria, cost, and performance over months of operation. Finally, there are the larger-scale pilot studies. These can consist of actual applications of the proposed treatment scheme to a small portion of the site.

The first step in performing a treatability study is to establish the goals and the treatability budget. These two factors determine the specific program, and they are often at odds with each other. The goal of the treatability study establishes the scope of the experimental

TABLE 10.1   Goals of Treatability Studies

Evaluate the overall feasibility of a process or scheme of processes
Establish the attainable level of treatment
Determine design criteria for process design
Estimate capital and operating cost of proposed process scheme
Determine the control parameters and their limits for optimizing performance
Evaluate material handling techniques or equipment
Confirmation of field performance and ultimate fate of contaminants
Evaluate performance problems
Provide for continued optimization performance during field cleanup

design. Determine if the goal is to obtain basic data or to simulate all or part of the field program. Typical goals of a treatability program are illustrated in Table 10.1. Generally, one goal is to determine if the bioremediation scheme will succeed in meeting cleanup limits at a specific location. A second goal is the development of data for design and operation. Another goal may be the collection of data for process optimization and confirmation of degradation. For any treatability program, these goals are much too broad to develop successful protocols. Specific issues must be formulated and prioritized. The questions that must be answered establish the treatability protocols.

If the appropriate biological system or its ability to meet cleanup limits is not known, the first objective is an evaluation of the overall feasibility of one or more alternate biological systems. Second, these evaluations must provide information that allows one to predict the attainable treatment level. Achieving the above objectives is a minimum and should result in a decision on the appropriateness of the evaluated treatment. If attainable treatment levels are satisfactory, additional treatability studies for developing design criteria may be necessary. They are usually conducted in a second-phase study. These studies may include the establishment of loading rates, hydraulic contact times, sludge retention times, bulk densities, mixing rates, etc. They should establish the bioremediation control parameters and their operational limits. For soil and slurry-phase bioremediation, evaluations may include material handling techniques and equipment. Finally, the treatability study should provide a means to predict the cost and duration of the treatment scheme.

Facilities used for treatability studies can have an important application during field bioremediation. They can be used to confirm field performance and fate of the hazardous compounds. With field remediation activities, it is almost impossible to perform an accurate mass or materials balance. The parallel operation of a pilot treatability program in conjunction with field operations can confirm the mass of chemicals degraded, volatilized, or solubilized in discharge water. Litigation on the actual fate of chemicals has significantly impacted on the ultimate duration and cost of bioremediation projects. The second benefit of running pilot facilities during field bioremediation is the ability to evaluate optimization procedures and performance problems. These may range from simple questions on the effect of additional nutrients, minerals, or electron donor supplements, to more complex interactions.

Experimental Design

Before goals of a treatability study are established, the forces that shape the treatability program must be defined. This requires evaluation of factors that are difficult to quantify, since they are based on inputs that may be poorly defined and change over time. These considerations include:

1. The goal of the regulatory agency
2. The goal of the client
3. The hazard level or health sensitivity of the contamination
4. The environmental sensitivity of the contaminated site
5. The political sensitivity of the contaminated site
6. The degree of flexibility with time schedules
7. The degree of flexibility with project budgets
8. The engineering confidence factor

Bases of design

After addressing the forces that shape the program, establish the bases on which protocols will be formulated. These include:

1. How one is to accomplish, quantify, and document the treatability study
2. Determining if a customized treatability protocol will be developed or if standardized protocols are appropriate
3. Determining if the treatability study will be conducted under laboratory optimal conditions, or under conditions that simulate those at the site
4. Determining the level of quality control to be applied to all test protocols and analytical data
5. Determining if analytical data will be collected to provide statistically significant data and, if so, to what level of confidence

6. Determining what analytical protocols will be applied to data collection

The treatability program can be a one-phase effort or divided into two or more phases. Multiple phases have the advantage of using results from the first phase to either change or confirm the proposed experimental design of the next phase. Multiple-phase programs, however, have the disadvantage of requiring significantly longer completion times. When the cleanup is conducted under the direction of a private client, the treatability program is usually performed in one phase because of the overall efficiency in cost and time. However, if there are doubts on the feasibility of a technology, multiple-phase approaches are preferred. Often the treatability program is dictated by the regulatory program.

Treatability studies conducted under the Comprehensive Environmental Response Compensation and Liability Act (CERCLA) have three phases. The first phase is remedy screening. This may be a bench-scale laboratory study. The second phase is remedy selection and can be either bench-scale or pilot in nature. These first two treatability studies are usually conducted under the Remedy Investigation/Feasibility Study (RI/FS) program. The goal is to determine if the technology can meet the cleanup goals. The third phase is conducted during the Remedial Design/Remedial Action (RD/RA) program. This level of treatability is pilot scale. Its goal is to establish design and operating parameters for optimization of the process.

There is no typical or best approach to the scope of each phase or its comprehensiveness. Each situation is unique to the needs of the client, requirements of regulatory officials, project schedule, and cost constraints. The goal is to design the most cost-effective treatability study program, provide the essential data for successful project design and operation, and maintain necessary schedules and budget. A treatability program can be viewed as a set of objectives with specific tests designed to achieve each objective (Table 10.2).

Developing protocols

Standardized protocols and guidance documents are becoming available. Examples include:

Guide for Conducting Treatability Studies under CERCLA: Biodegradation Remedy Selection, U.S. Environmental Protection Agency, 2nd and Final Draft, March 1993.

Guide for Conducting Treatability Studies under CERCLA: Aerobic Biodegradation Remedy Screening, U.S. Environmental Protection Agency, EPA/54012-91/-13A, July 1991.

TABLE 10.2   Typical Objectives of Bioremediation Treatability Studies

1. Evaluate the capability of indigenous microorganisms to degrade the target compounds
2. Evaluate the enhancement capability of seed microorganisms
3. Evaluate the optimum range for environmental parameters:

   Moisture
   pH
   Nutrients
   Trace minerals

4. Evaluate the need and effect of supplemental substrates and electron acceptors
5. Determine the feed and starvation cycle for primary substrates
6. Evaluate the need to provide supplemental electron donors
7. Evaluate the rate of degradation for target compounds under idealized laboratory conditions or modified to represent expected field response
8. Evaluate the expected duration of the bioremediation project
9. Determine the attainable level of treatment
10. Evaluate potential soil-water reactions and clogging potential of in situ treatment
11. Evaluate the potential for toxicity changes due to mixing, surfactants, or buildup of intermediates
12. Evaluate the degree of volatilization
13. Determine the cost-effectiveness of various optimization measures
14. Evaluate the monitoring frequency for process control
15. Evaluate the operational limits on process control parameters without significant decrease in performance

Standard protocols are designed to meet certain minimum standards throughout the country. However, standardization removes the flexibility to modify the protocol for site-specific conditions. Standardized protocols cannot be used to evaluate treatment problems encountered during the field program. More importantly, standard protocols limit the ability to evaluate the more sophisticated bioremediation processes.

Specific objectives are delineated to evaluate and successfully design a bioremediation project. These objectives are then supported by the development of detailed protocols. Most treatability studies require the design of a customized protocol. The protocols are designed to assure that the appropriate questions can receive an interpretable answer. A treatability study is only as good as its protocols. Protocols are stated in a step-by-step procedure, leaving nothing to another's interpretation. An example protocol to answer the question "Can indigenous microorganisms degrade a contaminant?" is presented at the end of this chapter.

Controls.   All treatability evaluations, regardless of their simplicity, require controls. The use of controls cannot be overemphasized and should never be omitted. Controls are used to establish the base with which the parameter under evaluation is to be compared. Without controls, one can never be certain of the meaning of any test results. Except for the variable being evaluated, controls are handled in identical manner to the test units. For example, a control to evaluate nonbiological conditions will contain the same contaminated media and all supplements and will be containerized, stored, mixed, incubated, and sampled identical to the test unit. The only difference is an additive to block or prevent microbial activity. Typical additives for this purpose are sodium azide or mercuric chloride. If the test is to evaluate a nutrient supplement, a control for evaluating the effect of nutrient supplements would be set up identical to the test unit except for no nutrient supplement. Microbial response and target compound removal for test units are then compared with the control.

Indigenous microorganisms. The capability of indigenous microorganisms to degrade a target compound can be determined with simple laboratory procedures. The first step is the collection of adequate soil or water samples. Tests are usually conducted using actual site soils or liquids. At a minimum, the contaminated media should be analyzed for the target compounds, microbial count, pH, and soil moisture content. The pH and moisture are optimized for either bacteria or fungi. Establishing nutrient levels is not necessary at this stage of the evaluation. The contaminated media are supplemented with all potential nutrients. Typical growth supplements are provided in Table 10.3. Either supplement is adequate, although supplement B provides more trace nutrients. Soil and groundwater typically contain these trace nutrients in adequate concentrations. The contaminated medium is incubated at temperatures typical for optimized degradation (20 to 30°C). Microbial growth and target compound concentrations are followed over intervals of the test duration.

Seed microorganisms. The use of seed microorganisms is important for some compounds. Evaluating the benefit of microbial seeding is conducted under test conditions essentially identical to the indigenous microbial studies. Appropriate controls will include the nonseeded test for establishing a baseline. Control tests would consist of a control with no biological activity, a control with indigenous organisms, and a control with indigenous organisms plus any added substances that the microbial seed is combined with. This may consist of wood chips or a liquid growth medium in which the seed is carried, since seed materials can contain significant organic matter. The seed organism may be inactivated by autoclaving or other techniques suitable to the solution. Data must distinguish between microbial response to the target compound and added organic matter. The term "essentially identical" is used above since the seed organism may have specific environmental requirements, pH, primary substrate, etc., that must be accommodated by the test protocol. This will, by

TABLE 10.3 Typical Growth
Supplements

Growth Supplement A*	
22.5 mg	$MgSO_4 \cdot 7H_2O$
27.5mg	$CaCl_2$
250 mg	$FeCl_3 \cdot 6H_2O$
25 mg	$NaMoO_4 \cdot 2H_2O$

Growth Supplement B	
35 mg	NH_4Cl
15 mg	KNO_3
75 mg	$K_2HPO_4 \cdot 3H_2O$
25 mg	$NaH_2HPO_4 \cdot H_2O$
10 mg	KCl
20 mg	$MgSO_4$
1.0 mg	$FeSO_4 \cdot 7H_2O$
5.0 mg	$CaCl_2$
0.05 mg	$ZnCl_2$
0.5 mg	$MnCl_2 \cdot 4H_2O$
0.05 mg	$CuCl_2$
0.001 mg	$CoCl_2$
0.001 mg	H_3BO_3
0.0004 mg	MoO_3

For an aqueous medium the above
is per liter.
*All buffer at pH 7.1 with 5 mM
potassium phosphate buffer.

necessity, provide for some differences between the seeded and
indigenous tests.

The techniques of augmenting soil or a bioreactor with specialized
microorganisms are dependent on the nature of the organism. The
procedures differ for aerobic bacteria, anaerobic bacteria, and fungi.
Each procedure is based on the growth needs and environmental
requirements of the organism. Inocula can be obtained from groundwater, sludge, or soil demonstrating effective degradation, from commercial suppliers, and from culture collection centers and research
laboratories. Many research centers have isolated organisms with
specialized capabilities and maintain these cultures in their facilities.

Organisms must be grown on an appropriate substrate before inoculation in the field. Bacteria are usually grown in liquid cultures.
They may be settled or centrifuged to provide high concentrations for
transportation. This is usually performed in the laboratory. For the
wood rot fungi, a nutrient fortified grain-sawdust mixture has been
used (Lamar, 1993). Wood chips provide an excellent physical and
substrate growth support for many fungi. The fungal inocula can be
stored and transported to the field in bags containing a microporous

filter to allow air exchange. The inocula are typically refrigerated during transportation, but not frozen.

The fugal inocula, nutrients, and wood chips are then tilled into the soil. Mixed ratios have consisted of 2.5 percent of inocula and wood chips to soil on a dry weight basis. Old wood chips may contain excessive indigenous fungal growth. If such growth is undesirable, it can be significantly reduced by fumigating the wood chips with methyl bromide prior to inoculation with the desired fungus. For fungal growth, moisture content should be approximately 20 percent weight basis. Aeration is provided by mechanical mixing such as tilling or with forced aeration.

Process optimization.   Protocols are developed to evaluate the effectiveness of primary substrates, supplemental nutrients, electron acceptors, and their mode of delivery. This testing focuses on the optimization of the treatment process. The ideal approach for evaluating the effects of environmental parameters, including nutrients, electron acceptors, and supplemental substrates or electron donors, is to hold all variables constant except the variable under evaluation. For example, a series of tests evaluating the effect of moisture is evaluated by holding all parameters constant and varying moisture in each of the test units. As many as 10 test units at different moisture levels may be evaluated, and these may be run in duplicates or more. This approach is necessary for research efforts. However, treatability studies seldom have the luxury of time or budget for this number of evaluations. Furthermore, previous information as presented in Chaps. 3 to 5 will usually provide a basic range for optimized performance. Thus treatability tests are performed with only partial and frequently mixed parameter variations to confirm expected optimized performance. One can fine-tune these variables during the actual field bioremediation activity.

In treatability studies, many chemicals have been used as the source of nutrients, minerals, and electron acceptors or donors. Typical additives are provided in Table 10.4. The importance of adding nutrients to field bioremediation projects isn't clear. Field results yield mixed conclusions, and the benefits demonstrated by nutrient addition in laboratory studies frequently are not realized during field performance. There are several possible explanations for this. The two most likely are a result of other factors that reduce the biodegradation rates and the potential for nutrient recycling. If other factors limit the rate of degradation, rate of oxygen delivery being the most likely, the benefits of nutrient addition may not be realized. A reduced rate of degradation provides for greater nutrient recycling than that experienced in laboratory studies. As a rule, don't add nutrients for in situ treatment unless it is definitely limiting the rate

TABLE 10.4 **Typical Additives for Treatability Studies**

Nutrients	Minerals	Electron acceptors	Electron donors
KH_2PO_4	$MgSO_4$	$MgSO_4$	Methanol
K_2HPO_4	$CaCl_2$	$MnSO_4$	Ferric citrate
KNO_3	$CuSO_4$	$CuSO_4$	Butyrate
$MgSO_4$	$FeCl_3$	$Zn(NO_3)_2$	Acetate
NH_4Cl	MoO_3	Acetate	Glucose sucrose
$(NH_4)_2SO_4$	$MnSO_4$	Hydrogen	Toluene
$Zn(NO_3)_2$	$NaMoO_4$	Carbon dioxide	
Agricultural fertilizers			

pH control	To achieve oxygen depletion	To block microbial activity
Lime	Thioglycollate	Sodium azide
NaOH	Cysteine hydrochloride ascorbic acid	Mercuric chloride
$NaHCO_3$		
Phosphoric acid		
HCl		
H_2SO_4		

of degradation and the reduced rate is considered significant (see Chap. 7).

Attainable treatment level. Probably the most important questions of any bioremediation system are what level of treatment can be attained and what is the expected duration of the project. These two questions can be evaluated only after a microbial system is selected and the optimized environmental conditions are established. Both questions can be addressed in the same test protocol. The rate of treatment is established by measuring the decrease in target compounds with time, and the test program is operated until the target compound is nondetectable or there is no further change in concentration.

The rate of degradation achieved in laboratory treatability studies is seldom achieved in field activities. This is particularly true for solid-phase, slurry-phase, and in situ bioremediation. Laboratory studies frequently demonstrate first- or second-order reaction rates, whereas many solid-phase and in situ bioremediation field activities have been reported as zero-order with respect to the contaminant. Field degradation rates for soil- and slurry-phase and in situ bioremediation are significantly less for several reasons. Common limiting factors are the heterogeneity of the media, inadequate mixing, desorption, and inadequate feed rate of a key reactant. This usually results from an oxygen-limiting reaction rate. Liquid-phase bioreac-

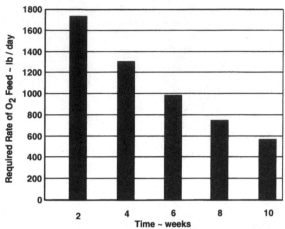

Assuming 100,000 pounds of ultimate oxygen demand and the mean
95 % confidence level for the degradation rate of naphthalene of
0.02 / day (Sherman, 1988)

Figure 10.1 Required rates of oxygen feed to optimize the
degradation of naphthalene in soil.

tors can be modeled with greater accuracy, since their process kinetics are better defined and easier to duplicate in the laboratory.

In many field situations it is impossible to deliver the rate of oxygen or other electron acceptor adequately and distribute it uniformly. It also may be impractical to design an oxygen delivery system to satisfy the demand for the initial rates of degradation. Oxygen demand rates decrease as bioremediation progresses, resulting in an overdesigned system. For example, the oxygen demand rates for oxidizing many hydrocarbons as represented by first-order kinetics drops to less than 50 percent within several months of a bioremediation program (Fig. 10.1).

If the attainable level is inadequate, then one should evaluate the potential reasons for lack of complete degradation. These reasons can include many factors related to the microbial-enzyme reaction system. These are discussed in detail in Chaps. 3 to 5. They are a function of the specific microbial system and reactor configuration. Some potential problems and considerations for their solution are summarized in Table 10.5.

Methodologies and Costs

Methodologies vary and reflect the budget and time schedule placed on the project. Most treatability studies are custom-designed. Designs are based on the specific needs and project scope. The testing protocols, number of control tests, analytical procedures, and level of quality con-

TABLE 10.5 Potential Reasons for Failure to Attain an Adequate Level of Biological Degradation

1. pH or moisture out of microbial desired range
   Solution:
   - Adjust pH and moisture as necessary
   - Reseeding with appropriate microorganisms may be necessary if a seed was initially used

2. Depletion of the enzyme inducer
   Solution:
   - Add a primary or analog substrate

3. Buildup of toxic intermediates
   Solution:
   - Provide environmental changes to support a greater variety of microbial species or use a reactor design that hydraulically removes these intermediates
   - Modify the microbial system to a group that can handle the toxic intermediates

4. Primary substrate inhibits the degradation of the target compound
   Solution:
   - Decrease field concentration of primary substrate
   - Use pulse feeding of primary substrate
   - Provide for separation of biomass growth and target compound degradation through reactor design

5. Incomplete anaerobic dehalogenation of compounds with mixed levels of halogenation
   Solution:
   - Check for electron acceptor inhibition by NO_3^- or SO_4^-
   - For compounds having a low degree of halogenation change to an aerobic mode with emphasis on methane, butane, propane, and phenol-degrading bacteria
   - Check for inadequate levels of electron donors. Evaluate various additives such as butanol, acetate, phenol, toluene

6. Partial or incomplete transformation of target compounds
   Solution:
   - Check for adequacy of sludge retention time
   - Check for short circuit in the bioreactor
   - Check for pH drift within the bioreactor as a result of the biodegradation
   - Check for incomplete mixing of chemical additives
   - Check for moisture drifting between additions for soil-phase systems
   - Check for inadequate transformation capacity
   - Reseeding with appropriate microorganisms may be necessary

7. Significant levels of adsorbed-phase target compounds
   Solution:
   - Utilize slurry-phase reactors
   - Evaluate surfactants and mixing levels for rendering the target compound more available

trol are usually a function of a project's environmental importance from potential health and ecological impacts. The overall cost of a project definitely influences the treatability budget. Budgeting $100,000 to $300,000 on treatability studies can save millions on a multimillion-dollar cleanup. However, little treatability can be supported on a $50,000 cleanup. The client and regulatory agencies are usually the driving force behind specific methodologies. However, the design engineer must never forget his or her purpose, achieving a successful cleanup without risking health or ecological impacts or exposing the client or employer to unneeded litigation. Never agree on a program that is compromised because of insufficient data. However, collect the data that are necessary for design and operation without overkill.

Costs for a treatability program can range from $10,000 to several hundred thousands of dollars. For large projects, $200,000 invested in a well-designed treatability study can save millions in final cleanup costs. Twenty million on field pilot studies reduced the cleanup cost for the French Limited Superfund site in Crosby, Tex., from over $125 million to less than $60 million.

A cost-effective treatability program will have flexibility and coordination between members of the treatability team. This team consists of those applying the treatment protocols, those evaluating the data, those performing the analytical analysis, those responsible for design of the protocols, and the project manager. Rigid programs that set a protocol to be performed with no provision for continued data evaluation and interaction between team members usually result in incomplete answers. The net result is inadequate data that compromise the field design, the need to repeat the test, or a failed treatment response because of unrecognized solutions. It isn't uncommon for a knowledgeable treatability team to request additional analysis to evaluate unexpected treatment responses. Unexpected treatment responses are the norm. If this wasn't the case, there would probably not be a need for the treatability study. Allowances for additional analytical cost and troubleshooting should be planned as a contingency in any treatability budget.

A significant portion of treatability costs is associated with analytical costs. Analytical expertise must be included in the design of the treatment protocols. The limitations of specific analytical methods must be evaluated in reference to a protocol's goal. During the treatability study, support from a chemical and biological laboratory is an important part of the program. The need for good analytical data on a routine basis is clearly recognized for process control. Analytical measurements also are required for emergency response to unexpected process performance. Results are frequently needed within 24 h to recognize problems and take corrective solutions.

Analytical cost usually represents 50 percent of the total treatability budget. The frequency of monitoring and level of quality assurance influence cost, but the most significant is the nature of the analysis. This ranges from simple dissolved oxygen measurements to GC/MS analysis. Monitoring costs can be controlled by recognizing when indicator parameters are adequate versus the measurement of specific compounds or organisms.

A significantly lower number of samples are required for a given confidence level when 95 percent of the target compounds have been removed versus only 50 percent. A high level of confidence on the data is necessary only near the end of the treatment program. Analytical analysis may include intermediate breakdown products. The importance of monitoring intermediates rests with knowledge of the transformation products of the parent compound. If more toxic components or inhibiting intermediates are generated, these should be monitored in any treatability and field installation. Monitoring intermediates also offers excellent proof that biodegradation is the removal mechanism.

Equipment

Treatability equipment varies from simple flasks to designed pilot facilities for scale-up (Table 10.6). Systems can use sealed reactors for prevention of volatilization and performing mass balances. Typical systems for solid-phase studies include soil pans, beakers, buckets or tubs, and enclosed reactors for establishing mass balances to small field pilots. An enclosed reactor for soil piles and composting studies that allows one to determine mass balances is illustrated in Fig. 10.2. The rate and composition of all gas and liquid discharges can be measured. Soil pans are usually used in laboratory studies. They may range in size from 1 to 100 lb of soil. Nutrients and the electron acceptor, usually oxygen, are added by mixing or tilling the soil several times a week. The frequency is a function of the oxygen uptake rate. Field plots provide more realistic information on potential field performance. These range in size from 1 or more to several hundred square yards.

Slurry-phase systems have ranged from beakers, serum bottles, buckets, 55-gal drums, 40,000-gal tanks to sectioning off a field lagoon. Liquid-phase treatability studies have included reaction flasks, respiration apparatuses, and pilot-scaled contact beds and suspended reactors (Figs. 10.3 to 10.5). In situ studies range from laboratory columns of several feet in length to the isolation of in situ field columns. Field pilot studies for in situ bioremediation can use cylinders to section off the soil under evaluation. A column of soil can be isolated by driving a cylinder into the soil sediment or sand. Large

TABLE 10.6 Equipment and Materials

Parameter	Soil columns	Field plots	Slurry reactors	Contained soil systems
Test systems	Lab and field cylinders	In-ground barriers Above-ground berms	Lab reactors Small tanks (lab and field)	Lined and bermed area in the field
Contaminant sampling	Small coring device	Split-spoon Shelby tube	Bailer sample port	Split-spoon Shelby tube
Moisture control	Sprinkler Upflow percolation Watering can	Sprinkler Subsurface irrigation Plastic covers	N/A	Sprinkler Watering can Plastic covers
Temperature measurement	Temperature probe Soil thermometer	Temperature probes	Thermometer Temperature probe	Temperature probe Soil thermometer
Nutrient addition (agricultural chemicals or other chemicals)	Pumps and sprinklers for dissolved nutrients	Shovel, rake, etc. Tractor Spreader or sprayer Sprinkler and irrigation system for dissolved nutrients	Metering pump Mix tank Hand addition	Trowel, shovel, rake, etc. Tractor Sprinkler and irrigation
Oxygen addition (for aerobic studies)	Forced aeration Oxygenated water injection system H_2O_2 injection system H_2O_2	Tractor and disk garden tiller Bioventing and forced aeration	Floating aerators Diffusers, H_2O_2 Injection	Trowel, hand tool Forced aeration
pH control	pH probe for soil dissolved in water Acid Base	pH probe for soil dissolved in water Lime Phosphoric acid	pH probe Acid Base Lime	pH probe for soil dissolved in water Acid Base Lime

SOURCE: Modified from EPA Final Draft, "Guide for Conducting Treatability Studies under CERCLA: Biodegradation Remedy Selections," March 1993.

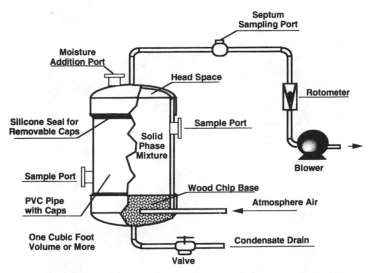

Figure 10.2 Schematic of pilot reactor for solid phase and compost biore-mediation treatability studies.

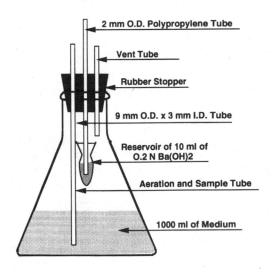

Figure 10.3 Reaction vessel for aerobic aquatic biodegradation shake-flask test standard protocol 40 CFR 796.310.

areas can be isolated with interlock sheet piling. For large projects a small portion of the site can be operated as a pilot program.

Protocols

Standardized protocols have not and probably will not be developed for all the potential media and metabolism modes. Media can be

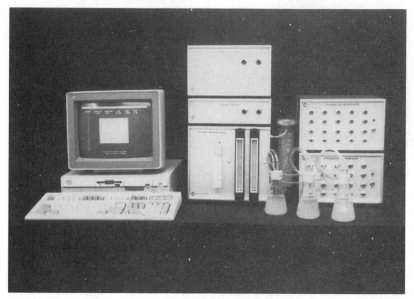

Figure 10.4 Computerized oxygen–carbon dioxide respirometer for bioremediation studies. (*Courtesy of Columbus Instruments International Corporation, Columbus, Ohio.*)

grouped as solids, slurries, sludges, liquids, and gases. The nature of the contacting bioreactor can vary significantly, as can the metabolism mode and specific microbial organism. Since each variation can influence the treatment protocol, standardized protocols are usually limited to testing the initial feasibility of biological degradation.

The U.S. EPA has adopted several standard protocols to examine biodegradability. These protocols can be found in the Code of Federal Regulations, Title 40 (40 CFR). A procedure for evaluating liquid aerobic biodegradation is set forth in 40 CFR 796.3100. This test measures the degree of mineralization by monitoring the amount of carbon dioxide produced. The contaminated water sample is inoculated with soil or sewage bacteria and incubated in a flask (Fig. 10.3). The bacteria are incubated in the dark in a CO_2-free atmosphere. An alkaline solution of $Ba(OH)_2$ is used to adsorb generated carbon dioxide. Samples are analyzed for dissolved organic carbon, and the $Ba(OH)_2$ from the reservoir is titrated to measure the amount of CO_2 evolved. A control flask without the contaminants is run with the test. The degree of mineralization is estimated from the difference between the extent of organic carbon disappearance and the formation of CO_2 in experimental flasks compared with the control.

A soil protocol is described in 40 CFR 796.3400 called "Inherent Biodegradability in Soil." This procedure also can be modified to

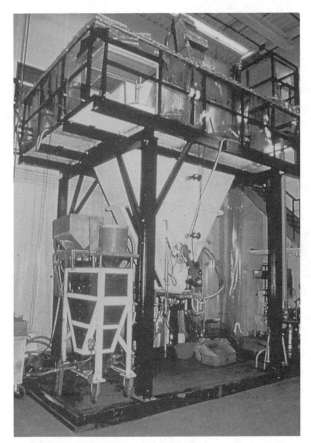

Figure 10.5   Pilot slurry-phase bioreactor using dual treatment through turbulent separation.

accommodate liquids and slurries. The protocol requires the use of radio-labeled ^{14}C target compounds. Obtaining radiological label compounds is costly and sometimes difficult. Analytical monitoring is also expensive. These factors place practical limitations on the usefulness of this protocol for most bioremediation projects. The protocol calls for the use of a standardized soil chosen to represent the type of soil present at a given site. This is a significant deficiency for real world response. The radio-labeled compounds are mixed into these standardized soils. Compounds are mineralized, yielding radio-labeled carbon dioxide that is trapped in an alkaline solution (Fig. 10.6). The amount of radioactive carbon in the alkaline solution is measured by liquid scintillation counting.

Procedures for aerobic and anaerobic microorganisms differ significantly. Anaerobic treatability studies and culture techniques require

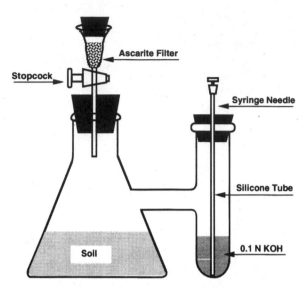

Figure 10.6 Reaction vessel for biodegradability in soil. (*standard protocol 40 CFR 796.3400.*)

atmospheres that are absent of oxygen. They typically use special gas mixtures, electron donors, and acceptors. Most culturing and material transfer operations are conducted in glove boxes. Microbial plating techniques are presented under microbial activity in Chap. 6. Treatability studies are conducted in sealed chambers designed to maintain anaerobic environments. They include glass desiccators or specialized jars (Mcintosh and Fildes).

Treatment containers and glove boxes are first purged of oxygen with a mixed gas of nitrogen, hydrogen, and carbon dioxide. A ratio of $N_a{:}H_2{:}O_2$ of 85:10:5 is typical. The transfer of all agents must be performed in an O_2-free glove box if the strict anaerobic organisms are to survive. Oxygen yields biologically produced hydrogen peroxide which is highly toxic to the strict anaerobes. Anaerobic metabolism will occur only under low redox potentials of -150 to -350 mV depending on the electron acceptor. The use of reducing agents such as thioglycollate or cysteine hydrochloride ascorbic acid may be required in conjunction with the oxygen-free environment. The redox can be monitored easily by the addition of a color indicator such as resazurin.

For bench-scale anaerobic studies, serum bottles have found acceptance as reaction vessels. A serum bottle protocol for anaerobic soils or sediment is available in 40 CFR 795.54, "Anaerobic Microbiological Transformation Rate Data for Chemicals in the Subsurface Environment." Serum bottles are small glass containers (150 mL) that can be completely sealed and sampled by a syringe. Each serum bottle is capped with a Teflon-lined septum and an aluminum seal. The anaerobic procedure uses 20 percent (weight per volume) soil and sediment slurries and strict anaerobic techniques. Methanogenic, sulfate-reducing, and denitrifying conditions can be investigated with this technique. The decrease in target compound concentration is followed by sacrificing a serum bottle for each data point. Gas samples also can be collected from the head space to monitor methane production.

A major limitation with the above procedures is that they use small batch reactors with a short time duration. This limits the usefulness of the test results. For example, reactors do not yield meaningful data on the amount of biomass versus contact time to achieve cleanup limits.

For suspended-growth reactors biomass is monitored by measuring mixed liquor volatile suspended solids (MLVSS). This is a gravimetric measurement of the amount of suspended and combustible organic matter. Microbial mass is first filtered from the sample, dried, weighed, and then combusted and reweighed. It is assumed that all combusted organic matter is biomass. This technique is good for bioreactor systems, since biomass makes up the majority of the organic matter. However, most small-scale batch treatability protocols, such as the standardized flask test, do not accumulate the biomass quantity for measuring weight differences. For these systems, microscope counting is frequently used. Microscopic counting and plate count enumerations provide only an indication of increases or decreases in total or specific microbial populations. They do not provide exact counts but only relative data to the control or previous measurement.

A much larger number of microorganisms exist in a reactor or on contaminated soil than can be cultured on growth media. An example of the task involved is illustrated by one site contaminated with aviation fuel. Approximately 60 aerobic microorganisms, including more than 40 bacteria, 6 actinomycetes, and 10 fungi, were identified (Swindell, 1988). These 100+ organisms may represent only a fraction of actual organisms. Only 1 to 10 percent of the microbes in soil can be detected by typical laboratory culture techniques (Swindell, 1988). Second, the organisms counted may not be the active degraders, whereas changes in biomass weight respond to degradation of the energy source.

Most standard procedures cannot be used to generate design criteria or aid in process optimization. This requires the use of pilot systems that are based on design principles, either batch or continuous flow. They require operation over extended time with more monitoring and attention to process variables.

The design principles of liquid-phase biological reactors are well developed. This includes both suspended and fixed-film reactors. Mathematical models based on reactor kinetics and mass transfer theory are adequately developed. One can design pilot studies for liquid-phase reactors with a high degree of confidence when moving to full-scale facilities. Design variables include such parameters as:

- Sludge age or solids retention time
- Food to microorganism ratio
- Hydraulic detention time
- Recycle ratio

Solid- and slurry-phase pilot studies are more difficult to scale. Limitations include the heterogeneity of soil and sludges, the inability to evaluate the application of materials handling equipment on small volumes, and the inability to distribute nutrients and other additives with the degree of efficiency achieved in small pilot facilities. Solid-phase treatability results are dependent on soil size distribution and the size distribution selected for the study. Treatability studies on soils with a higher percentage of fines yield slower degradation rates than coarser soils for identical compounds and conditions (Pedersen, 1992), whereas slurry reactors yield results that are independent of the soil size distribution.

Small-scale solid-phase studies yield unrealistic rates of degradation. Field systems cannot employ the mixing energy per unit weight of material as achieved on small systems. Thus field activities do not result in the homogeneity, the rate of mass transfer, and the control of moisture and pH typical of pilot studies. For solid phase, pilot field studies using field plots of several square yards are recommended over bench-scale studies.

Typical pilot soil plots have been constructed of wood and galvanized steel. Sizes range from test units of 1 to several square yards by 2 to 4 ft in depth. Multiple soil test units are used for duplication and controls. The soil plots are usually isolated from surface soil using polyethylene tarpaulin or similar liners. The plots may contain a leachate collection system that is overlaid with sand. The soil to be studied may be screened before application in the plots. Soil depths are typically 1 to 2 ft. Each plot is usually covered with polyethylene to prevent excessive moisture changes. Tensiometers can be used to monitor moisture level. Each plot is tilled at a designated frequency with garden tilling equipment.

Slurry reactors can be used to evaluate the ability to degrade compounds by solid-phase applications. Since slurry reactors optimize degradation, they should not be used for obtaining the rate of degradation. Slurry reactors optimize transfer rates and contact between the microorganisms and waste constituents. It is also inappropriate to use slurry reactors to evaluate microbial systems that prefer solid-phase growth over liquid-phase. Fungi are important degraders in solid-phase systems, including soil piles, composting, and some land farming activities. Many fungi cannot compete in slurry-phase reactors since they favor bacteria.

Typical slurry treatability reactors consist of glass reactors of about 5- to 55-gal drums. Each is equipped with baffles and mechanical mixing. Mixing should not be performed with air sparging unless this will be the field design. The type of aeration system significantly affects volatilization (see Chap. 8). For most slurry treatment applications, volatilization must be minimized. Second, air alone will not provide necessary energy input for field mixing of slurries. Pilot slurry reactors are usually loaded with 20 to 30 percent solids by weight, and some aeration equipment will not function well under this solids loading.

Pilot studies for in situ treatment are the most difficult to design. It is impossible to model the heterogeneity of the subsurface because of the hydraulic, chemical, and biological variations. For large projects, it is advisable to use a small portion of a site as a field pilot study after laboratory evaluations have been successfully completed.

Developing a protocol for in situ bioremediation requires attention to the chemical, physical, biological, and hydraulic characteristics of the aquifer. First, a comprehensive site evaluation is necessary to understand the subsurface. Laboratory treatability should attempt to use and duplicate subsurface conditions as closely as possible. The base growth medium should approximate the characteristics of the contaminated groundwater. It is best to use actual groundwater with necessary testing supplements.

Liquid-phase degradation tests do not represent the microbial subsurface conditions. If liquid biodegradation testing is desired, it is important to include sediment. By using a slurry of actual soil the microbial species may be more realistic. The major part of subsurface microbial biomass is attached to groundwater sediments (approximately 2×10^7 cells per gram) and only a minor part suspended in groundwater (approximately 2×10^6 cells per mL). The ratio of biomass on sediment to biomass in groundwater yields ratios of 20:1 and 67:1 (Holm, 1992). Sediment-attached biomass is not uniformly distributed. The sediment-attached biomass is found primarily in the fine-particle fraction. Approximately 67 percent of all bacteria in groundwater sediment is attached to the silt and clay fraction, though this sediment fraction may account for a small fraction of the total sediment by weight.

An important evaluation for in situ bioremediation is the potential for permeability decreases in the subsurface (see Chap. 7). The complexities of soil, water, and biochemical reactions make it impossible to predict the degree or even potential for clogging. An in situ treatability study must examine permeability changes. Undisturbed soil samples as collected from the contaminated formation are used for this evaluation. If significantly different soil groups or redox conditions exist at the site, each should be evaluated. The number of core samples is a function of the heterogeneity of the site as well as the redundancy desired for quality control. Actual groundwater with all anticipated supplements as based on slurry testing should be pumped through the undisturbed cores. If hydrogen peroxide or other electron acceptors are used, their depletion rate per soil length should be established.

The column sizing must allow appropriate development of mass transfer zones. For low hydraulic conductivity these zones are small, but with the higher-conductivity sands and gravels, they can be of several feet. The mass transfer zone height is a function of flow rate, the diffusion coefficients, and sorption affinities of the supplements. A column length of several feet should be used for most in situ studies. This can be composed of smaller columns connected in series. This column length should provide for the development of transfer zones so the effect of chemical variations across these zones can be evaluated for clogging potential. The column diameter should be at least 10 times the medium soil particle diameter. The column width is important to prevent column wall effects. A clear plastic column is preferred for visual observations. An upflow system has been found to prevent unsaturated flows that may occur from downflow channeling.

The column should contain ports taped at intervals for measurement of pressure drop. The change in pressure is recorded over time to evaluate the effect of the injection solution on permeability. Visual observations are recorded and if necessary clogging deposits analyzed for chemical and biological composition. Column testing should be conducted for a minimum of 4 weeks for evaluating clogging before making any changes in feed solution. Short-term tests will not provide an adequate time for assessing precipitation problems or the growth of fouling bacteria such as iron bacteria. Precipitation reactions are not immediate. For example, the kinetics of calcium phosphate precipitation suggest that a precursor phase is transformed upon aging before a stable precipitate is formed (Aggarwal, 1991).

Although most in situ studies use soil disturbed as little as possible, biological degradation can be evaluated in homogenized soil. Homogenized soil will optimize the delivery of nutrients and obtain more uniformity in degradation rates. These soils should be compacted to yield hydraulic conductivity similar to that of the undisturbed soil. If cleanup limits cannot be achieved under these conditions, in situ bioremediation is probably not feasible.

All pilot programs require monitoring for process control and treatment response. Process control parameters must be monitored at an adequate frequency. Monitoring frequency is a function of the reaction rates and the stability of the biochemical system. Some parameters may be monitored several times a day and others at weekly intervals. This stability is influenced by the water and soil chemistry and the amount of biomass. For example, pH change may be weakly or strongly buffered. The greater the biomass or sludge retention time the greater the stability to shock loads. However, fixed biofilm systems usually contain significantly higher biomass levels than liquid-phase systems and processes can rapidly change from aerobic to anaerobic. For high-organic-content solids, such as sewage supplemented soil composting, this change can occur within 15 min or less.

For most treatment systems, a monitoring frequency of 2 to 3 times a day is adequate for all process parameters except for pH and oxygen. Automation for these parameters is recommended. Monitoring frequency for solid-phase and in situ systems is difficult to judge until the treatability study is underway. Oxygen transfer is significantly lower and frequently limits the rate of degradation. In situ response is influenced by the injection rate and groundwater flow rates. Monitoring should be initiated at short intervals and extended as based on data results.

The key parameter to track in treatability studies is the target compound or compounds. The best method is to measure the specific compounds of concern. This will usually require gas chromatography, gas chromatography and mass spectrometry, or high-performance liquid chromatography. Direct compound quantification reduces ambiguous results and aids troubleshooting during a treatability study. Finally, direct quantification is frequently the only acceptable measurement of the regulatory agency. The most cost-effective monitoring is usually a combination of procedures that uses indirect methods coupled with direct measurement of the parent compounds and specific degradation products. Indirect monitoring can be applied at a higher frequency with periodic direct monitoring checks. Knowledge about the biochemistry of the metabolic pathways as discussed in Chaps. 4 and 5 will aid in establishing appropriate monitoring methods. Indirect methods for measuring total target compounds include biochemical oxygen demand (BOD), chemical oxygen demand (COD), dissolved organic carbon (DOC), and total organic carbon (TOC), modified oil and grease analyses, and the various total petroleum hydrocarbon procedures. Monitoring of degradation gases of CO_2 and CH_4 is a good indicator when coupled with mass balances of oxygen consumed and stoichiometric yields.

Example Protocol

Test protocols should contain a brief but precise statement on the objective of the protocol. The tasks to be accomplished should be stated and then the procedures necessary to perform these tasks. These procedures must be detail (cookbook) steps that include the preparation of solutions and each operation of the experiment. The requirements for analytical analysis, quality control, and data handling need to be specified. In brief, any person should be able to use the protocol and completely duplicate the experiment. An example is provided below.

Test protocol for evaluating the capability of indigenous microorganisms

A. Objectives

1. To evaluate the ability of indigenous microorganisms to degrade nonvolatile contaminants in soil samples. This protocol is to study the removal of contaminants under ideal conditions by indigenous microorganisms. It is not the intent to provide realistic field conditions since there is a basic question on the ability of indigenous organisms to degrade the target compounds in question.
2. To provide ideal degradation conditions the soil samples will be treated in a slurry reaction vessel. This provides optimized dispersion of nutrients and promotes solubilization and bioavailability of the target chemicals.

B. Tasks

1. Prepare standard solutions for the treatability evaluation.
2. Obtain aseptic soil samples and optimize conditions for degradation.
3. Set up five reaction vessels and conduct the test for a duration to be determined by the project manager. Test duration is based on several factors that will be delineated during the testing program, such as rate of degradation, level attained, and/or no further removal.
4. Collect samples at designated intervals during the test.
5. Provide analysis according to analytical methods specified below.
6. Submit weekly data results and a final report that includes all data and records.
7. Submit quality control data for analytical methods as specified in the analytical methods procedures.

C. Procedure—preparation of stock solutions

1. Use water of the highest quality, glass still, containing less than 0.01 mg/L copper, free of chlorine and chloramines, and free of organic material.

2. Prepare individual stock solutions of nutrient supplement by dissolving the given chemicals in separate 1-L containers.
   a. Dissolve 22.5 g $MgSO_4 \cdot 7H_2O$ in water and dilute to 1 L. Label as $MgSO_4$ stock solution, date and initial.
   b. Dissolve 27.5 g anhydrous $CaCl_2$ in water and dilute to 1 L. Label as $CaCl_2$ stock solution, date and initial.
   c. Dissolve 0.25 g $FeCl_3 \cdot 6H_2O$ in water and dilute to 1 L. Label as $FeCl_3$ stock solution, date and initial.
   d. Dissolve 25 gm $NaMoO_4 \cdot 2H_2O$ in water and dilute to 1 L. Label as $NaMoO_4$ stock solution, date and initial.
3. Prepare a potassium phosphate buffer solution at pH 7.2 as follows:
   a. Dissolve 8.5 g KH_2PO_4, 21.75 g K_2HPO_4, 33.4 g $Na_2HPO_4 \cdot 7H_2O$, and 1.7 g NH_4Cl in 500 mL of water and dilute to 1 L. Label as pH 7.2 buffer, date and initial.
   b. Check the solution for a pH of 7.2 with a calibrated pH meter.
4. Autoclave all solutions to provide aseptic stocks to prevent the inoculation of nonindigenous organisms.
5. Prepare a 1N solution of NaOH and HCl. Utilize standard concentrated stocks (commercially available) and follow dilution directions.
6. Prepare a 1.5-mole solution of sodium azide (NaN_3). Dissolve 1 g of NaN_3 in 500 mL of water and dilute to 1 L. Label as stock NaN_3 solution, date and initial.

D. Procedure—treatability study

1. Prepare five slurry reactors for evaluating the degradation of the target compound or compounds. Three reactors will be used as test reactors and two as the control.
2. Use 2-L vessels for reactors; 2-L flasks can be used and closed with cotton plugs. Sterilize clean flasks in an autoclave or hot air oven. Using aseptic procedures fill each reactor with 500 mL of sterilized distilled water.
3. To each reactor add 1 mL of the stock solution of:

   $MgSO_4 \cdot 7H_2O$

   $CaCl_2$

   $FeCl_3 \cdot 6H_2O$

   $NaMoO_4 \cdot 2H_2O$

   Phosphate buffer

4. Mix contents and to each reactor add 500 g of aseptic collected soil from the contaminated site.
5. Label three reactors as test 1, test 2, and test 3. Label the fourth and fifth reactors as control 1 and control 2.

6. Take an initial sample of (volume)[1] from each reactor for time zero characterization. Label each sample and all others according to the following code:
Reactor Number—Date—Time—Parameter to Be Analyzed
The first four sample labels will read as follows:

Test 1-01/01/94-0:00—Parameter[2]

Test 2-01/01/94-0:00—Parameter[2]

Test 3-01/01/94-0:00—Parameter[2]

Control 1-01/01/94-0:00—Parameter[2]

Control 2-01/01/94-0:00—Parameter[2]

One sample is to be collected for each parameter of:

- Measurement of target compound (specify the compounds or method)
- Microbial count
- pH
- Dissolved oxygen

[1]NOTE: The volume to be sampled is dependent on the nature of the target compound and analytical procedures. The protocol must provide the volume to be collected for each parameter. Analytical requirements may determine the reactor size.

[2]NOTE: Specify the parameter or parameters to be analyzed.

7. To block biological activity in the control reactors add 10 mL of stock sodium azide solution to yield 15 mM. This solution is added to reactors labeled control 1 and control 2 only.

8. Measure the pH of each zero time sample immediately and if necessary (dependent on the pH and buffer capacity of the soil sample) adjust pH with the stock solution of caustic or acid to pH 7.0 to 7.2.

To adjust pH, perform a titration on the sample collected for pH measurement with either acid or base. Calculate the amount needed to lower or raise the pH to 7.0 to 7.2, correcting for sample volume versus the reactor volume. Add that amount of acid or base to each reactor while mixing. After addition and mixing for at least 10 min, measure pH again and record value.

9. Measure the initial dissolved oxygen and if necessary aerate to saturation, 9 to 10 mg/L.

10. Place each reactor in an incubator shaker or other appropriate device for mixing and temperature control. Incubate while mixing at 20 to 30°C (temperature depends on the protocol design).

11. Collect samples as designated below. After collection of samples, submit for *immediate* analysis of:

Dissolved oxygen

pH

Microbial count

Target compound

Preserve with (Specify the preservative. This depends on the nature of the target compound), or refrigerate at 4°C. Refrigerated samples must be analyzed within 24 h.

12. Collect samples at the following frequency and label accordingly:

1 day (24 h)[3]	Reactor Number—Date—24:00[3]—Parameter
4 days (96 h)	Reactor Number—Date—96:00—Parameter
8 days (192 h)	Reactor Number—Date—192:00—Parameter
14 days (336 h)	Reactor Number—Date—336:00—Parameter
21 days (504 h)	Reactor Number—Date—504:00—Parameter
28 days (672 h)	Reactor Number—Date—672:00—Parameter
42 days (1008 h)	Reactor Number—Date—1008:00—Parameter

NOTE: Use actual hours and minutes if different from designation. Total test duration to be established by project manager.

E. Analytical methods

As part of this protocol, the required analytical methods must be clearly referenced by a standard number or name or provided in detail.

F. Quality control

As part of this protocol, the minimum acceptable quality control must be designated.

G. Data recording and reporting

As part of this protocol, provide the data reporting requirements. Indicate the degree of detail including submission of laboratory log records.

Definition of Hazardous Waste

Many chemical compounds are toxic and can be considered as a hazard. A hazardous waste, however, is defined by regulations. The regulations for hazardous waste have been promulgated by the U.S. EPA. This is an ever-changing definition, and at present EPA has proposed the addition of over 300 chemicals and chemical categories. A chemical may or may not be a hazardous waste since the definition depends on factors other than the chemical structure.

The classification of a waste as hazardous is sometimes complex. For most wastes, however, the following will serve as a guide for the Resource Conservation and Recovery Act (RCRA) hazardous and solid waste definitions.

An RCRA hazardous waste is identified as follows:

- A solid waste is a hazardous waste if it is not excluded from regulation as a hazardous waste and meets one of the following conditions:

  - It is listed in Subpart D of 40 CFR Part 261.

  - It exhibits a characteristic defined in Subpart C of 40 CFR Part 261.

  - It is a mixture containing a listed hazardous waste and a non-hazardous solid waste.

  - It is derived from storage, treatment, or disposal of a hazardous waste.

- Materials that do not satisfy any of the above conditions are not hazardous, regardless of how dangerous they actually are.

The RCRA hazardous waste is so classified:

  - Because it is listed.

  - Because it has the defined characteristics.

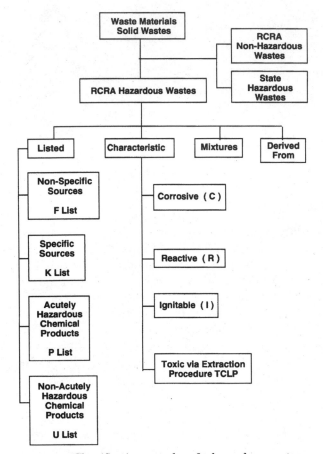

Figure A.1 Classification procedure for hazardous waste.

- Because it is derived from a designated source.
- Because of its mixture.

Figure A.1 provides an overall diagram of the classification approach.

A solid waste is defined as follows:

- A solid waste is any material that is discarded or disposed. A solid waste does not need to be solid; it may be semisolid, liquid, or even a contained gaseous material.
- A material is discarded when it is:
    - Abandoned (e.g., disposed of, burned or incinerated, or treated before or in lieu of being abandoned).

- Recycled (in a manner constituting disposal, burned for energy recovery, reclaimed, or accumulated speculatively).
- Considered inherently waste-like (like F020-F023, F026, and F028).

- EPA has developed product-like and waste-like distinctions for determining which recycled materials are wastes and which are raw materials.

There are exceptions and exclusions to the definition of solid waste as provided in 40 CFR 261.2. These exceptions are:

- Materials used or reused as ingredients in an industrial process to make a product, provided the materials are not being reclaimed.
- Materials used or reused as effective substitutes for commercial products.
- Materials returned to the original process from which they are generated, without first being reclaimed.
- Domestic sewage and any mixture of domestic sewage and other wastes that pass through a sewer system to a public operated treatment works (POTW).
- Industrial wastewater discharges regulated under the Clean Water Act.
- Irrigation return flow.
- Nuclear materials or by-products as defined by the Atomic Energy Act of 1954.
- Mining materials that are not removed from the ground during the extraction process.
- Pulping liquors reclaimed in a pulping liquor recovery furnace and reused in the pulping process (unless speculatively accumulated).
- Spent sulfuric acid used to produce a virgin sulfuric acid (unless speculatively accumulated).
- Secondary materials that are reclaimed and returned to the original process in which they were generated and are being reused in the process.

Listings of Hazardous Waste

Some chemicals are listed by name as a hazardous waste. Any solid waste is listed as a hazardous waste if it is named on one of the following lists:

- Commercial chemical products, P and U wastes [40 CFR 261.33(e) and (f)].

- Specific source wastes, K wastes (40 CFR 261.32).

- Nonspecific source wastes, F wastes (40 CFR 261.31).

The commercial chemical product listings (P and U wastes) only cover products that are generically identified. Benzene is so listed, and discarded benzene would be classified as a hazardous waste. The actual manner in which the chemical was used affects the classification. When dealing with mixtures of discarded chemicals, the definition may not be clear. Experts in environmental law are frequently needed to review potential interpretations.

The K wastes are from specific sources that are listed as hazardous wastes. Regulations 40 CFR 261.32 identify waste discharges from specific industries. Again, many of these decisions on the classification of a hazardous waste are difficult to interpret. There also are the wastes from nonspecific sources, called the F wastes. The following wastes as generated by industry are listed:

- Spent solvents

- Electroplating wastes

- Dioxin-containing wastes

Finally, EPA has identified four characteristics of a hazardous waste (Subpart C of 40 CFR 261). If the waste exhibits any of these characteristics, it is classified as a hazardous waste. These are:

- Ignitability

- Corrosivity

- Reactivity

- Toxicity.

Each is defined as follows:

Ignitability

- A liquid material is ignitable if it has a flash point lower than 140°F.

- Solid materials and contained gaseous materials also may be ignitable, but different standards apply to such materials:

  - Is solid material capable under standard temperature and pressure of causing fire through friction, adsorption of moisture, or spontaneous chemical changes and, when ignited, burns so vigorously and persistently that it creates a hazard?

- Is the substance an ignitable compressed gas as defined in 40 CFR 173.300?
- Is the substance an oxidizer as defined in 49 CFR 173.151?

Corrosivity

- A liquid material is corrosive if any one of the following criteria is satisfied:

  - It has a pH less than or equal to 2.0.
  - It has a pH greater than or equal to 12.5.
  - It is capable of corroding steel at a rate of more than $\frac{1}{4}$ in/year at a temperature of 55°C.

- Solid and gaseous materials are never considered corrosive.

Reactivity

- A material is reactive if:

  - It is explosive.
  - It reacts violently with water.
  - It generates toxic gas when exposed to water or other liquids that are moderately acidic or alkaline.

Toxicity

As determined by a defined laboratory extraction procedure, if one or more compounds are extracted at a concentration above a specified limit, the material is a hazardous waste because of its characteristics.

Mixture and "Derived from" Rules

Any mixture of a listed hazardous waste with a solid waste is defined as a hazardous waste. Any solid waste derived from the treatment, storage, or disposal of a listed hazardous waste is defined as a hazardous waste.

Hazardous Substances List and Contaminant Group Codes

Compounds	Code
Organic	
Halogenated volatiles	A
Halogenated semivolatiles	B
Nonhalogenated volatiles	C
Nonhalogenated semivolatiles	D
Organic pesticides and herbicides	E
Dioxins and furans	F
PCBs	G
Polynuclear aromatics (PNAs)	H
Solvents	I
Benzene-toluene-ethylbenzene-xylene (BTEX)	J
Organic cyanide	K
Organic corrosives	L
Inorganic	
Heavy metals	M
Nonmetallic toxic elements (As, F)	N
Radioactive metals	O
Asbestos	P
Inorganic cyanides	Q
Inorganic corrosives	R
Miscellaneous	
Explosives and propellants	S
Organometallic pesticides and herbicides	T
Organometallic compounds	U

Hazardous Substances List and Contaminant Group Codes (Continued)

Organic Compounds

CAS No.	Compound	Code	CAS No.	Compound	Code
83329	Acenaphthene	D,H	133062	Capstan	B
208968	Acenaphthylene	D,H	63252	Carbaryl	E
75070	Acetaldehyde	C	1563662	Carbofuran	E,F
67641	Acetone	C,I	71150	Carbon disulfide	C
75058	Acetonitrile	C,K	56235	Carbon tetra-	A
98862	Acetophenone	D		chloride	
591082	Acetyl-2-thiourea,1	D	78196	Carbophenothione	E
107028	Acrolein	C	75876	Chloral	A
79061	Acrylamide	D	57749	Chlordane	E
79107	Acrylic acid	C,L	106478	Chloroaniline, p	B
107131	Acrylonitrile	C	108907	Chlorobenzene	A
124049	Adipic acid	L	67663	Chloroform	A
116063	Aldicarb	E	74873	Chloromethane	A
309002	Aldrin	E	107302	Chloromethyl	A
107186	Allyl alcohol	E		methyl ether	
62533	Aniline	D,I,L	106898	Chloromethyl-	E
120127	Anthracene	D,H		oxirane, 2-	
1912249	Atrasine	E	91587	Chloronaphtha-	B
2642719	Azinphos-ethyl	E		lene, 2	
86500	Azinphos-methyl	E	95578	Chlorophenol, 2-	B
151564	Aziridine	C	59507	Chloro-3-methyl-	B
71432	Benzene	C,I,J		phenol, 4-	
98884	Benzene carbonyl	B	2921882	Chloropyrifos	E
	chloride		218019	Chrysene	D,H
92875	Benzidine	D	56724	Coumaphos	E
205992	Benzofluoranthene,	H	8021394	Creosote	H
	3,4		108394	Cresol, m-	D
65850	Benzoic acid	D,L	106445	Cresol, p-	D
100470	Benzonitrile	A,C,I	98828	Cumene	C,I
95169	Benzothiazole, 1,2	D,I	21725462	Cyanazine	E
50328	Benzo(a)pyrene	D,H	110827	Cyclohexane	C,I
206440	Benzo(j,k)fluorene	H	108941	Cyclohexanone	C
207089	Benzo(k)fluor-	D,H	72548	DDD	E
	anthene		72559	DDE	E
56553	Benzo(a)anthra-	D,H	50293	DDT	E
	cene		78488	DEF	C,E
100447	Benzyl chloride	A	333415	Diazinon	E
117817	Bis(2-ethylhexyl)	D	132649	Dibenzofuran	D
	phthalate		53703	Dibenz(a,h)anthra-	D,H
111911	Bis(2-chloro-ethoxy)	B		cene	
	methane		124481	Dibromochloromethane	A
111444	Bis(2-chloro-ethyl)-	B	106934	Dibromoethane, 1,2-	A
	ether		96128	Dibromo-3-chloro-	A
542881	Bis(chloromethyl)-	B		propane, 1,2-	
	ether		1918009	Dicamba	E
75274	Bromodichloro	A	95501	Dichlorobenzene, 1,2-	B
	methane		541731	Dichlorobenzene, 1,3-	B
74964	Bromomethane	A	106467	Dichlorobenzene, 1,4-	B
1689845	Bromoxynil	E	91941	Dichlorobenzidine, 3,3-	B
106990	Butadiene, 1,3	C	75718	Dichlorodifluoro-	A
71363	Butanol	C		methane	
85687	Butylbenzyl	D	75343	Dichloroethane, 1,1-	A
	phthalate		107062	Dichloroethane, 1,2-	A
94826	Butyric acid, 4-	C,L			
	(2,4-dichlorop)				

Hazardous Substances List and Contaminant Group Codes (Continued)

Organic Compounds

CAS No.	Compound	Code	CAS No.	Compound	Code
75354	Dichloroethene, 1,1-	A	122145	Fenitrothion	E
156592	Dichloro-ethylene, *cis*-1,2	A	86737	Fluorene	D,H
			50000	Formaldehyde	C
156605	Dichloroethylene, *trans*-1,2-	A	64186	Formic acid	L
			110009	Furan	F
120832	Dichlorophenol, 2,4-	B	98011	Furfural	I,C
94757	Dichlorophenoxy acetic acid, 2-	L	765344	Glycidylaldehyde	C
			76448	Heptachlor	E
78875	Dichloropropane, 1,2-	A	1024573	Heptachlor epoxide	E
542756	Dichloropropene, 1,3-	A	118741	Hexachlorobenzene	B
62737	Dichlorvos	E	87683	Hexachlorobutadiene	B
115322	Dicofol	E	60873	Hexachlorocyclo-hexane, alpha-	E
60571	Dieldrin	E			
84662	Diethyl phthalate	D	60873	Hexachlorocyclo-hexane, beta-	E
111466	Diethylene glycol	D,I			
1660942	Diisopropylmethyl phosphonate	D	60873	Hexachlorocyclo-hexane, delta-	E
60515	Dimethoate	E	77474	Hexachlorocyclo-pentadiene	B
119904	Dimethoxy-benzidine, 3,3-	D			
			67721	Hexachloroethane	B
105679	Dimethyl phenol, 2,4-	D	70304	Hexachlorophene	B
13113	Dimethyl phthalate	D	110543	Hexane	C,I
77781	Dimethyl sulfate	C	1689834	Ioxynil	E
99650	Dinitrobenzene, 1,3-	D	78831	Isobutanol	C
51285	Dinitrophenol, 2,4-	D	78591	Isophorone	D
121142	Dinitrotoluene, 2,4-	D	143500	Kepone	E
606202	Dinitrotoluene, 2,6-	D	58899	Lindane	E
88857	Dinoseb	E	121755	Malathion	C,E
123911	Dioxane, 1,4-	C	108316	Maleic anhydride	E
78342	Dioxathion	E	123331	Maleic hydrazide	E
122667	Diphenylhydra-zine, 1,2-	D,H	126987	Methacrylonitrile	C
			67561	Methanol	C
85007	Diquat	E	16752775	Methomyl	E
298044	Disulfoton	C,E	72435	Methoxychlor	E
330541	Diuron	E	79221	Methyl chloro-carbonate	L
84742	Di-*n*-butyl phthalate	D			
117840	Di-*n*-octyl phthalate	D	78933	Methyl ethyl ketone	C
115297	Endosulfan	E	108101	Methyl isobutyl ketone	C,I
959988	Endosulfan	I	80626	Methyl methacrylate	C
33212659	Endosulfan II	E	101144	Methylene bis (2-chloroaniline)	B
1031078	Endosulfan sulfate	E			
145733	Endothall	E	75092	Methylene chloride	A
77208	Endrin	E	23855	Mirex	E
7421934	Endrin aldehyde	E	91203	Naphthalene	D,H
563122	Ethion	E	100016	Nitroaniline, *p*-	D
141786	Ethyl acetate	C	98953	Nitrobenzene	D
100414	Ethyl benzene	C,J	100027	Nitrophenol, 4-	D
75003	Ethyl chloride	A,I	1116547	Nitrosodiethano-lamine, *n*-	D
60297	Ethyl ether	C			
107211	Ethyl glycol	I	55185	Nitrosodiethyl-amine, *n*-	D
110805	Ethylene glycol monoethyl ether	C,I			
759944	Ethylpropylthio-carbamate, S-	E	62759	Nitrosodimethyl-amine, *n*-	D

Hazardous Substances List and Contaminant Group Codes (Continued)

Organic Compounds

CAS No.	Compound	Code	CAS No.	Compound	Code
86306	Nitrosodiphenyl-amine, *n-*	D	7934	Tetrachloroethane, 1,1,2,2	A
930552	Nitrosopyrrolidine, *n-*	D	127184	Tetrachloroethene	E
924163	Nitroso-di-*n*-butyl-amine, *n-*	D	58902	Tetrachlorophenol, 2,3,4,6	F,I
615532	Nitroso-di-*n*-methyl-urethane, *n-*	D	3689245	Tetraethyldihiopyro-phosphate	E
99990	Nitrotoluene,4-	D	109999	Tetrahydrofuran	C,J
56382	Parathion, ethyl	G	137268	Thiram	D
298000	Parathion, methyl	B	108883	Toluene	E
1336363	PCBs	B	584849	Toluene diisocyanate	E
608935	Pentachlorobenzene	B	8001352	Toxaphene	A
76017	Pentachloroethane	B	93721	TP, 2,4,5-	B
82688	Pentachloronitro-benzene	D,H	75252	Tribromomethane	A
87865	Pentachlorophenol	D	120821	Trichlorobenzene, 1,2,4-	A
85018	Phenanthrene	D	71556	Trichloroethane, 1,1,1-	A
108952	Phenol	E			
139662	Phenyl sulfide	C,E	79005	Trichloroethane, 1,1,2-	A
62384	Phenylmercuric acetate	E	79016	Trichloroethylene	B
298022	Phorate	E	75694	Trichlorofluoro-methane	B
75445	Phosgene	E			
13171216	Phosphamidion	D,E	933788	Trichlorophenol, 2,3,5-	B
7803512	Phosphine	D	95954	Trichlorophenol, 2,4,5-	B
85449	Phthalic anhydride	D,H	88062	Trichlorophenol, 2,4,6-	L
23950585	Pronamide	C,I	609198	Trichlorophenol, 3,4,5-	A,I
129000	Pyrene	D,H	93765	Trichlorophen-oxyacetic acid, 2-	E
110861	Pyridine	D			
91225	Quinoline	E	933788	Trichloro-1,2,2-trifluoroethane	B
108463	Resorcinol	E,H			
299843	Ronnel	C	27323417	Triethanolamine	C
57249	Strychnine	F	126727	Tris(2,3-dibromo-propyl)phosphate	A
100425	Styrene	B			
1746016	TCDD	A,E,I	108054	Vinyl acetate	E
95943	Tetrachlorobenzene, 1,2,4,5	A	75014	Vinyl chloride	C,J
			81812	Warfarin	C,J
630206	Tetrachloroethane, 1,1,1,2	A	108383	Xylene, *m-*	C,J
			95476	Xylene, *o-*	C,J
			106423	Xylene, *p-*	C,J

Inorganic Compounds

CAS No.	Compound	Code	CAS No.	Compound	Code
7429905	Aluminum	M	7778543	Calcium hypochlorite	M
20859738	Aluminum phosphide	M	1333820	Chromic acid	M,R
7440360	Antimony	M	7440473	Chromium	M
7440382	Arsenic	M		Chromium (III)	M
1327533	Arsenic trioxide	M		Chromium (IV)	M
1303339	Arsenic trisulfide	M	7440484	Cobalt	M
7440393	Barium	M	7440508	Copper	M
542621	Barium cyanide	M,Q	544923	Copper cyanide	M,Q
7440417	Beryllium	M	7720787	Ferrous sulfate	M
7440439	Cadmium	M	7439896	Iron	M
13765190	Calcium chromate	M	7439921	Lead	M

Hazardous Substances List and Contaminant Group Codes (Continued)

Inorganic Compounds

CAS No.	Compound	Code	CAS No.	Compound	Code
7439965	Manganese	M	7681494	Sodium fluoride	M
7439976	Mercury	M	1310732	Sodium hydroxide	M,R
7440020	Nickel	M	1314325	Thallic oxide	M
7718549	Nickel chloride	M	7440280	Thallium	M
10102440	Nitrogen dioxide	R	563688	Thallium acetate	M
7789006	Potassium chromate	M	6533739	Thallium carbonate	M
151508	Potassium cyanide	M,Q	7791120	Thallium chloride	M
506616	Potassium silver cyanide	M,Q	10102451	Thallium nitrate	M
			12039520	Thallium selenide	M
7783008	Selenious acid	M,R	7446186	Thallium (I) sulfate	M
7782492	Selenium	M	7440291	Thorium	M
7440224	Silver	M	1314621	Vanadium pentoxide	M
506649	Silver cyanide	M,Q	7440666	Zinc	M
7440235	Sodium	M	557211	Zinc cyanide	M,Q
26628228	Sodium azide	M	1314847	Zinc phosphide	M
7775113	Sodium chromate	M	7733020	Zinc sulfate	M
143339	Sodium cyanide	M,Q			

Explosives and Propellants

CAS No.	Compound	Code	CAS No.	Compound	Code
7664417	Ammonia	S	302012	Hydrazine	S
131748	Ammonium picrate	S	55630	Nitroglycerin	S
7773060	Ammonium sulfamate	S	99990	Nitrotoluene, 4-	S
460195	Cyanogen	S	26628228	Sodium azide	M,S
2691410	Cyclotetramethyl-enetetranitramine	S	99354	Trinitrobenzene, 1,3,5-	S
			118967	Trinitrotoluene	S

Organometallic Compounds

CAS No.	Compound
630104	Selenourea
78002	Tetraethyl lead

Calculating Process Requirements

The design of a bioremediation project requires estimates on the quantity of materials that must be delivered to a bioreactor or to the subsurface for in situ treatment. These calculations form the basis for sizing process facilities, including piping, pumps, emissions control, chemical storage, and the cost of a project. The total quantity and rates of delivery are projected for such materials as:

Electron acceptor

Electron donors

Primary substrates

pH control

Nutrient supplement

Estimating the material needs for bioremediation is based on the same principles used for all chemical reactions. The ratio of reactants is dictated by the stoichiometric equation. For bioremediation this requires a balanced redox equation.

Mass Requirements for Electron Acceptor and Nutrients

Given the amount of contaminant per mass of soil or volume of groundwater, the total quantity of reactants can be calculated from a balance reaction. Although design estimates are frequently based on typical reactant ratios, there are advantages to writing a balance reaction for generating these ratios. First, estimates can be made on supplement needs for which rule-of-thumb ratios are unavailable. Second, it is frequently desirable to estimate the amount of a by-product, such as CO_2, during bioventing. Third, the reactant ratios vary with the metabolic mode selected and the nature of the nutrients. Finally, the ability to develop the stoichiometric equation provides insight on discrepancies between treatability studies and field response. Discrepancies may result from significant natural oxygen demand, an underestimation of the original organic load, or lower utilization rates that can result from poor distribution of supplements.

Estimating process requirements involves several steps. First, field data provide the type and concentration of contaminants and nutrients available. The second step is to convert field data to a mass loading of contaminants. This is typically done by using concentration contours as illustrated in Figs. 6.1 and 6.6. Any contaminant plume varies significantly in concentration. As a result, a plume must be subdivided into zones for which mean concentrations are estimated to calculate mass contaminant loads per zone. The extensiveness of this calculation is related to the accuracy desired and limited by the adequacy of the field data. The delivery of chemicals for in situ bioremediation is not uniform throughout a site but is based on the localized mass load within selected zones. For example, nutrient addition may not be necessary for a significant portion of a plume but may be vital for the heart of the contamination area. A large plume, such as the 3000-ft-long plume illustrated in Fig. 6.1, can be divided into three treatment zones.

Once the contaminant mass is estimated, the third step is to develop the ratio of electron acceptor and nutrients for bioremediation. This requires the writing of the balanced redox equation. Since the redox reactions consist of two parts, an oxidation reaction and a reduction reaction, the electron flow must be balanced. Free electrons cannot exist in solution. No more electrons can be donated than those received. After the electron flow is balanced, the two half reactions can be combined for a mass balance.

Balancing redox reactions

The technique for coupling and balancing two halves of a redox reaction is illustrated by the simple inorganic reaction frequently dealt with in wastewater treatment. The oxidation of nitrite to nitrate is reviewed below.

The first step is to identify the principal reactants and the species that are undergoing oxidation and reduction, that is, donating or receiving an electron. The second step is to write the balanced half reactions for the oxidation of nitrate. These half reactions are provided in many handbooks. The oxidation half reaction is given as follows:

$$NO_2^- + 2\,OH^- \rightarrow NO_3^- + H_2O + 2e$$

For this half reaction, two electrons are generated when nitrite is oxidized to nitrate. These electrons are the result of nitrogen going from a +3 state to a +5 state. The third step is to write the reduction half reaction and identify the electron acceptor. Since this is an aerobic respiration reaction, the electron acceptor must be molecular oxygen. The half reaction for the reduction of oxygen is

$$O_2 + H_2O + 4e \rightarrow 4\ OH^-$$

Molecular oxygen is reduced through the acceptance of four electrons. These two half reactions are individually balanced, but these cannot be combined until the electron flow is balanced. The oxidation of nitrate is generating two electrons and the reduction of oxygen is requiring four electrons. The next step is to balance the electron flow by multiplying the oxidation reaction components by 2. This multiplication step results in four electrons generated, which matches the four to be accepted. The two half reactions can now be summed, yielding

$$2\ NO_3^- + O_2 + 4\ OH^- + 2\ H_2O \rightarrow 2\ NO_3^- + 2\ H_2O + 4\ OH^-$$

Note that all species on the left side of the equal sign are summed for both half reactions, as are all components on the right side. This equation can be simplified by canceling like species on each side of the equation, resulting in

$$2\ NO_2^- + O_2 \rightarrow 2\ NO_3^-$$

and dividing by 2, yields the final equation

$$NO_2^- + 1/2\ O_2 \rightarrow NO_3^-$$

Each molecule of nitrite requires one-half molecule of oxygen for this redox reaction to proceed.

The development of stoichiometric equations for the breakdown of organic compounds uses the same principles, but they are not as straightforward as above. The total reaction must include the organic species being oxidized, the electron donor being reduced, and the major nutrients for cell growth. There are four common electron acceptors and two common forms of nitrogen for nutrients. To make the writing of the stoichiometric equation easy, a generalized approach for organic biodegradation with each potential electron acceptor and two nutrient sources has been developed (McCarty, 1987).

The overall stoichiometric equation is a result of the summation of half reactions for the organic substrate, the selected electron acceptor, and the biomass synthesis reaction. Table B.1 provides the half reactions for several electron acceptors and two cell synthesis equations.

TABLE B.1   Generic Half Reactions for Organic Redox Reactions

---

Half Reaction of Electron Donor ~ H_D

$$1/Z\,(C_aH_bO_cN_d) + (2a - c/Z\,(H_2O)) = a/Z(CO_2) + d/Z(NH_3) + H^+ + e^-$$

where $Z = 4a + b - 2c - 3d$: a, b, c, and d represent the average number of atoms for C, H, O, and N, respectively, in the organic contaminant

---

Half Reaction of Electron Acceptor ~ H_A

Aerobic: When oxygen is the electron acceptor:

$$1/4\,O_2 + H^+ + e^- = 1/2\,H_2O$$

Anaerobic: When nitrate is the electron acceptor:

$$1/6\,NO_3^- + H^+ + 5/6e^- = 1/12\,N_2 + 1/2\,H_2O$$

When sulfate is the electron acceptor:

$$1/8\,SO_4^{-2} + H^+ + e^- = S^{-2} + 1/2\,H_2O$$

When carbon dioxide is the electron acceptor:

$$1/8\,CO_2 + H^+ + e^- = 1/8\,CH_4 + 1/4\,H_2O$$

Cell Synthesis Equation ~ C_S

When ammonia is the nitrogen source:

$$1/4\,CO_2 + 1/20\,NH_3 + H^+ + e^- = 1/20\,C_5H_7O_2N + 2/5\,H_2O$$

When nitrate is the nitrogen source:

$$5/28\,CO_2 + 1/28\,NO_3^- + 29/28\,H^+ + e^- = 1/28\,C_5H_7O_2N + 11/28\,H_2O$$

NOTE: An approximate composition of cellular structure is $C_5H_7O_2N$. The phosphorus need for microbial growth is approximately one-sixth of that for nitrogen
SOURCE: McCarty, 1987.

Ammonia is the nitrogen source for one cell synthesis equation and nitrate is the source for the second. The overall reaction can be given in general terms by

$$H_D + f_eH_A + f_sC_s \tag{B.1}$$

where H_D = half reaction for the organic compound oxidation, electron donor
H_A = half reaction for the electron acceptor
C_s = reaction that provides nutrient requirements for biomass synthesis

The cell synthesis reaction equates the nutrient demand to the amount of biomass that will be produced. During the degradation of the organic compound, a portion of that energy yields cellular growth. A factor is included in the reaction for distribution of this energy between biomass synthesis and other needs. These factors are represented by

$$f_e = \text{fraction of organic oxidized for energy}$$

$$f_s = \text{fraction associated with conversion to microbial cells}$$

where

$$f_e + f_s = 1 \qquad\qquad \text{(B.2)}$$

For aerobic systems, the f_s factor for energy distribution is found to range between 0.12 and 0.6 (Table B.2). The slower the reaction (the harder the compound is to degrade) the smaller the value of f_s. The energy fraction going to cell synthesis for anaerobic systems is much lower than for aerobic systems. Thus the amount of biomass produced is significantly less.

Example problem

Estimate the electron acceptor and nutrient needs for bioremediation of soil contaminated with 7000 mg/kg of hydrocarbons having concentrations given below.

n-Pentane	C_5H_{12}	1200 mg/kg
Xylene	C_8H_{10}	800 mg/kg
Benzene	C_6H_6	980 mg/kg
Toluene	C_7H_8	400 mg/kg
2,2-Dimethylheptane	C_9H_{20}	830 mg/kg
Benzene 1,2-diol	$C_6H_6O_2$	1470 mg/kg
Butyric acid	$C_4H_8O_2$	1320 mg/kg
		7000 mg/kg

Step 1. Since most contaminated sites are a result of a mixture of organic compounds, it is easier to write one redox reaction for the

TABLE B.2   Factors for Energy Distribution for Biomass Synthesis in Redox Reactions

Electron acceptor	f_s values for Eq. (B.1)
O_2	0.12–0.60 (mean 0.5)
NO_3	0.1–0.5
SO_4	0.04–0.2
CO_2	0.04–0.2

SOURCE: McCarty, 1987.

TABLE B.3   Estimating the Average Chemical Structure of Mixed Contaminants

Chemical	Mass proportion by atomic weight*			Fraction of total mass as based on concentration†	Total molecular atomic weight contribution‡		
	C (1)	H (2)	O (3)	(4)	C (5)	H (6)	O (7)
C_5H_{12}	60	12	0	1.2/7	10.3	2.1	0
C_8H_{10}	96	10	0	0.8/7	11.0	1.1	0
C_6H_6	72	6	0	0.98/7	10.1	0.8	0
C_7H_8	84	8	0	0.4/7	4.8	0.5	0
C_9H_{20}	108	20	0	0.83/7	12.8	2.4	0
$C_6H_6O_2$	72	6	32	1.47/7	15.1	1.3	6.7
$C_4H_8O_2$	48	8	32	1.32/7	9.1	1.5	6.0
Total atomic mass contribution					73.2	9.7	12.7
Average number of atoms (mass/atomic weight)					6.1	10	0.8

Answer. $C_{6.1}H_{10}O_{0.8}$; use $C_6H_{10}O$ as the overall structure of the mixture of contaminants.

*Atomic weight times number of atoms of each.
†Average concentration of the compound divided by the total concentration of all contaminants.
‡Columns (1) × (4), (2) × (4), and (3) × (4).

mean chemical structure rather than a reaction for each compound present.

The above mixture is represented by a single chemical structure for writing a stoichiometric equation. Determining the average chemical structure is illustrated in Table B.3. After the overall chemical structure $(C_6H_{10}O)$ is obtained, the balance stoichiometric equation is obtained by substituting according to the equations of Table B.1.

Step 2.   Determine the most appropriate electron acceptor and metabolism mode for destruction of the above hydrocarbon mixture.

Answer.   Oxygen, aerobic (the reason for this selection is presented in Chap. 4).

Step 3.   Select an appropriate source for the nutrient nitrogen.

Answer.   Either ammonia or nitrate can be used. For this example, ammonia nitrogen is selected.

Ammonia is readily available to microorganisms, but it does create an additional oxygen demand. For aerobic systems, some ammonia will be oxidized to nitrite and nitrate by indigenous bacteria. Oxygen demands will be greater than estimated from the reaction for the sole oxidization of the contaminants. Nitrate, however, can serve as an electron acceptor which may not be desired for some bioremediation systems (see Chaps. 3 and 4).

Step 4. Select an appropriate energy factor for the conversion of hydrocarbon to cellular mass.

Answer. Hydrocarbons oxidize relatively fast, so a valve for f_s of 0.5 is selected.

Step 5. Write the appropriate half reactions and the equation for cell growth using the yield fractions f_e and f_s (see Table B.1):

$$Z = 4a + b - 2c - 3d \qquad \text{(B.3)}$$

for $C_6H_{10}O$: $a = 6$, $b = 10$, $c = 1$, $d = 0$.

Answer. $Z = 4(6) + 10 - 2(1) = 32$.

For H_D one obtains by substitution:

$$1/32\ C_6H_{10}O + 11/32\ H_2O = 6/32\ CO_2 + H^+ + e^- \qquad \text{(B.4)}$$

For $f_e H_A$ one obtains by substitution

$$1/8\ O_2 + 1/2\ H^+ + 1/2\ e^- = 1/4\ H_2O \qquad \text{(B.5)}$$

For $f_s C_s$ one obtains by substitution:

$$1/8\ CO_2 + 1/40\ NH_3 + 1/2\ H^+ + 1/2\ e^-$$

$$= 1/40\ C_5H_7\ O_2\ N + 1/20\ H_2O \qquad \text{(B.6)}$$

Step 6. Write the overall reaction by summation of half reactions and cell growth equation (sum H_D, $f_e H_A$, and $f_s C_s$). Sum all quantities on the left side of the equal sign and then all quantities on the right side:
Left side:

$$1/32\ C_6H_{10}O + 11/32\ H_2O + 1/8\ O_2 + 1/2\ H^+$$

$$+\ 1/2\ e^- + 1/8\ CO_2 + 1/40\ NH_3 + 1/2\ H^+ + 1/2\ e^- = \qquad \text{(B.7)}$$

Right side:

$$6/32\ CO_2 + H^+ + e^- + 1/4\ H_2O$$

$$+\ 1/40\ C_5H_7O_2N + 2/10\ H_2O \qquad \text{(B.8)}$$

Answer. The stoichiometric equation is, for NH_3 as the nutrient nitrogen source:

$$C_6H_{10}O + 4\ O_2 + 0.8\ NH_3 = 2\ CO_2 + 0.8\ C_5H_7O_2N + 3.4\ H_2O \qquad (B.9)$$

If NO_3 was selected as the nutrient nitrogen source, one obtains

$$C_6H_{10}O + 4\ O_2 + 0.57\ NO_3$$

$$= 3.1\ CO_2 + 0.57\ C_5H_7O_2N + 3.3\ H_2O \qquad (B.10)$$

Step 7. Calculate molecular weights to establish mass ratios.

Answer. Molecular weights:

$$C_6H_{10}O = 98$$

$$O_2 = 32$$

$$NH_3 \text{ as } N^* = 14^*$$

Step 8. Determine mass ratios:

$$C_6H_{10}O{:}O_2{:}N{:}P^\dagger{:}CO_2 \qquad (B.11)$$

$$1{:}4(32/98){:}0.8(14/98){:}1/6(0.8)(14/98){:}2(44/98)$$

or

$$1{:}1.31{:}0.114{:}0.019{:}09$$

For every pound of contaminant degraded 1.31 lb of oxygen, 0.114 lb of nitrogen, 0.019 lb of phosphorus are required, and 0.9 lb of CO_2 are produced.

Step 9. Calculate the pounds of oxygen, ammonia, and phosphorus necessary to bioremediate 3000 yd^3 of soil contaminated with 7000 mg/kg of hydrocarbons. Using a soil density of 115 lb/ft^3 yields 65,205 lb of hydrocarbons to be biodegraded. The chemical supply needs are:

$$\text{Oxygen: } (65{,}205)(1.31) = 85{,}420 \text{ lb}$$

$$\text{Nitrogen: } (65{,}205)(0.114) = 7430 \text{ lb}$$

$$\text{Phosphorus } (65{,}205)(0.019) = 1240 \text{ lb}$$

The amount of nitrogen and phosphorus available in the contaminated medium must be established so the net balance can be delivered to the system.

---

*Nutrients are sold in terms of nitrogen (N).
†Phosphorus biological requirement is approximately one-sixth of nitrogen.

Delivery rates for electron acceptor and nutrients

The above calculations establish the total mass needs for remediating the given hydrocarbon contamination. The next engineering require- ment is to determine the rate that these materials must be supplied to the contaminated soil. It is the rate of delivery that determines the sizing of equipment such as chemical feed pumps, water injection or application rates, and gas handling rates. To calculate delivery rates one needs information on rate of treatment and a decision on the desired rate of cleanup. Providing equipment for the highest possible rate that the biological system can perform may be impractical over the last half of the project's life. These engineering decisions are dis- cussed in Chaps. 7 through 10. A method to estimate delivery rate is provided in the following example.

Rate of degradation

Degradation rates are frequently based on laboratory treatability studies. If these data are unavailable, published data can be used to estimate rates. For many studies, degradation is expressed as a first- order rate equation. The following example uses the first order kinet- ic expression; however, many field systems follow a zero-order degra- dation rate as discussed in Chaps. 7 and 8. The reaction order believed most appropriate for the selected treatment should be used. Bioventing systems usually use a zero-order rate as illustrated in the example for bioventing. The first-order expression is given by

$$\frac{dL}{dt} = -kL \tag{B.12}$$

or

$$L_t = Le^{-kt} \tag{B.13}$$

where L = original concentration of organic compounds
L_t = concentration after degradation for time t in days
k = first-order degradation rate constant, unit per day

Example problem

The application of Eq. (B.12) is demonstrated in the following problem.
Oxygen is to be supplied to treat soil contaminated with naphtha- lene. Assume that the contaminant load will require a total mass of 100,000 lb of oxygen for bioremediation. Estimate the delivery rate of oxygen necessary for maximum naphthalene degradation. Since the actual oxygen supply rate varies with time, assume that delivery

rates will be changed every 2 weeks in the field. Therefore, a biweekly delivery rate (a constant rate over 2-week intervals) can be calculated.

From data provided in Chap. 8 (Figs. 8.1 and 8.2), the mean half-life for biodegradation of naphthalene is 35 days. The first-order rate constant is calculated by using Eq. (B.13):

$$0.5 = e^{-k(35)} \tag{B.14}$$

and $k = 0.02$/day

The rate of degradation dL/dt is given by Eq. (B.12.) Since the ratio of hydrocarbon mass to required oxygen mass is constant, this equation also provides the rate at which oxygen and nutrients are utilized. The utilization rate is calculated from the respective mass ratio times the rate of organic compound degradation.

The rate of oxygen or nutrient utilization is given by

$$\frac{dN}{dt} = -k(R_N)L_t \tag{B.15}$$

where dN/dt = pounds of oxygen, nitrogen, or phosphorus required per day
R_N = mass ratio of the oxygen, nitrogen, or phosphorus from Eq. (B.11)

To calculate the biweekly rate of oxygen required, establish the amount of oxygen demand at the midpoint of each biweekly period, for example, at 7 days, 21 days, etc. Table B.4 illustrates the calculation procedure, yielding an oxygen demand rate of 1740 lb/day during the first 2 weeks and 570 lb/day for the ninth and tenth week. After 10 weeks, the oxygen delivery rate is less than one-third the initial requirement. Because of the rapid decrease in degradation rate, equipment capacity for a project is usually designed at less than the maximum rate. Variable pumping rates or multiple units can be used to adjust for the changing needs. If the oxygen supply rate is designed at 1000 lb/day, the rate of bioremediation would follow a zero-order degradation rate until the field requirement reached the 1000 lb/day level. For the above problem, this would extend the cleanup time by approximately 2 weeks.

Equation (B.13) also can be used to estimate cleanup time. For cleanup to a level of 100 ppm approximately 99 percent removal of naphthalene and oxygen demand must be achieved. Substituting 0.99 for L_t/L in Eq. (B.13) yields 230 days for t.

Several precautions must be recognized when calculating a cleanup time by this technique. First, determine if the rate order and rate constant are suitable for the field program. These points are dis-

TABLE B.4 Solving for Oxygen Delivery Rates

Biweek period (1)	t, days (2)	$-kt$ (3)	PAH fraction remaining, L_t/L (4)	Total remaining oxygen requirement, lb (5)	Maximum rate of oxygen update, lb/day (6)
0	0			100,000	
1	7	−0.14	0.869	86,900	1,740
2	21	−0.42	0.657	65,700	1,319
3	35	−0.70	0.49	49,000	990
4	49	−0.98	0.375	37,531	750
5	63	−1.26	0.284	28,365	570

Given $(R_N)L$ for oxygen = 100,000 lb
Column (3) = column (2) in days times the first-order rate constant (0.02/day)
Column (4) = solution to Eq. (B.13) for L_t/L
Column (5) = column (4) times 100,000 lb
Column (6) = solution to Eq. (B.15)

cussed in Chap. 10. Laboratory data frequently yield first-order kinetics when field degradation responds as a zero-order equation.

Degradation rates also are slower in the field because sites contain mixed organic compounds. The degradation rate constant is significantly less for some chemicals (Fig. 8.1 and Table 9.1). This results in a field rate constant that decays with time.

Second, microorganisms do not respond immediately. A lag period occurs before any significant degradation, and utilization of electron acceptor occurs. Lags of several weeks to several months have been experienced in the field. Lag periods are typical and result from several factors. The most common lag period is the time required for a small population of bacteria to grow before enough specific degraders accumulate to cause measurable increase in chemical transformation. Many factors can affect the growth of small populations. Indigenous protozoa have been shown to extend lag periods as a result of their predator activities (Spain, 1990). Lags can be influenced by the lack of nutrients or the buildup of specific degraders for the more difficult to degrade compounds. Lags in achieving mineralization can be influenced by the buildup of toxic intermediates and the corresponding lag period of the microbial species required to degrade the intermediates.

Extensive biodegradation periods (a year or more) have been noted after the addition of chlorinated aromatic compounds to soil. These long lag periods probably result from the lack of an indigenous organism to degrade the compound. The lag period ends as a result of evolved genetic exchange or mutation. In some cases, the appropriate enzymes are not induced or growth is somehow inhibited. Extended lags can be resolved by controlled laboratory studies to establish potential causes.

Calculations based on treatability studies are good for estimating cleanup times and sizing equipment, but actual feed rates must be adjusted according to field data.

Bioventing example problem

The oxygen delivery rate for bioventing is based on in situ respiration testing as discussed in Chap. 6. The rate of change in oxygen concentration as a percent of the gas phase is measured for an injected slug of air. Field data are interpreted by zero-order kinetics and converted to a rate constant (percent oxygen per hour) as illustrated in Fig. 6.3.

This rate of oxygen uptake, given that the oxygen uptake in the control or background test is not significant, can be used to predict a cleanup rate and project duration. The rate of biodegradation K_B can be related to the oxygen uptake rate K_O as

$$K_B = \frac{K_O \, A \, D_O \, C}{100}$$

where K_B = biodegradation rate, mg/kg/day
K_O = oxygen uptake rate, percent oxygen/day
A = volume of air/mass of soil, L/kg
D_O = density of oxygen gas, L/kg (1330 mg/L at 1 atm, 20°C)
C = mass ratio of hydrocarbon to oxygen required for mini-
malization

The volume of air per mass of soil is a function of the soil structure, porosity , and moisture as discussed in Chaps. 3 and 6. The air volume can be calculated from density, specific volume, moisture content, and specific gravity of minerals using the phase diagram relationships for unsaturated soils. For soil with a bulk density of 1440 kg/m and an air void to total volume of 0.3, the value of A is calculated as follows:

$$A = (0.3)\,(1000 \text{ L air/m}^3 \text{ of soil})/1440 \text{ kg m}^3$$

$$A = 0.21 \text{ L air/kg of soil}$$

Using the mass ratio for hydrocarbon degradation in Eq. (B.11), hydrocarbon:O_2 of 1:1.3, and a K_O of 7.0 percent/day, the biodegradation rate is

$$K_B = (7.0 \text{ percent/day})\,(0.21 \text{ L/kg})\,(1330 \text{ mg/L})\,(1/1.3)$$

$$K_B = 15 \text{ mg/kg/day}$$

This rate assumes that air can be uniformly delivered to the subsurface. Any heterogeneity significantly decreases this degradation rate, and safety factors of 2 to 3 have been applied, yielding a hydrocarbon degradation rate of 5 to 7.5 mg/kg/day.
For a hydrocarbon spill averaging 7000 mg/kg, the estimated cleanup time is

$$\frac{7000 \text{ mg/kg}}{5 \text{ mg/kg/day}} = 1400 \text{ days, or 3.8 years}$$

to

$$\frac{7000 \text{ mg/kg}}{7.5 \text{ mg/kg/day}} = 933 \text{ days, or 2.6 years}$$

The oxygen and air requirement is based on the optimum rate of degradation of 15 mg/kg/day. Using the mass ratio of oxygen per unit mass of hydrocarbon the oxygen requirement is

$$15 \text{ mg/kg/day } (1.3) = 19.5 \text{ mg } O_2/\text{kg/day}$$

or

$$0.0195 \text{ lb } O_2/1000 \text{ lb soil/day}$$

For 3000 yd^3 at a density of 115 lb/ft^3, 9.315×10^6 lb of soil is contaminated. This requires the delivery of

$$(0.0195 \text{ lb } O_2/1000 \text{ lb/day}) (9.315 \times 10^6 \text{ lb}) = 182 \text{ lb } O_2/\text{day}$$

Given the density of air at 0.08 lb air/ft^3 and that oxygen is 21 percent, the required airflow rate is

$$(182 \text{ lb } O_2/\text{day})/(0.08 \text{ lb/ft}^3) (0.21) = 10,833 \text{ ft}^3/\text{day or } 7.5 \text{ ft}^3/\text{min}$$

The number of required wells for air injection or extraction is based on the radius of influence as established from field data on pressure drop versus distance from the test well. As with all in situ bioremediation projects, these calculations provide a prediction of performance. Field operation must be based on data collected from monitoring stations during active bioremediation.

References

Adriaens, P., and Focht, D. D., "Continuous Coculture Degradation of Selected Polychlorinated Biphenyl Cogeners by *Acinetobacter Spp.* in an Aerobic Reactor System," *Environ. Sci. Technol.,* Vol. 24, pp. 1042–1049, 1990.

Adriaens, P., Kohler, H. P. E., Kohler-Staub, D., and Focht, D. D., "Bacterial Dehalogenation of Chlorobenzoates and Coculture Biodegradation of 4,4′-dichlorobiphenyl," *Appl. Environ. Microbiol.,* Vol. 55, No. 4, pp. 887–892, 1989.

Aggarwal, P. K., Means, J. L., and Hinchee, R. E., "Formulation of Nutrient Solutions for In-Situ Bioremediation," *In-Situ Bioremediation, Applications and Investigations for Hydrocarbon and Contaminate Site Remediation,* Hinchee, R. E., and Olfenbuttel, R. F. (Eds.), Butterworth-Heinemann, Boston, Mass., pp. 51–66, 1991.

Aharonson, N., Katan, J., Aurdou, E., and Yarden, O., "The Role of Fungi and Bacteria in the Enhanced Degradation of the Fungicide Carbendazim and the Herbicide Diphenamid," *Enhanced Biodegradation of Pesticides in the Environment,* Racke, K. D., and Coats, J. R. (Eds.), American Chemical Society, Washington, D.C., pp. 113–128, 1990.

Ahring, B. K., and Westermann, P., "Kinetics of Butyrate, Acetate, and Hydrogen Metabolism in a Thermophilic, Anaerobic, Butyrate-Degrading Triculture," *Appl. Environ. Microbiol.,* Vol. 53, pp. 434–339, 1987.

Aiba, S., Humphrey, A. E., and Millis, N. F., *Biochemical Engineering,* Academic Press, New York, 1965.

Albrecht, C. R., and Balmer, A. G., "Large-Scale Composting Using the Aerated Static Pile Method," *Proceedings of the International Conference on Composting of Solid Wastes and Slurries,* Edward Stentiford (Ed.), University of Leeds, Leeds, England, pp. 42–57, 1983.

Alexander, M., *Introduction to Soil Microbiology,* 2d Edition, John Wiley & Sons, New York, 1977.

Alvarez-Cohen, L. M., and McCarty, P. L., "TCE Transformation by a Mixed Methanotrophic Culture-Effects of Toxicity, Aeration and Reductant Supply," *Appl. Environ. Microbiol.,* Vol. 57, pp. 228–235, 1991.

Alvarez-Cohen, L. M., and McCarty, P. L., "Two-Stage Dispersed-Growth Treatment of Halogenated Aliphatic Compounds by Cometabolism," *Environ. Sci. Technol.,* Vol. 25, pp. 1387–1393, 1991a.

Anderson, J. P. E., "Herbicide Degradation in Soil: Influence of Microbial Biomass," *Soil Biol. Biochem.,* Vol. 16, pp. 483–489, 1984.

Andrews, J. F., "A Mathematical Model for the Continuous Culture of Microorganisms Utilizing Inhibitory Substrates," *Biotechnol. Bioeng.,* Vol. 10, pp. 707–723, 1968.

Arvin, E., Jensen, B., Godsy, E. M., and Grbić-Galić, D., "Microbial Degradation of Oil and Creosote Related Aromatic Compounds under Aerobic and Anaerobic Conditions," presented at the *International Conference on Physicochemical and Biological Detoxification of Hazardous Wastes,* Wu, X. C. (Ed.), Technomic Publication, Lancaster, Pa., Vol. II, pp. 828–847, 1988.

Aslanzadeh, J., and Hedrick, H. G., "Search for Mirex-Degrading Soil Microorganisms," *Soil Sci.,* Vol. 139, pp. 369–374, 1985.

Babcock, R. W., Kyoung, S. R., Hsieh, C. C., and Stenstrom, M. K., "Development of an Off-Line Enricher-Reactor Process for Activated Sludge Degradation of Hazardous Waste," *Water Environ. Res.,* Vol. 54, pp. 782–791, 1992.

Baek, N. H., and Jaffe, P. R., "Anaerobic Mineralization of Trichloroethylene," presented at the *International Conference on Physicochemical and Biological Detoxification of Hazardous Wastes,* Wu, X. C. (Ed.), Technomic Publication, Lancaster, Pa., Vol. II, pp. 772–782, 1988.

Bahr, J. M., "Analysis of Non-Equilibrium Desorption of Volatile Organics During Field Test of Aquifer Decontamination," *J. Contam. Hydrol.,* Vol. 4, pp. 205–222, 1989.

Bailey, J. E., and Ollis, D. F., *Biochemical Engineering Fundamentals,* 2d Edition, McGraw-Hill Book Company, New York, 1986.

Barnhart, M. J., and Meyers, J. M., "Pilot bioremediation tells all about petroleum contaminated soil," *Pollution Engineering,* Vol. 21, pp. 110–112, 1989.

Bayona, J. M., and Solanas, A. M., "Isolation and Characterization of a Fluorene-Degrading Bacterium: Identification of Ring Oxidation and Ring Fission Products," *Appl. Environ. Microbiol.,* Vol. 53, pp. 2910–2917, 1992.

Berg, J. D., Eikum, A. S., Eggen, T., and Selfor, H., "Cold-Climate Bioremediation: Composting and Groundwater Treatment Near the Arctic Circle at a Coke Works," *Hazardous Materials Control / SUPERFUND '91,* HMCRI, pp. 321–325, 1991.

Bergman, T. J., Greene, J. M., and Davis, T. R., "In-Situ Slurry-Phase Bioremediation Case with Emphasis on Selection and Design of a Pure Oxygen Dissolution System," *Proceedings of HMC / SUPERFUND 1992 National Conference,* Washington, D.C., pp. 430–443, December 1992.

Berry, D. F., Madsen, E. L., and Bollag, J. M., "Conversion of Indole to Oxindole under Methanogenic Conditions," *Appl. Environ. Microbiol.,* Vol. 53, pp. 180–182, 1987.

Bioremediation Report, King Publishing Group, Washington, D.C., 1993.

Bitton, G., and Gerba, C. P., *Groundwater Pollution Microbiology,* John Wiley & Sons, New York, p. 23, 1984.

Bleam, R. D., and Cauthray, M. K., "Microbial Polynuclear Aromatic Hydrocarbon Degradation—Review and Case Histories," presented at the *International Conference on Physicochemical and Biological Detoxification of Hazardous Wastes,* Wu, X. C. (Ed.), Technomic Publication, Lancaster, Pa., Vol. II, pp. 867–882, 1988.

Block, R. N., Clark, J. P., and Bishop, M., "Biological Remediation of Petroleum Hydrocarbons," *Proceedings of the 6th National Conference on Hazardous Wastes and Hazardous Materials,* The Hazardous Material Control Research Institute, pp. 315–318, April 12–14, 1989.

Bollag, J. M., "Decontaminating Soil with Enzymes," *Environ. Sci. Technol.,* Vol. 26, No. 10, pp. 1876–1881, 1992.

Bollag, J. M., et al., *Soil Sci. Soc. Am. J.,* Vol. 44, pp. 52–56, 1980.

Borow, H. S., and Kinsella, J. V., "Bioremediation of Pesticides and Chlorinated Phenolic Herbicides—Above Ground and In-Situ Case Studies," *Proceedings of HMCRI, 10th National Conference,* Washington, D.C., pp. 325–331, 1989.

Bouwer, H., "Elements of Soil Science and Groundwater Hydrology," *Groundwater Pollution Microbiology,* Bitton, G., and Gerba, C. (Eds.), John Wiley & Sons, New York, 1984.

Bouwer, E. J., Rittmann, B. E., and McCarty, P. L., "Anaerobic Degradation of Halogenated 1- and 2-Carbon Organic Compounds," *Environ. Sci. Technol.,* Vol. 15, pp. 596–599, 1981.

Bouwer, E. J. and Wright, J. P., "Transformations of Trace Halogenated Aliphatics in Anoxic Biofilm Columns," *J. Contam. Hydro.,* Vol. 2, pp. 155–169, 1988.

Bouwer, E. J., and McCarty, P. L., *Proceedings of the ASCE Environmental Engineering Specialty Conference,* Atlanta, Ga., July 8–10, pp. 196–202, 1981.

Brady, N. C., *The Nature and Property of Soils,* 8th Edition, Macmillan Publishing Co., 1974.

Braus-Stromeyer, S. A., Hermann, R., Cook, A. M., and Leisinger, T., "Dichloromethane as a Sole Carbon Source for an Acetogenic Mixed Culture and Isolation of a Fermentative Dicholoromethane-Degrading Bacterium," *Appl. Environ. Microbiol.,* Vol. 59, pp. 3790–3797, 1993.

Britton, L. N., "Microbial Degradation of Aliphatic Hydrocarbons," *Microbial Degradation of Organic Compounds,* Gibson, D. T. (Ed.), Marcel Dekker, New York, pp. 89–130, 1984.

Brodkorb, T. S., and Legge, R. L., "Enhanced Biodegradation of Phenanthrene in Oil-Tar Contaminated Soils Supplemented with *Phanerochaete chrysosporium,*" *Appl. Environ. Microbiol.,* Vol. 53, pp. 3117–3121, 1992.

Bromilow, R. H., Briggs, G. G., Williams, M. R., Smelt, J. H., Tuinstra, L., and Traag, W. A., "The Role of Ferrous Ions in the Rapid Degradation of Oxamyl, Methomyl, and Aldicarb in Anaerobic Soils," *Pestic. Sci.,* Vol. 17, pp. 535–547, 1986.

Brown, R. A., "Oxygen Sources for In-Situ Bioreclamation," *In-Situ Bioremediation of Groundwater and Contaminated Soils, Water Pollution Control Federation 63rd Annual Conference,* Washington, D.C., pp. 99–118, October, 1990.

Brown, R., and Crosbie, J., "Oxygen Sources for In-Situ Bioremediation," *Proceedings 10th National Conference HMCRI,* Washington, D.C., 1989.

Brown, S. C., Grady, C. P. L., and Tabak, H. H., "Biodegradation Kinetics of Substituted Phenolics: Demonstration of a Protocol Based on Electrolytic Respirometry," *Water Res.,* Vol. 24, No. 7, pp. 853–861, 1990.

Brox, G., "Bioslurry Treatment," *Proceedings Applied Bioremediation 1993,* Fairfield, N.J., Oct. 25–26, 1993.

Brubaker, G., "Screening Criteria for In-Situ Bioreclamation of Contaminated Aquifers," *Proceedings Sixth National RCRA Superfund Conference,* New Orleans, La., p. 319, 1989.

Brunner, W., Sutherland, F. H., and Focht, D. D., "Enhanced Biodegradation of Polychlorinated Biphenyls in Soil by Analog Enrichment and Bacterial Inoculation," *J. Environ. Qual.,* Vol. 14, pp. 324–328, 1985.

California Department of Health Services, "The California Site Migration Decision Tree," Department of Health Services, Toxic Substances Control Division, 1985.

Caramagno, D., and Cheremisinoff, R. N., "Design of Groundwater Injection Systems for the Remediation of Groundwater at Hazardous Waste Sites," *Encyclopedia of Environmental Control Technology,* Vol. 4, Paul N. Cheremisinoff (Ed.), 1991.

Castro, C. E., and Belser, N. O., "Biodehalogenation Reductive Dehalogenation of the Biocides Ethylene Dibromide, 1,2-Dibromo-3-chloropropane, and 2,3-Dibromobutane in Soil," *Environ. Sci. Technol.,* Vol. 2, pp. 779–783, 1968.

Cerniglia, C. E., "Microbial Metabolism of Polycyclic Aromatic Hydrocarbons," *National Center for Toxicological Research,* Vol. 30, pp. 31–65, 1984.

Chang, Hou-Min, Joyce, T. W., Kick, T. K., and Huynh, V. B., "Process of Degrading Chloro-Organics by White-Rot Fungi," United States Patent No. 4,554,075 (Nov. 19, 1985).

Chang, Chow-Feng, "Effects of Reactor Configuration on the Performance of Static-Bed Submerged Media Anaerobic Reactors," Dissertation, Iowa State University, 1988.

Chaudhry, G. R., and Cortez, L., "Degradation of Bromacil by a *Pseudomonas sp.,*" *Appl. Environ. Microbiol.,* Vol, 54, pp. 2203–2207, 1988.

Chresand, T. J., and Anderson, M. J., "Biological Treatment of Phenolic Process Water at a Wood Treating Site," Report, Biotrol, Inc., Chaska, Minn., 1990.

Chu, K. H., and Jewell, W. J., "Treatment of Tetrachloroethylene with Anaerobic Attached Film Process," *J. Environ. Eng.,* Vol. 120, No. 1, pp. 58–71, 1994.

Clark, R. R., Chian, E. S. K., and Griffin, R. A., "Degradation of Polychlorinated Biphenyls by Mixed Microbial Cultures," *Appl. Environ. Microbiol.,* Vol. 37, pp. 680–685, 1979.

Colberg, P. J. S., "Degradation of Chlorinated Aromatic Compounds under Sulfate-Reducing Conditions," *EPA Symposium on Bioremediation of Hazardous Wastes,* Falls Church, Va., April 1991.

Conner, J. R., "Case Study of Soil Venting," *Pollut. Eng.,* Vol. 7, pp. 74–78, 1988.

Cookson, J. T., Jr., and Leszczynski, J. E., "Restoration of a Contaminated Drinking Water Supply Aquifer, *Proceedings of the 4th National Outdoor Action Conference on Aquifer Restoration, Groundwater Monitoring, and Geophysical Methods,* National Groundwater Association, Las Vegas, Nev., 1990a.

Cookson, J. T., and Murray, C. M., "Fundamental Studies for Developing Operational Strategies to Minimize Odors Associated with Sewage Sludge Composting," *Water Pollution Control Federation 63rd Annual Conference,* Washington, D.C., October 1990b.

Cookson, J. T., "Volatilization of VOCs During the Treatment of Groundwater in Activated Sludge Basins," Unpublished report prepared for Baker Chemical Co., New Jersey, June 1987.

Cripps, C., Bumpus, J. A., and Aust, S. D., "Biodegradation of Azo and Heterocyclic Dyes by *Phanerochaete chrysosporium,*" *Appl. Environ. Microbiol.,* Vol. 56, pp. 1114–1118, 1990.

Dague, R. R., and Pidapart, S. R., "Anaerobic Sequencing Batch Reactor Treatment of Swine Wastes," *46th Purdue Industrial Waste Conference Proceedings,* Lewis Publishers, Chelsea, Mich., pp. 751–759, 1992.

Davis, M. W., Glaser, J. A., Evans, J. W., and Lamar, R. T., "Field Evaluation of the Lingin-Degrading Fungus *Phanerochaete sordida* to Treat Creosote-Contaminated Soil," *Environ. Sci. Technol.,* Vol. 27, pp. 2572–2576, 1993.

Dean, N., and Kremer, F., "Advancing Research for Bioremediation," *Environ. Prot.,* Vol. 3, pp. 19–23, 1992.

Deever, W. R., and White, R. C., "Composting Petroleum Refinery Sludges," Texaco, Inc., Port Arthur, Tex., 1978.

Department of the Air Force, *Test Plan and Technical Protocol for a Field Treatability Test for Bioventing,* Air Force Center for Environmental Excellence, May 1992.

Dhawale, S. W., Dhawale, S. S., and Dean-Ross, D., "Degradation of Phenanthrene by *Phanerochaete chrysosporium* Occurs under Ligninolytic as well as Nonligninolytic Conditions," *Appl. Environ. Microbiol.,* Vol. 53, pp. 3000–3006, 1992.

DiStefano, T. D., Gossett, J. M., and Zinder, S. H., "Reductive Dechlorination of High Concentrations of Tetrachloroethane to Ethane by an Anaerobic Enrichment Culture in the Absence of Methanogenesis," *Appl. Environ. Microbiol.,* Vol. 57, pp. 2287–2292, 1991.

Dolfing, J., and Tiedje, J. M., "Growth Yield Increase Linked to Reductive Dechlorination in a Defined 3-Chlorobenzoate Degrading Methanogenic Coculture," *Arch. Microbiol.,* Vol. 149, pp. 102–105, 1987.

Dolfing, J., "Gibbs Free Energy of Formation of Halogenated Aromatic Compounds and Their Potential Role as Electron Acceptors in Anaerobic Environments," *Environ. Sci. Technol.,* Vol. 26, pp. 2213–2218, 1992.

Dupont, R. R., Doucette, W. J., and Hinchee, R. E., "Assessment of In-Situ Bioremediation Potential and the Application of Bioventing at a Fuel-Contaminated Site," *Bioventing and Vapor Extraction: Uses and Applications in Remediation Operations Seminar,* sponsored by Air and Waste Management Association, Apr. 15, 1992.

Dwyer, D. F., Krumme, M. L., Boyd, S. A., and Tiedje, J. M., "Kinetics of Phenol Biodegradation by an Immobilized Methanogenic Consortium," *Appl. Environ. Microbiol.,* Vol. 52, pp. 345–351, 1986.

Edwards, D. A., Luthy, R. G., and Liu, Z., "Solubilization of Polycyclic Aromatic Hydrocarbons in Micellar Nonionic Surfactant Solutions," *Environ. Sci. Technol.,* Vol. 25, pp. 127–133, 1991.

Ely, D. L., and Heffner, D. A., "Process for In-Situ Biodegradation of Hydrocarbon Contaminated Soil," U.S. Patent No. 4,765,902, 1988.

English, C. W., and Loehr, R. C., "Degradation of Organic Vapors in Unsaturated Soils," *Bioremediation Fundamentals and Effective Application, Proceedings of the 3rd Annual Symposium at the Gulf Coast Hazardous Substance Research Center,* pp. 65–74, February 1991.

Ensign, S. A., Hyman, M. R., and Arp, D. J., "Cometabolic Degradation of Chlorinated Alkenes by Alkene Monooxygenase in a Propylene Grown *Xanthobacter* Strain," *Appl. Environ. Microbiol.,* Vol. 53, pp. 3038–3046, 1992.

ENTEC Directory of Environmental Technology, Earthscan Publications and Lewis Publishers/CRC Press, Ann Arbor, Mich., 1993.

Fair, G. M., Geyer, J. C., and Okun, D. A., *Water and Wastewater Engineering,* Vol. 2, *Water Purification and Wastewater Treatment and Disposal,* John Wiley & Sons, pp. 24–26, 1968.

Finnemore, E. J., "Estimation of Groundwater Mounding beneath Septic Drain Fields," *Groundwater,* Vol. 31, No. 6, pp. 884–889, 1993.

Finnemore, E. J., and Hantzsche, N. N., "Groundwater Mounding Due to On-Site Sewage Disposal," *J. Irrig. Drainage Eng.,* ASCE, Vol. 109, No. 2, pp. 199–210, 1983.

Focht, D. D., "Performance of Biodegradative Microorganisms in Soil: Xenobiotic Chemicals as Unexploited Metabolic Niches," *Environmental Biotechnology,* Omenn, G. S. (Ed.), Plenum Press, New York, pp. 15–30, 1987.

Focht, D. D., and Brunner, W., "Kinetics of Biphenyl and Polychlorinated Biphenyl Metabolism in Soil," *Appl. Environ. Microbiol.,* Vol. 50, pp. 1058–1063, 1985.

Fox, P., Suidan, M. T., and Bandy, J. T., "Comparison of Attachment Media Types in Anaerobic Expended-Bed Reactors," presented at the *International Conference on Physicochemical and Biological Detoxification of Hazardous Wastes,* Wu, X. C. (Ed.), Technomic Publication, Lancaster, Pa., Vol. II, pp. 803–814, 1988.

Fredlund, D. G., and Rahardjo, H., *Soil Mechanics for Unsaturated Soils,* John Wiley & Sons, New York, 1993.

French Limited Task Group and ERT, a Resource Engineering Company, *In-Situ Biodegradation Demonstration Report,* Vols. I, II, *French Limited Site,* Revised Feb. 1, 1988.

Gale, E. F., *The Chemical Activities of Bacteria,* Academic Press, London, 1952.

Gälli, R., "Biodegradation of Dichloromethane in Waste Water Using a Fluidized Bed Bioreactor," *Appl. Microbiol. Biotechnol.,* Vol. 27, pp. 206–213, 1987.

Gälli, R., and McCarty, P. L., "Kinetics of Biotransformation of 1,1,1-trichloroethane by *Clostridium sp.* Strain TCAIIB," *Appl. Environ. Microbiol.,* Vol. 55, pp. 845–851, 1989.

Gaudy, A. F., and Gaudy, E. T., *Elements of Bioenvironmental Engineering,* Engineering Press, Inc., San Jose, Calif., 1988.

Gibson, D. T., and V. Subramanian, "Microbial Degradation of Aromatic Hydrocarbons," *Microbial Degradation of Organic Compounds,* Gibson, D. T. (Ed.), Marcel Dekker, New York, pp. 181–252, 1984.

Gibson, S. A., and Suflita, J. M., "Anaerobic Biodegradation of 2,4-5-Trichlorophenoxyacetic Acid in Samples from a Methanogenic Aquifer: Stimulated by Short Chain Organic Acids and Alcohols," *Appl. Environ. Microbiol.,* Vol. 56, pp. 1825–1832, 1990.

Glaser, J. A., "The Development of Water and Soil Treatment Technology Based on the Utilization of a White-Rot, Wood Rotting Fungus," presented at the *International Conference on Physicochemical and Biological Detoxification of Hazardous Wastes,* Wu, X. C. (Ed.), Technomic Publication, Lancaster, Pa., Vol. II, pp. 633–644, 1988.

Godsy, E. M., and Grbić-Galić, D., "Biodegradation Pathways for Benzothiophene in Methanogenic Microcosms," Mallard, G. E., and Ragone, S. E. (Eds.), *U.S. Geological Survey Toxic Substances Hydrology—Proceedings of the 4th Technical Meeting,* Phoenix, Ariz., Sept. 26–30, pp. 3–8, 1988.

Goltz, M. W., *Water Resources Report,* Vol. 27, pp. 554–556, 1991.

Grady, C. P. L., Aichinger, G., Cooper, S. F., and Naziruddin, M., "Biodegradation Kinetics for Selected Toxic/Hazardous Organic Compounds," Environmental Systems Engineering, Clemson University, S.C., unpublished paper.

Grbić-Galić, D., "Microbial Degradation of Homocyclic and Heterocyclic Aromatic Hydrocarbons under Anaerobic Conditions," Unpublished report, Department of Civil Engineering, Environmental Engineering and Science, Stanford University, 1990.

Grbić-Galić, D., and Vogel, T. M., "Transformation of Toluene and Benzene by Mixed Methanogenic Cultures," *Appl. Environ. Microbiol.,* Vol. 53, pp. 254–260, 1987.

Grbić-Galić, D., "Anaerobic Microbial Transformation of Aromatic Hydrocarbons and the Relevance to Regeneration of Contaminated Ground Waste Aquifers," Stanford University, 1988.

Griffin, D. M., "Water Potential as a Selective Factor in the Microbial Ecology of Soils," *Water Potential Relations in Soil Microbiology,* Soil Science Society of America, pp. 141–151, 1980.

Grifoll, M., Casellas, M., Bayona, J. M., and Solanas, A. M., "Isolation and Characterization of a Fluorene-Degrading Bacterium: Identification of Ring

Oxidation and Ring Fission Products," *Appl. Environ. Microbiol.,* Vol. 58, pp. 2910–2917, 1992.

Gunner, H. B., and Zuckerman, B. M., *Nature,* Vol. 217, p. 1183, 1968.

Haber, L., Allen, L. N., Zhao, S., and Hanson, R. S., "Methylotrophic Bacteria: Biochemical Diversity and Genetics," *Science,* Vol. 221, pp. 1147–1153, 1983.

Hagedorn, C., "Microbiological Aspects of Groundwater Pollution Due to Septic Tanks," *Groundwater Pollution Microbiology,* Bitton, G., and Gerba, C. P. (Eds.), John Wiley & Sons, New York, p. 181, 1984.

Haggblom, M. M., Rivera, M. D., and Young, L. Y., "Anaerobic Degradation of Chloroaromatic Compounds under Different Reducing Conditions," *EPA Symposium on Bioremediation of Hazardous Wastes,* Falls Church, Va., April 1991.

Hakwala, F. S., Lewandowski, G. A., and Sofer, S. S., "Design of Toxic Waste Treatment Bioreactor: Viability Studies of Microorganisms Entrapped in Alginate Gel," presented at the *International Conference on Physicochemical and Biological Detoxification of Hazardous Wastes,* Wu, X. C. (Ed.), Technomic Publication, Lancaster, Pa., Vol. II, pp. 587–599, 1988.

Hallbeck, L., and Pedersen, K., "Autotrophic and Mixotrophic Growth of *Gallionella ferruginea,*" *J. Gen. Microbiol.,* Vol. 137, pp. 2657–2661, 1991.

Harding-Lawson Associates, personal communications, 1990.

Harmsen, J., "Possibilities and Limitations of Landfarming for Cleaning Contaminated Soils," *On-Site Bioremediation: Processes for Xenobiotic and Hydrocarbon Treatment,* Hinchee, R. E., and Offenbuttel, R. E. (Eds.), pp. 255–272, Butterworth-Heinemann, Boston, Mass., 1991.

Hartman, S., deBoni, J. A. M., Tramper, J., and Luyben, K. C. A. M., "Bacterial Degradation of Vinyl Chloride," *Biotechnol. Lett.,* Vol. 1, pp. 383–388, 1985.

Harvey, R. W., Sinith, R. L., and LeBlanc, D. R., "Transport of Microspheres and Indigenous Bacteria through a Sandy Aquifer: Results of Natural- and Forced-Gradient Tracer Experiments," *Environ. Sci. Technol.,* Vol. 23, No. 1, pp. 51–56, 1989.

Haug, W., Schmidt, A., Nortemann, B., Hempel, D. C., Stolz, A., and Knackmuss, H. J., "Mineralization of the Sulfonated Azo Dye Mordant Yellow 3 by a 6-Aminonaphthalene-2-Sulfonate-Degrading Bacterial Consortium," *Appl. Environ. Microbiol.,* Vol. 57, pp. 3144–3149, 1991.

Heitkamp, M. A., et al., "Microbial Metabolism of Polycyclic Aromatic Hydrocarbons Isolation and Characterization of Pyrene-Degrading Bacterium," *Appl. Environ. Microbiol.,* Vol. 54, pp. 2549–2555, 1988.

Heitkamp, M. A., and Cerniglia, C. E., "Mineralization of Polycyclic Aromatic Hydrocarbons by a Bacterium Isolated from Sediment below an Oil Field," *Appl. Environ. Microbiol.,* Vol. 54, pp. 1612–1614, 1988a.

Heitkamp, M. A., Freeman, J. P., Miller, D. W., and Cerniglia, C. E., "Pyrene Degradation by a Mycobacterium sp.: Identification of Ring Oxidation and Ring Fission Products," *Appl. Environ. Microbiol.,* October 1988b.

Henry, S. M., and Grbić-Galić, D., "Effect of Mineral Media on Trichloroethylene Oxidation by Aquifer Methanotrophs," *Microb. Ecol.,* Vol. 20, pp. 151–169, 1990.

Henry, S. M., and Grbić-Galić, D., "Influence of Endogenous and Exogenous Electron Donors and Trichloroethylene Oxidation Toxicity on Trichloroethylene Oxidation by Methanotrophic Cultures from a Groundwater Aquifer," *Appl. Environ. Microbiol.,* Vol. 57, pp. 236–244, 1991.

Henry, S. M., and Grbić-Galić, D., "Inhibition of Trichloroethylene Oxidation by the Transformation Intermediate Carbon Monoxide," *Appl. Environ. Microbiol.,* Vol. 57, pp. 1770–1776, 1991a.

Herman, D. C., and Costerton, J. W., "Starvation-Survival of a p-Nitrophenol-Degrading Bacterium," *Appl. Environ. Microbiol.,* Vol. 59, pp. 340–343, 1993.

Hernandez, B. S., Higson, F. K., Kondrat, R., and Focht, D. D., "Metabolism of and Inhibition of Chlorobenzoates in *Pseudomonas putida P111,*" *Appl. Environ. Microbiol.,* Vol. 57, pp. 3361–3366, 1991.

Herrling, B., and Baurmann, S. W., "Hydraulic Circulation System for In-Situ Bioreclamation and/or In-Situ Remediation of Strippable Contamination," *In-Situ Bioreclamation: Applications and Investigations for Hydrocarbon and Contaminated*

Site Remediation, Hinchee, R. E., and Offenbuttel, R. F. (Eds.), Butterworth-Heinemann, Stoneham, Mass., pp. 173–195, 1991.

Hickey, R. F., Wagner, D., and Mazewski, G., "Treating Contaminated Groundwater Using a Fluidized-Bed Reactor," *Remediation,* Vol. 1, pp. 447–460, 1991.

Hickman, G. T., Novak, J. T., Morris, M. S., and Rebhun, M., "Effects of Site Variations on Subsurface Biodegradation Potential," *J. Water Pollut. Control Fed.,* Vol. 61, pp. 1564–1575, 1989.

Hicks, P. M., Hicks, R., and Cortin, F., "In-Situ Bioremediation of Soils and Groundwater Contaminated with Petroleum Hydrocarbons," *Proceedings of the Hazardous Waste/Groundwater Symposia, Water Environment Federation 65th Annual Conference and Exposition,* Vol. V, pp. 305–312, Sept. 20–24, 1992.

Hicks, R. J., "Above Ground Bioremediation: Practical Approaches and Field Experiences," *Proceedings Applied Bioremediation 1993,* Fairfield, N.J., Oct. 25–26, 1993.

Hinchee, R. E., and Ong, S. E., "A Rapid In-Situ Respiration Test for Measuring Anaerobic Biodegradation Rates of Hydrocarbons in Soil," *J. Air Waste Manage. Assoc.,* Vol. 42, pp. 1305–1312, 1992.

Hinchee, R. E., and Downey, D. C., "The Role of Hydrogen Peroxide Stability in Enhanced Bioreclamation," *Proceedings of the NWWA/API Conference on Petroleum Hydrocarbons and Organic Chemicals in Groundwater,* Houston, Tex., pp. 715–722, 1988.

Hinchee, R. E., Miller, R. N., and DuPont, R. R., "Enhanced Biodegradation of Petroleum Hydrocarbons: An Air-Based In-Situ Process," *Biological Processes: Innovative Hazardous Waste Treatment Technology Series,* Freeman, H. M., and Sferra, P. R. (Eds.), Technomic Publishing, Lancaster, Pa., pp. 177–183, 1991.

Hinchee, R. E., "Bioventing: Principles, Applications and Case Studies," Training Program of International Network for Environmental Training, Potomac, Md., 1993b.

Hogan, J. A., Toffoli, G. R., Miller, F. C., Hunter, J. V., and Finstein, M. S., "Composting Physical Model Demonstration: Mass Balance of Hydrocarbons and PCBs," presented at the *International Conference on Physicochemical and Biological Detoxification of Hazardous Wastes,* Wu, X. C. (Ed.), Technomic Publication, Lancaster, Pa., Vol. II, pp. 742–758, 1988.

Holm, P. E., Nielsen, P. H., Albrechtsen, H.-J., and Christensen, T. H., "Importance of Unattached Bacteria and Bacteria Attached to Sediment in Determining Potentials for Degradation of Xenobiotic Organic Contaminants in an Aerobic Aquifer," *Appl. Environ. Microbiol.,* Vol. 58, pp. 3020–3026, 1992.

Hong, S. N., Krichten, D. J., and Bowers, D. L., "Application of Anaerobic Selector Technology in Activated Sludge Systems," presented at the *62nd Annual Conference of the Water Pollution Control Federation,* San Francisco, Calif., October 1989.

Hopkins, G. D., Munakata, J., Semprini, L., and McCarty, P. L., "Trichloroethylene Concentration Effects on Pilot Field-Scale In-Situ Groundwater Bioremediation by Phenol-Oxidizing Microorganisms," *Environ. Sci. Technol.,* Vol. 27, pp. 2542–2547, 1993b.

Hopkins, G. D., Semprini, L., and McCarty, P. L., "Microcosm and In-Situ Field Studies of Enhanced Biotransformation of Trichloroethylene by Phenol-Utilizing Microorganisms," *Appl. Environ. Microbiol.,* Vol. 59, pp. 2277–2285, 1993a.

Horvath, R. S., "Microbial Co-Metabolism and the Degradation of Organic Compounds in Nature," *Bacteriol. Rev.,* Vol. 36, No. 2, pp. 146–155, 1972.

Howard, P. H., *Handbook of Environmental Fate and Exposure Data for Organic Chemicals,* Vol. I, Lewis Publishers, Chelsea, Mich., 1989.

Huang, S., and Wu, Y. C., "Cometabolism of Trichloroethylene with Sugar Using Two Stage Upflow Activated Carbon Fluidized Beds," presented at the *International Conference on Physicochemical and Biological Detoxification of Hazardous Wastes,* Wu, X. C. (Ed.), Technomic Publication, Lancaster, Pa., Vol. II, pp. 783–802, 1988.

Hutchins, S. R., and Wilson, J. T., "Laboratory and Field Studies on BTEX Biodegradation in a Fuel-Contaminated Aquifer under Denitrifying Conditions," *In-Situ Bioremediation Applications and Investigations for Hydrocarbon and*

Contaminated Site Remediation, Hinchee, R. E,, and Olfenbutter, R. F. (Eds.), pp. 157–172, Butterworth-Heinemann, Boston, Mass., 1991.

Hutzler, N. J., Baillod, C. R., and Schaepe, P. A., "Biological Reclamation of Soils Contaminated with Pentachlorophenol," *Proceedings of the 6th National Conference on Hazardous Wastes and Hazardous Materials,* Apr. 12–14, 1989, New Orleans, La., The Hazardous Material Control Research Institute, pp. 361–365.

INET, *Bioventing: Principles, Applications and Case Studies,* training program manual, instructor R. E. Hinchee, International Network for Environmental Training, Inc., Potomac, Md., pp. 7.1–7.26, 1993.

Jacob, G. S., Garbow, J. R., Halles, L. E., Kimack, N. M., Kishore, G. M., and Schaefor, J., "Metabolism of Glyphosate in *Pseudomonas sp.* strain LB," *Appl. Environ. Microbiol.,* Vol. 54, pp. 2953–2958, 1988.

Jagnow, G., Haider, K., and Ellwardt, P., "Anaerobic Dechlorination and Degradation of Hexachlorocyclohexane Isomers by Anaerobic and Facultative Anaerobic Bacteria," *Arch. Microbiol.,* Vol. 115, pp. 285–292, 1977.

Janssen, D. B., Van den Wijngaard, A. J., vanderWaarde, J. J., and Oldenhuis, R., "Biochemistry and Kinetics of Aerobic Degradation of Chlorinated Aliphatic Hydrocarbons," *On-Site Bioreclamation: Processes for Xenobiotic and Hydrocarbon Treatment,* Hinchee, R. E., and Offenbuttel, R. F. (Eds.), Butterworth-Heinemann, Boston, Mass., pp. 92–112, 1991.

Jensen, H. L., *Can. J. Microbiol.,* Vol. 3, p. 165, 1957.

Jerger, D. E., "Slurry-Phase Treatment of Wood Preserving Wastes: A Practical Success Story," *Proceedings Applied Bioremediation 1993,* Fairfield, N.J., Oct. 25–26, 1993.

Jimeno, A., Bermudez, J., Canovas-Diaz, M., Manjon, A., and Iborra, J. L., "Methanogenic Biofilm Growth Studies in an Anaerobic Fixed-Film Reactor," *Process Biochem.,* pp. 55–60, April, 1990.

Johnson, P. C., Kemblowski, M. W., and Colhart, J. D., "Quantitative Analysis for the Cleanup of Hydrocarbon-Contaminated Soils by In-Situ Soil Venting," *Groundwater,* Vol. 28, No. 3, May–June 1990.

Jury, W. A., Winer, A. M., Spencer, W. F., and Focht, D. D., "Transport and Transformations of Organic Chemicals in the Soil-Air-Water Ecosystem," *Reviews of Environmental Contamination and Toxicology,* Springer-Verlag, New York, Vol. 99, pp. 120–164, 1987.

Karlsson, H., and Bitto, R., "New Horizontal Wellbore System for Monitor and Remedial Wells," *Proceedings of Hazardous Materials Control Research Institute 11th National Conference,* pp. 357–362, Washington, D.C., 1990.

Kataoka, N., Tokiwa, Y., Tanaka, Y, Fujiki, K., Taroda, H., and Takeda, K., "Examination of Bacterial Characteristics of Anaerobic Membrane Bioreactors in Three Pilot-Scale Plants for Treating Low-Strength Wastewater by Application of the Colony-Forming-Curve Analysis Method," *Appl. Environ. Microbiol.,* Vol. 58, pp. 2751–2757, 1992.

Kaufman, D. D., "Degradation of Pentachlorophenol in Soil and by Soil Microorganisms," *U.S. EPA Proceedings of a Symposium Held in Pensacola, FL,* Plenum Press, New York, pp. 27–39, June 1977.

Kearney, P. C., and Kaufman, D. D., "Microbial Degradation of Some Chlorinated Pesticides," *Degradation of Synthetic Organic Molecules in the Biosphere,* National Academy of Sciences, pp. 166–196, 1972.

Kennedy, D. W., Aust, S. D., and Bumpus, J. A., "Comparative Biodegradation of Alkyl Halide Insecticides by the White Rot Fungus *Phanerochaete chrysosporium* (BKM-F-1767)," *Appl. Environ. Microbiol.,* Vol. 56, pp. 2347–2353, 1990.

Kharak, Y. K., Gunter, W. D., Aggarwal, P. K., et al., "SOLMINEQ.88: A Computer Program for Geochemical Modeling of Water-Rock Interactions," *U.S. Geological Survey, Water Resources Investigations,* Report 88-4227, 1988.

Kindzierski, W. B., Gray, M. R., Fedorak, P. M., and Hrudy, S. E., "Activated Carbon and Synthetic Resins as Support Material for Methanogenic Phenol-Degrading Consortia: Comparison of Surface Characteristics and Initial Colonization," *Water Environ. Res.,* Vol. 65, pp. 766–775, 1992.

Klecka, G. M., and Maier, W. J., "Kinetics of Microbial Growth on Pentachlorophenol," *Appl. Environ. Microbiol.,* Vol. 49, pp. 46–53, 1985.

Kohler, H.-P. E., Kohler-Staub, D., and Focht, D. D., "Degradation of 2-Hydroxybiphenyl and 2,2'-Dihydroxy biphenyl by *Pseudomonas sp.* Strain HBP1," *Appl. Environ. Microbiol.*, 54, pp. 2683–2688, 1988.

Kohler, H.-P. E., Kohler-Staub, D., and Focht, D. D., "Cometabolism of Polychlorinated Biphenyls: Enhanced Transformation of Aroclor 1254 by Growing Bacterial Cells," *Appl. Environ. Microbiol.*, 54, pp. 1940–1945, 1988a.

Kshirsagar, D. G., Parhad, N. M., Muthal, P. L., and Shivaraman, N., "Microbial Detoxification of Malathion from Wastewater," presented at the *International Conference on Physicochemical and Biological Detoxification of Hazardous Wastes,* Wu, X. C. (Ed.), Technomic Publication, Lancaster, Pa., Vol. II, pp. 695–787, 1988.

Lamar, R. T., Evans, J. W., and Glaser, J. A., "Solid-Phase Treatment of a Pentachlorophenol-Contaminated Soil Using Lignin-Degrading Fungi," *Environ. Sci. Technol.*, Vol. 27, pp. 2566–2571, 1993.

Lamar, R. T., Glaser, J. A., and Kirk, T. K., "Fate of Pentachlorophenol (PCP) in Sterile Soils Inoculated with the White-Rot *Basidiomycete Phanerochaete Chrysosporium:* Mineralization, Volatilization and Depletion of PCP," *Soil Biol. Biochem.*, Vol. 22, No. 4., pp. 433–440, 1990b.

Lamar, R. T., Larsen, M. J., and Kirk, T. K., "Sensitivity to and Degradation of Pentachlorophenol by *Phanerochaete spp.*," *Appl. Environ. Microbiol.*, Vol. 56, pp. 3519–3526, 1990a.

Lamar, R. T, Evans, J. W., and Glaser, J. A., "Solid-Phase Treatment of a Pentachlorophenol-Contaminated Soil Using Lignin-Degrading Fungi," *Environ. Sci. Technol.*, Vol. 27, pp. 2566–2571, 1993.

Lamar, R. T. and Dietrich, D. M., "In-Situ Depletion of Pentachlorophenol from Contaminated Soil by *Phanerochaete sp.*," *Appl. Environ. Microbiol.*, Vol. 56, pp. 3093–3100, 1990.

Langkilde, J., Christensen, T. H., Skov, B., and Foverskov, A., "Redox Zones Downgradient of Organic Compounds," *In-Situ Bioreclamation: Applications and Investigations for Hydrocarbon and Contaminated Site Remediation,* Hinchee, R. E., and Offenbuttel, R. (Eds.), Butterworth-Heinemann, Stoneham, Mass., pp. 363–376, 1991.

Langseth, D. E., "Hydraulic Performance of Horizontal Wells," *Proceedings of the Hazardous Materials Control Research Institute, 11th National Conference,* pp. 398–408, Washington, D.C., 1990.

Lawes, B. C., "Soil-Induced Decomposition of Hydrogen Peroxide," *In-Situ Bioreclamation: Applications and Investigations for Hydrocarbon and Contaminated Site Remediation,* Hinchee, R. E. and Offenbuttel, R. F. (Eds.), Butterworth-Heinemann, Stoneham, Mass., pp. 143–156, 1991.

Lawrence, A. W., and McCarty, P. L., "A Unified Basis for Biological Treatment Design and Operation," *J. Sanit. Eng. Div., ASCE,* Vol. 96, 1970.

Lee, M. D., Thomas, J. M., Border, R. C., Bedient, P. B., Ward, C. H., and Wilson, J. T., *CRC Critical Reviews in Environmental Control—Biorestoration of Aquifers Contaminated with Organic Compounds,* Report, Rice University, Vol. 18, 1988.

Lenke, H., Dieper, D. H., Bruhn, C., and Knackmuss, H. J., "Degradation of 2,4-Dinitrophenol by Two *Rhodococcus erythropolis* Strains HL 24-1 and HL 24-2," *Appl. Environ. Microbiol.*, Vol. 53, pp. 2928–2932, 1992.

Lewandowski, G. A., Armenante, P. M., and Pak, D., "Reactor Design for Hazardous Waste Treatment Using a White Rot Fungus," *Water Res.*, Vol. 24, pp. 75–82, 1990.

Lightfoot, E. N., Thorne, P. S., Jones, R. L., Hansen, J. L., and Romine, R. R., "Laboratory Studies on Mechanisms for the Degradation of Aldicarb, Aldicarb Sulfoxide, and Aldicarb Sulfone," *Environ. Toxicol. Chem.*, Vol. 6, pp. 377–394, 1987.

Looney, B. B., Kaback, D. S., and Corey, J. C., "Field Demonstration of Environmental Restoration Using Horizontal Wells," *Bioventing and Vapor Extraction: Uses and Applications in Remediation Operations,* Air and Waste Management Association, pp. 104–118, April 1992.

Lynkilde, J., Christensen, T. H., Skov, B., and Foverskov, A., "Redox Zones Downgradient of a Landfill and Implications for Biodegradation of Organic Compounds," *In-Situ Bioreclamation, Applications and Investigations for*

Hydrocarbon and Contaminated Site Remediation, Hinchee, R. E., and Offenbuttel, R. F. (Eds.), Butterworth-Heinemann, Stoneham, Mass., pp. 363–376, 1991.

MacRae, I. C., Raghu, K., and Bautista, E. M., "Anaerobic Degradation of the Insecticide Lindane by *Clostridium sp.,*" *Nature,* Vol. 221, pp. 859–860, 1969.

Mahaffey, W. R., Compeau, G., Nelson, M., and Kinsella, J., "Development of Strategies for Bioremediation of PAH's and TCE," *In-Situ Bioremediation of Groundwater and Contaminated Soils,* WPCF Annual Conference, Washington, D.C., pp. 3–47, October 1990.

Mahaffey, W. R., Gibon, D. T., and Cerniglia, C. E., "Bacterial Oxidation of Chemical Carcinogens: Formation of Polycyclic Aromatic Acids from Benz[a]anthracene," *Appl. Environ. Microbiol.,* Vol. 54, No. 10, pp. 2415–2423, 1988.

Maloney, S. E., Maule, A., and Smith, A. R. W., "Microbial Transformation of the Pyrethroid Insecticides Permethrin, Deltamethrin, Fastac, Fenvalerate and Fluvalinate," *Appl. Environ. Microbiol.,* Vol. 54, pp. 2874–2876, 1988.

Marks, T. S., Allpress, J. D., and Maule, A., "Dehalogenation of Lindane by a Variety of Porphyrins and Corrins," *Appl. Environ. Microbiol.,* Vol. 55, pp. 1258–1261, 1989.

Mashayekhi, M., "Full-Scale Experience with the Fixed-Film Immobilized Cell Bioreactor," *Appl. Bioremediation 93,* Fairfield, N.J., Oct. 27–29, 1993.

Mayer, K. P., Grbić-Galić, D., Semprini, L., and McCarty, P. L., "Degradation of Trichloroethylene by Methanotrophic Bacteria in a Laboratory Column of Saturated Aquifer Material," *Water Sci. Technol.,* Vol. 20, No. 11/12, pp. 175–178, 1988.

Mays, M. K., Sikora, L. J., Hatton, J. W., and Lucia, S. M., "Composting as a Method for Hazardous Waste Treatment," *Proceedings of the 10th National Conference of the Hazardous Materials Control Research Institute,* Nov. 27–29, 1989, Washington, D.C., pp. 298–300.

McCarty, P. L., Semprini, L., Robert, P. V., and Hokins, G., "Biostimulation of Methanotrophic Bacteria to Transform Halogenated Alkenes for Aquifer Restoration," *In-Situ Bioremediation of Groundwater and Contaminated Soils,* WPCF Annual Conference, Washington, D.C., pp. 80–96, October 1990.

McCarty, P. L., Semprini, L., Dolan, M. E., et al., "In-Situ Methanotrophic Bioremediation for Contaminated Groundwater at St. Joseph, Michigan," *On-Site Bioremediation: Processes for Xenobiotic and Hydrocarbon Treatment,* Hinchee, R. E., and Offenbuttel, R. F. (Eds.), Butterworth-Heinemann, Stoneham, Mass., pp. 16–40, 1991.

McCarty, P. L., "Bioengineering Issues Related to In-Situ Remediation of Contaminated Soils and Groundwater," *Environmental Biotechnology,* Omenn, G. S. (Ed.), Plenum Press, New York, pp. 143–162, 1987.

McKenna, E., "Biodegradation of Polynuclear Aromatic Hydrocarbon Pollutants by Soil and Water Microorganisms," Final Report, Project No. A-073-ILL, University of Illinois, Water Resources Center, Urbana, Ill., 1976.

Melcer, H., and Nutt, S. G., "Removal of Organic Trace Contaminants During Biological Treatment of Coal Liquefaction Process Condensates," *International Conference on Physicochemical and Biological Detoxification of Hazardous Wastes,* Wu, X. C. (Ed.), Technomic Publication, Lancaster, Pa., Vol. II, pp. 815–827, 1988.

Mercer, J. W., "Basics of Pump and Treat Groundwater Remediation Technology," EPA/600/8-90/003. Kerr Engineering Research Lab.

Metcalf & Eddy, Inc., *Wastewater Engineering Treatment, Disposal and Reuse,* McGraw-Hill Book Company, New York, pp. 359–442, 1991.

Mihelcic, J. R., and Luthy, R. G., "The Potential Effects of Sorption Processes on the Microbial Degradation of Hydrophobic Organic Compounds in Soil-Water Suspensions," presented at the *International Conference on Physicochemical and Biological Detoxification of Hazardous Wastes,* Wu, X. C. (Ed.), Technomic Publication, Lancaster, Pa., Vol. II, pp. 708–721, 1988.

Mikesell, M. D., and Boyd, S., "Reductive Dechlorination of the Pesticides 2,4-D, 2,4,5-T, and Pentachlorophenol in Anaerobic Sludges," *J. Environ. Qual.,* Vol. 14, pp. 337–340, 1985.

Mikesell, M. D., Olsen, R. H., and Kukor, J. J., "Redox Zones Downgradient of Organic Compounds," *In-Situ Bioreclamation: Applications and Investigations for*

Hydrocarbon and Contaminated Site Remediation, Hinchee, R. E., and Offenbuttel, R. (Eds.), Butterworth-Heinemann, Stoneham, Mass., pp. 351–362, 1991.

Miller, G. P., Portier, R. J., Hoover, D. G., Friday, D. D., and Sicard, J. L., "Biodegradation of Chlorinated Hydrocarbons in an Immobilized Bed Reactor," *Environ. Prog.,* Vol. 9, No. 3, pp. 161–164, 1990.

Morris, P. J., Mohn, W. W., Quensen III, J. F., Tiedje, J. M., and Boyd, S. A., "Establishment of a Polychlorinated Biphenyl-Degrading Enrichment Culture with Predominantly meta Dechlorination," *Appl. Environ. Microbiol.,* Vol. 53, pp. 3088–3094, 1992.

Mueller, J. G., Lantz, S. E., Thomas, R. L., Kline, E. L., Chapman, P. J., Middaugh, D. P., Pritchard, P. H., Colvin, R. J., Rozich, A. P., and Ross, D., "Aerobic Biodegradation of Creosote," EPA Symposium on Bioremediation of Hazardous Wastes, Falls Church, Va., April 1991.

Mueller, J. G., Chapman, P. J., Blattmann, B. O., and Pritchard, P. H., "Isolation and Characterization of a Fluoranthene-Utilizing Strain of *Pseudomonas paucimobilis, Appl. Environ. Microbiol.,* Vol. 56, No. 4, pp. 1079–1086, 1990.

Nelson, J. A., Clark, M. L., Jones, G. R., Andrews, S. O., and Porter, R. J., "Field Evaluation of a Packed Bed Immobilized Microbe Reactor for the Continuous Biodegradation of Still Bottoms Wastewater," presented at the *International Conference on Physicochemical and Biological Detoxification of Hazardous Wastes,* Wu, X. C. (Ed.), Technomic Publication, Lancaster, Pa., Vol. II, pp. 600–618, 1988.

Nelson, M. J. K., Montgomery, S. O., O'Neill, E. J., and Pritchard, P. H., "Aerobic Metabolism of Trichloroethylene by a Bacterial Isolate," *Appl. Environ. Microbiol.,* Vol. 52, pp. 383–384, 1986.

Norcross, K. L., Irvine, R. L., and Herzbaun, P. H., "SBR Treatment of Hazardous Wastewater-Full-Scale Results," *Biotechnology for Degradation of Toxic Chemicals in Hazardous Wastes,* Schulze, R. J., Smith, E. D., Bandy, J. T., Wu, X. C., and Basikeo, J. V. (Eds.), Noyes Data Corp., Park Ridge, N.J., 1985.

O'Brien, G. J., Reich, R. A., Salata, L. S., Feibes, M. H., McManus, C. N., and Health, W. H., "Carbon Columns vs the Pact Process for Priority-Pollutant Removal," *Water Environ. Technol.,* Vol. 72–75, pp. 238–240, 1990.

Ohisa, N., Yamaguchi, M., and Kurihara, N., "Lindane Degradation by Cell-Free Extracts of *Clostridium rectum,"Arch. Microbiol.,* Vol. 125, pp. 221–225, 1980.

Ostendorf, D. W., and Kambel, D. H., "Vertical Profiles and Near Surface Traps for Field Measurement of Volatile Pollution in the Subsurface Environment," *Proceedings of the National Water Well Association Conference on New Techniques for Quantifying the Physical and Chemical Properties of Heterogeneous Aquifers,* Dallas, Tex., 1989.

O'Sullivan, D., "New Studies Pinpoint Pathway of B_{12} Biosynthesis," *Chem. Eng. News,* Vol. 69, No. 5, 1991.

Palumbo, A. V., Eng, W., Boerman, P. A., Strandberg, G. W., Donaldson, T. L., and Herber, S. E., "Effects of Diverse Organic Contaminants on Trichloroethylene Degradation by Methanotrophic Bacteria and Methane-Utilizing Consortia," *On-Site Bioreclamation Processes for Xenobiotic and Hydrocarbon Treatment,* Hinchee, R. E., and Offenbuttel, R. F. (Eds.), Butterworth-Heinemann, Stoneham, Mass., pp. 77–91, 1991.

Paone, D. A. M., Brodacs-Irwin, K., and Sommerfield, T. A., "Novel Biotreatment Method for Toxic Industrial Wastes," presented at the 65th Annual Conference and Exposition, Water Environmental Federation, New Orleans, La. (September 20–24, 1992).

Parkhurst, D. L., Thorstenson, D. C., and Plummer, L. N., "PHREEQE—A Computer Program for Geochemical Calculations," U.S. Geological Survey, *Water Res. Invest.,* pp. 80–96, 1980.

Pasti-Grigsby, M. B., Paszczynski, A., Groszczyuski, S., Crawford, D. L., and Crawford, R. L., "Influence of Aromatic Substitution Patterns on Azo Dye Degradability by *Streptomyces spp.* and *Phanerochaete chrysosporium, Appl. Environ. Microbiol.,* Vol. 58, No. 11, pp. 3605–3613, 1992.

Pedersen, S. F. and Neufeld, R. D., "Studies on Degradation of Hexachlorobiphenyl in Soil," presentation at the *65th Annual Conference and Exposition,* Water Environment Federation, New Orleans, La., Sept. 20–24, 1992.

Pereira, W. E., et al., "Anaerobic Microbial Transformations of Azarenes in Groundwater at Hazardous Waste Sites," *Chemical Quality of Water and the Hydrologic Cycle,* Averett, R., and McKnight, D. (Eds.), Chap. 7, 1987.

Phelps, T. J., Niedzvelski, J. J., Schram, R. M., Herbes, S. E., and White, D. C., Biodegradation of Trichloroethylene in Continuous-Recycle Expanded-Bed Bioreactors," *Appl. Environ. Microbiol.,* Vol. 56, pp. 1702–1709, 1990.

Pipke, R., and Amrhein, N., "Isolation and Characterization of a Mutant of *Arthrobacter sp.* strain GLP-1 Which Utilizes the Herbicide Glyphosate as Its Sole Source of Phosphorus and Nitrogen," *Appl. Environ. Microbiol.,* Vol. 54, pp. 286–287, 1988.

Plummer, L. N., Busby, J. F., Lee, R. W., and Hanshaw, B. B., "Geochemical Modeling of the Madison Aquifer in Parts of Montana, Wyoming and South Dakota," *Water Res.,* Vol. 26, pp. 1981–2014, 1990.

Piotrowski, M. R., Benyhill, K. E., and Vernon, J. L., "Closed-Loop In-Situ Bioremediation of a Leaking Underground Storage Tank Site," *Proc. Appl. Bioremediation '93,* Fairfield, N.J., 1993.

Quensen, J. F., Boyd, S. A., and Tiedje, J. M., "Dechlorination of Four Commercial Polychlorinated Biphenyl Mixtures (Aroclors) by Anaerobic Microorganisms from Sediments," *Appl. Environ. Microbiol.,* Vol. 56, pp. 2360–2369, 1990.

Racke, K. D., and Coats, J. R., "Enhanced Biodegradation of Insecticides on Midwestern Corn Soils," *Enhanced Biodegradation of Pesticides in the Environment,* Racke, K. D., and Coats, J. R. (Eds.), American Chemical Society, Washington, D.C., pp. 68–81, 1990.

Racke, K. D., "Implementation of Enhanced Biodegradation for the Use and Study of Pesticides in the Soil Environment," *Enhanced Biodegradation of Pesticides in the Environment,* Racke, K. D. and Coats, J. R. (Eds.), American Chemical Society, Washington, D.C., pp. 270–281, 1990a.

Rainwater, K., Mayfield, M. P., Heintz, C. E., and Claborn, B. J., "Cyclic Vertical Water Table Movement for Enhancement of In-Situ Biodegradation of Diesel Fuel," *Biological Processes—Innovative Hazardous Waste Technology Series,* Freeman, H. M., and Sferra, P. R. (Eds.), Vol. 3, Technomic Publication, Lancaster, Pa., pp. 123–130, 1991.

Rainwater, K., and Scholze, R. J., "In-Situ Biodegradation for Treatment of Contaminated Soil and Groundwater," *Biological Processes—Innovative Hazardous Waste Treatment Technology Series,* Freeman, H. M., and Sferra, P. R. (Eds.), Technomic Publication, Lancaster, Pa., Vol. 3, pp. 107–122, 1991.

Ramanand, K., Sharmila, M., and Sethunathan, N., "Mineralization of Carbofuran by a Soil Bacterium," *Appl. Environ. Microbiol.,* Vol. 54, pp. 2129–2133, 1988.

Rasche, N., Hyman, M. R., and Arp, D. J., "Biodegradation of Halogenated Hydrocarbon Fumigants by Nitrifying Bacteria," *Appl. Environ. Microbiol.,* Vol. 56, pp. 2568–2571, 1990.

Reed, J. P., Kremer, R. J., and Keaster, A. J., "Spectrophotometric Methodologies for Predicting and Studying Enhanced Degradation," *Enhanced Biodegradation of Pesticides in the Environment,* Racke, K. D., and Coats, J. R. (Eds.), American Chemical Society, Washington, D.C., pp. 240–247, 1990.

Reineke, W., "Microbial Degradation of Halogenated Aromatic Compounds," *Microbial Degradation of Organic Compounds,* Gibson, D. T. (Ed.), Marcel Dekker, New York, pp. 319–360, 1984.

Reinhard, M., Barbash, J. E., and Kunzle, J. M., "Abiotic Dehalogenation Reactions of Haloaliphatic Compounds in Aqueous Solution," presented at the *International Conference on Physicochemical and Biological Detoxification of Hazardous Wastes,* Wu, X. C. (Ed.), Technomic Publication, Lancaster, Pa., Vol. II, pp. 722–741, 1988.

Rhee, G. Y., Sokul, R. C., Bush, B., and Bethoney, C. M., "Long-Term Study of the Anaerobic Dechlorination of Aroclor 1254 with and without Biphenyl Enrichment," *Environ. Sci. Technol.,* Vol. 27, pp. 714–719, 1993.

Roberts, P. V., Hopkins, G. D., MacKay, M., and Semprini, L., "A Field Evaluation of In-Situ Biodegradation of Chlorinated Ethenes: Part 1, Methodology and Field Site Characterization," *Groundwater,* Vol. 28, No. 4, pp. 591–604, 1990.

Roberts, P. V., Semprini, L., Hopkins, G. D., Grbić-Galić, D., McCarty, P. L., and Reinhard, M., "In-Situ Aquifer Restoration of Chlorinated Aliphatics by Methanotrophic Bacteria," Department of Civil Engineering, Stanford University, and U.S. EPA EPA/600/2-89/033, 1989.

Rochkind, M. L., Blackburn, J. W., and Sagler, G. S., "Microbial Decomposition of Chlorinated Aromatic Compounds," U.S. EPA, Office of Research and Development, EPA/600/2-86/090, 1986.

Rogers, J. E., Hale, D., Howard, W., Bryant, F., and Shiu-Mei, H., "Anaerobic Degradation of Chlorinated Aromatic Compounds," EPA Symposium on Bioremediation of Hazardous Wastes, Falls Church, Va., April 1991.

Rose, W. W., and Mercer, W. A., "Fate of Pesticides in Composted Agricultural Wastes," National Canners Association, Washington, D.C., Vol. 27, 1986.

Ross, D., Maziarz, T. P., and Bourguin, A. W., "Bioremediation of Hazardous Waste Sites in the USA—Case Histories," *HMCRI 9th National Conference,* Washington, D.C., The Hazardous Materials Control Research Institute, pp. 395–397, 1988.

Saber, D. L., and Crawford, R. L., "Isolation and Characterization of Flavobacterium Strains That Degrade Pentachlorophenol," *Appl. Environ. Microbiol.,* Vol. 50, pp. 1512–1518, 1985.

Saez, P. B., and Rittman, B. E., "Biodegradation Kinetics of 4-Chlorophenol, an Inhibitory Cometabolite," *Res. J. Water Pollut. Control Fed.,* Vol. 63, pp. 838–847, 1991.

Sawyer, C. J., "Contingency Plans," *Protecting Personnel at Hazardous Waste Sites,* Martin, W. F., and Levine, S. P. (Eds.), Butterworth-Heinemann, Boston, Mass., pp. 373–392, 1994.

Saxena, A., Zhang, R., and Bollag, J. M., "Microorganisms Capable of Metabolizing the Herbicide Metolachlor," *Appl. Environ. Microbiol.,* Vol. 53, pp. 390–396, 1987.

Saxon, A., Zhang, R., and Bollag, J. M., "Microorganisms Capable of Metabolizing the Herbicide Metolachlor," *Appl. Environ. Microbiol.,* Vol. 53, pp. 390–396, 1987.

Schafer, W., and Kizelbach, W., "Numerical Investigation into the Effects of Aquifer Heterogeneity on In-Situ Bioremediation," *In-Situ Bioreclamation: Applications and Investigations for Hydrocarbon and Contaminated Site Remediation,* Hinchee, R. E., and Offenbuttel, R. F. (Eds.), Butterworth-Heinemann, Stoneham, Mass., pp. 196–226, 1991.

Schnitzer, M., and Khan, S. U., *Humic Substances in the Environment,* Marcel Dekker, New York, p. 196, 1972.

Segar, R. L., DeWys, S. L., and Speitel, G. E., "Feeding Strategies That Sustain TCE Cometabolism in Sequencing Biofilm Reactors," *Water Environment Federation 65th Annual Conference,* New Orleans, La., September 1992.

Semprini, L., Grbić-Galić, D., McCarty, P. L., and Roberts, P. V., "Methodologies for Evaluating In-Situ Bioremediation of Chlorinated Solvents," U.S. Environmental Protection Agency EPA/600/R-92/042, March 1992.

Semprini, L., Robert, P. V., Hopkins, G. D., and McCarty, P. L., "Field Evaluation of Aquifer Restoration by Enhanced Biotransformation," presented at the *International Conference on Physicochemical and Biological Detoxification of Hazardous Wastes,* Wu, Yeun C. (Ed.), Technomic Publication, Lancaster, Pa., Vol. II, p. 955, 1988.

Semprini, L., Hopkins, G. D., Roberts, P. V., and McCarty, P. L., "In-Situ Biotransformation of Carbon Tetrachloride, Freon-113, Freon-11, and 1,1,1-TCA under Anoxic Conditions," *On-Site Bioreclamation Processes for Xenobiotic and Hydrocarbon Treatment,* Hinchee, R. E., and Offenbuttel, R. F. (Eds.), Butterworth-Heinemann, Boston, Mass., pp. 41–58, 1991.

Semprini, L., Roberts, P. V., Hopkins, G. D., and McCarty, P. L., "A Field Evaluation of In-Situ Biodegradation Methodologies for Aquifer Contaminated with Chlorinated Aliphatic Compounds: Part 2: The Results of Biostimulation and Biotransformation Experiments," Unpublished report, Environmental Engineering and Science, Department of Civil Engineering, Stanford University, 1989.

Sewell, G. W., Gibson, S. A., and Russell, H. H., "Anaerobic In-Situ Treatment of Chlorinated Ethenes," *Workshop on In-Situ Bioremediation of Groundwater and Contaminated Soils,* WPCF Water Pollution Control Federation Hazardous Wastes

Committee, Water Pollution Control Federation, Annual Conference, pp. 68–79, Washington, D.C., Oct. 7–11, 1990.

Shamat, N. A., and Maier, W. J., "Kinetics of Biodegradation of Chlorinated Organics," *J. Water Pollut. Control Fed.,* Vol. 52, No. 8, pp. 2158–2166, 1980.

Shelton, D. R., "Mineralization of Diethylthiophosphoric Acid by an Enriched Consortium from Cattle Dip," *Appl. Environ. Microbiol.,* Vol. 54, pp. 2572–2573, 1988.

Sherman, D. F., Loehr, R. C., and Newhauser, E. F., "Development of Innovative Biological Techniques for the Bioremediation of Manufactured Gas Plant Sites," presented at the *International Conference on Physicochemical and Biological Detoxification of Hazardous Wastes,* Wu, X. C. (Ed.), Technomic Publication, Lancaster, Pa., Vol. II, pp. 1035–1045, 1988.

Shields, M., "Treatment of TCE and Degradation Products Using *Pseudomonas cepacia,*" *Symposium on Bioremediation of Hazardous Wastes: EPA's Biosystems Technology Development Program Abstracts,* Falls Church, Va., Apr. 16–18, 1991.

Sikora, L. J., Kaufman, D. D., Ramirez, M. A., and Willson, G. B., "Degradation of Pentachlorophenol and Pentachloronitroabenzene on a Laboratory Composting. In *Proceedings of the Wight Annual Research Symposium,* Cincinnati, Ohio, EPA-600/9-82/002, pp. 372–381, 1982.

Simmons, K. E., et al., *Environ. Sci. Technol.,* Vol. 23, pp. 115–121, 1989.

Singer, G. M., Yu, X., Liu, C. H., and Rosen, J. D., "Extraction and Photodegradation of Polycyclic Aromatic Hydrocarbons from Coal Tar Contaminated Soil," *International Conference on Physicochemical and Biological Detoxification of Hazardous Wastes,* Wu, Yeun C. (Ed.), Technomic Publication, Lancaster, Pa., pp. 327–336, 1988.

Skipper, H. D., "Enhanced Biodegradation of Carbanothioate Herbicides in South Carolina," *Enhanced Biodegradation of Pesticides in the Environment,* Racke, K. D. and Coats, J. R. (Eds.), American Chemical Society, Washington, D.C., pp. 37–53, 1990.

Slater, H. J., and Lovatt, D., "Biodegradation and the Significance of Microbial Communities," *Microbial Degradation of Organic Compounds,* Gibson, David T. (Ed.), Marcel Dekker, New York, pp. 439–485, 1984.

Smelt, J. H., Dekker, A., Leistra, M., and Houx, N. W. H., "Conversion of Four Carbamoyloximes in Soil Samples from Above and Below the Water Table," *Pestic. Sci.,* Vol. 14, pp. 173–181, 1983.

Smith, R. L., and Duff, J. H., "Denitrification in a Sand and Gravel Aquifer," *Appl. Environ. Microbiol.,* Vol. 54, pp. 1071–1078, 1989.

Smith, A. E., and Lafond, G. P., "Effects of Long-Term Phenoxyalkanoic Acid Herbicide Field Applications on the Rate of Microbial Degradation," *Enhanced Biodegradation of Pesticides in the Environment,* Racke, K. D., and Coats, J. R. (Eds.), American Chemical Society, Washington, D.C., pp. 13–22, 1990.

Snoeyink, V. L., and Jenkins, D., *Water Chemistry,* John Wiley & Sons, New York, 1980.

Song, S., and Dague, R. R., "Laboratory Studies and Modeling of the Anaerobic Sequencing Batch Reactor," *Water Environment Federation 65th Annual Conference and Exposition,* New Orleans, La., September 1992.

Sontakke, S., Beaumier, D., Beaulieu, C., Greer, C., and Samson, R., "Continuous Culture Adaptation of *Pseudomonas putida* to Growth on Phenol as Sole Carbon Source," presented at the *International Conference on Physicochemical and Biological Detoxification of Hazardous Wastes,* Wu, X. C. (Ed.), Technomic Publication, Lancaster, Pa., Vol. II, pp. 619–632, 1988.

Spadaro, J. T., Gold, M. H., and Renganathan, V., "Degradation of Azo Dyes by the Lignin-Degrading Fungus *Phanerochaete chrysosporium,*" *Appl. Environ. Microbiol.,* Vol. 58, pp. 2397–2401, 1992.

Spain, J. C., "Microbial Adaptation in Aquatic Ecosystems," *Enhanced Biodegradation of Pesticides in the Environment,* Racke, K. D., and Coats, J. R. (Eds.), American Chemical Society, Washington, D.C., pp. 181–189, 1990.

Spiker, J. K., Crawford, D. L., Carford, R. L., "Influence of 2,4,6-Trinitrotoluene (TNT) Concentration on the Degradation of TNT in Explosive-Contaminated Soils by the White Rot Fungus *Phanerochaete chrysosporium,*" *Appl. Environ. Microbiol.,* Vol. 53, pp. 3199–3202, 1992.

Stanier, R. Y., Ingraham, J. L., Wheelis, M. L., and Painter, P. R., *The Microbial World,* 5th Edition, Prentice-Hall, Englewood Cliffs, N.J., 1986.

Staps, J. J. M., "International Evaluation of In-Situ Biorestoration of Contaminated Soil and Groundwater," *Third International Meeting of the NATO/CCMS Pilot Study-Demonstration of Remedial Action Technologies for Contaminated Land and Groundwater,* Montreal, Canada, Nov. 6–9, 1989.

Stegmann, R., Lotter, S., and Heerenklage, J., "Biological Treatment of Oil-Contaminated Soils in Bioreactors," *On-Site Bioremediation: Processes for Xenobiotic and Hydrocarbon Treatment,* Hinchee, R. E., and Offenbuttel, R. E. (Eds.), pp. 188–208, Butterworth-Heinemann, Stoneham, Mass., 1991.

Stensel, H. D., Strand, S. E., Richard, S. C., and Treat, T. P., "Toxicity Effects of Methanotrophic Biotransformation of Trichloroethylene," *Water Environment Federation 65th Annual Conference and Exposition,* New Orleans, La., September 1992.

Stentinson, M. K., Hahn, W., and Skovronek, H. S., "Site Demonstration of Biological Treatment of Groundwater by Biotrol, Inc. at a Wood Preserving Site in New Brighton, MN," *Biological Processes—Innovative Hazardous Waste Technology Series,* Vol. 3, Freeman, H. M., and Sferra, P. R. (Eds.), Technomic Publication, Lancaster, Pa., pp. 163–168, 1991.

Stirling, L. A., Watkinson, R. J., and Higgins, I. J., *Proc. Soc. Gen. Microbiol.,* Vol. 4., p. 28, 1976.

Storms, G. E., "Oxygen Dissolution Technologies for Bioremediation Applications," Praxair Technology, Tarrytown, N.Y., 1993.

Strand, S. E., Woodrich, J. V., and Stensel, D. H., "Biodegradation of Chlorinated Solvents in a Sparged, Methanotrophic Biofilm Reactor," *Res. J. Water Pollut. Control Fed.,* Vol. 63, pp. 859–867, 1991.

Strand, S. E., Bjelland, M. D., and Stensel, H. D., "Kinetics of Chlorinated Hydrocarbon Degradation by Suspended Cultures of Methane-Oxidizing Bacteria," *Res. J. Water Pollut. Control Fed.,* Vol. 62, No. 2, March/April 1990.

Strou, H. F., Mahaffey, W., and Bourguin, A. W., "Development of an In-Situ Bioremediation System for a Creosote-Contaminated Site," presented at the *International Conference on Physicochemical and Biological Detoxification of Hazardous Wastes,* Wu, X. C. (Ed.), Technomic Publication, Lancaster, Pa., Vol. II, pp. 919–936, 1988.

Stumm, W., and Morgan, J. J., *Aquatic Chemistry: An Introduction Emphasizing Chemical Equilibria in Natural Waters,* 2d Edition, John Wiley & Sons, New York, 1981.

Suflita, J. M., Robinson, J. A., and Tiedje, J. M., "Kinetics of Microbial Dehalogenation of Haloaromatic Substrates in Methanogenic Environments," *Appl. Environ. Microbiol.,* Vol., 45, pp. 1466–1473, 1983.

Suidan, M. T., Nath, R., and Schroeder, A. T., "Treatment of CERCLA Leachates by Carbon-Assisted Anaerobic Fluidized Beds," *EPA Symposium on Bioremediation of Hazardous Wastes,* Falls Church, Va., April 1991.

Sutherland, J. B., Selby, A. L., Freeman, J. P., Evans, F. E., and Cerniglia, C. E., "Metabolism of Phenanthrene by *Phanerochaete chrysosporium,*" *Appl. Environ. Microbiol.,* Vol. 57, pp. 3310–3316, 1991.

Sutton, P. M., "Engineered Systems for Biotreatment-Adsorption of Hazardous Wastes," presented at the *International Conference on Physicochemical and Biological Detoxification of Hazardous Wastes,* Wu, X. C. (Ed.), Technomic Publication, Lancaster, Pa., Vol. I, pp. 53–74, 1988.

Swindell, C. M., Landon-Arnold, S., and Flynn, D. I., "Characterization of Indigenous Microbial Degraders of Aviation Fuel Components from a Contaminated Site," presented at the *International Conference on Physicochemical and Biological Detoxification of Hazardous Wastes,* Wu, X. C. (Ed.), Technomic Publication, Lancaster, Pa., Vol. II, pp. 977–991, 1988.

Tabak, H. H., "Development and Application of a Multilevel Respirometric Protocol to Determine Biodegradability and Biodegradation Kinetics of Toxic Organic Pollutant Compounds," *On-Site Bioreclamation Processes for Xenobiotic and Hydrocarbon Treatment,* Hinchee, R. E., and Offenbuttel, R. F. (Eds.), pp. 324–340, 1991.

Tiedje, J. M., and Stevens, T. O., "The Ecology of an Anaerobic Dechlorinating Consortium," *Environmental Biotechnology*, Omenn, G. S. (Ed.), Plenum Press, New York, pp. 3–14, 1987.

Tiedje, J. M., and Alexander, M., "Enzymatic Cleavage of the Ether Bond of 2,4-dichloro-phenoxyacetate," *J. Agr. Food Chem.*, Vol. 17, pp. 1080–1084, 1969.

Tokuz, R. X., "Rotating Biological Contactor Treatment of Hazardous Wastes Containing Chlorinated Phenols," presented at the *International Conference on Physicochemical and Biological Detoxification of Hazardous Wastes*, Wu, X. C. (Ed.), Technomic Publication, Lancaster, Pa., Vol. II, pp. 552–566, 1988.

Troy, M. A., and Jerger, D. E., "Evaluation of Laboratory Treatability Study Data for the Full-Scale Bioremediation of Petroleum Hydrocarbon Contaminated Soils," *Proceedings Applied Bioremediation 93*, Fairfield, N.J., Oct. 25–26, 1993.

Trudgill, P. W., "Microbial Degradation of the Alicyclic Ring: Structural Relationships and Metabolic Pathways," *Microbial Degradation of Organic Compounds*, Gibson, D. T. (Ed.), Marcel Dekker, New York, pp. 131–180, 1984.

Tsien, H. C., Bousseau, A., Hanson, R. S., and Wackett, L. P., "Biodegradation of Trichloroethylene by *Methylosinus trichosporium*," *Appl. Environ. Microbiol.*, Vol. 55, pp. 3155–3161, 1989.

Tuhela, L., Smith, S. A., and Tuouinen, O. H., "Microbiological Analysis of Iron-Related Biofouling in Water Wells and a Flow-Cell Apparatus for Field and Laboratory Investigations, *Groundwater*, Vol. 31, No. 6, pp. 982–988, 1993.

Turco, R. F., and Konopka, H. E., "Response of Microbial Populations to Carbofuran in Soils Enhanced for its Degradation," *Enhanced Biodegradation of Pesticides in the Environment*, Racke, K. D., and Coats, J. R. (Eds.), American Chemical Society, Washington, D.C., pp. 154–166, 1990.

Uchiyama, H., Nakajima, T., Yagi, O., and Nakahara, T., "Role of Heterotrophic Bacteria on Complete Mineralization of Trichloroethylene by *Methylocystis sp.* Strain M," *Appl. Environ. Microbiol.*, Vol. 53, pp. 3067–3071, 1992.

Unterman, R., Bedard, D. L., Brennan, M. J., Bopp, L. H., Mondello, F. J., Brooks, R. E., Mobley, D. P., McDermott, J. B., Schwartz, C. C., and Dietrich, D. K., "Biological Approaches for PCB Degradation," *Reducing Risks from Environmental Chemicals through Biotechnology*, G. S. Omenn, et al., Plenum Press, New York, 1988.

U.S. Department of Agriculture, *Soil: The 1957 Year Book of Agriculture*, U.S. Department of Agriculture, Government Printing Office, p. 54, 1957.

U.S. Department of the Air Force, Air Force Center for Environmental Excellence, *Test Plan and Technical Protocol for a Field Treatability Test for Bioventing*, Environmental Services Office, May 1992.

U.S. Environmental Protection Agency, "Nitrate for Biorestoration of an Aquifer Contaminated with Jet Fuel," U.S. EPA, Robert S. Kerr Environmental Laboratory, Ada, Okla., NTIS Report PB91-164285, 1991.

U.S. Environmental Protection Agency, "Process Design Manual for Sludge Treatment and Disposal," Municipal Environmental Research Laboratory, EPA 625/1-79-011, pp. 4–8, September 1979.

U.S. Environmental Protection Agency, "Composting of Municipal Wastewater Sludges," Office of Research and Development, Cincinnati, Ohio, EPA 625/4-85-014, August 1985.

U.S. Environmental Protection Agency, "Summary Report, In-Vessel Composting of Municipal Wastewater Sludge," EPA 625/8-89/016, September 1989.

U.S. Environmental Protection Agency, "Toxic Release Inventory," Magnetic Tape Number P.B. 89-186-118, NTIS, or TRI Data Base, National Library of Medicine, Bethesda, Md., June 1989.

U.S. Environmental Protection Agency, "1991 Bioremediation Field Projects," Bioremediation in the Field, EPA 540/2-91/027, No. 4, pp. 7–8, December 1991a.

U.S. Environmental Protection Agency, "Summary Report, Sequencing Batch Reactors," Center of Environmental Research Information, EPA 625/8-86/001, 1986.

U.S. Environmental Protection Agency, *Bioremediation in the Field*, No. 8, EPA/540/N-93/001, May 1993.

U.S. Environmental Protection Agency, *Engineering Bulletin Slurry Biodegradation*, Center for Environmental Research, Cincinnati, Ohio, EPA/540/2-90/016.

U.S. Environmental Protection Agency, *Permit Guidance Manual on Unsaturated Zone Monitoring for Hazardous Waste Land Treatment Units,* Environmental Monitoring Systems Lab., EPA/530-SW-86-040, p. 5, 1986.

U.S. Environmental Protection Agency, "MINTEQA$_2$ An Equilibrium Metal Speciation Model, User's Manual," EPA600/3-87/012, 1988.

U.S. Environmental Protection Agency, "Emerging Technology—Pilot Scale Demonstration of a Two-Stage Methanotrophic Bioreactor for Biodegradation of Trichloroethane in Groundwater," EPA 540/S-93/505, October 1993.

U.S. Environmental Protection Agency, *Cleaning Up the Nation's Waste Sites: Markets and Technology Trends,* Office of Solid Waste and Emergency Response, Washington, D.C., EPA 542-R-92-012, April 1993.

U.S. House of Representatives, *Deep Pockets: Taxpayer Liability for Environmental Contamination,* Majority Staff Report of the Subcommittee on Oversight and Investigations of the Committee on Natural Resources, Government Printing Office, Washington, D.C., July 1993.

Valocchi, A. J., and Roberts, P. V., "Attenuation of Ground-Water Contaminant Pulses," *J. Hydraul. Eng.,* Vol. 109, No. 12, pp. 1665–1682, 1983.

Van den Wijngaard, A. J., Wind, R. D., and Janssen, D. B., "Kinetics of Bacterial Growth on Chlorinated Aliphatic Compounds," *Appl. Environ. Microbiol.,* Vol. 59, pp. 2041–2048, 1993a.

Van den Wijngaard, A. J., VanderKleij, R. G., Doornweerd, R. E., and Janssen, D. B., "Influence of Organic Nutrients and Cocultures on the Competitive Behavior of 1,2-Dichloroethane-Degrading Bacteria," *Appl. Environ. Microbiol.,* Vol. 59, pp. 3400–3405, 1993b.

Vanelli, T., Chapman, P., and Hooper, A. B., "Degradation of Halogenated Aliphatic Compounds by the Ammonia-Oxidizing Bacteria *Nitrosomonas europaea,*" *EPA Symposium on Bioremediation of Hazardous Wastes,* Falls Church, Va., April 1991.

Vanelli, T., Logan, M., Arciero, D. M., and Hooper, A. B., "Degradation of Halogenated Aliphatic Compounds by the Ammonia-Oxidizing Bacterium *Nitrosomonas europaea,*" *Appl. Environ. Microbiol.,* Vol. 56, No. 4, pp. 1169–1171, 1990.

Venkataramani, E. S., and Ahlert, R. C., "Microbial Remediation of Landfill Liquids," *Biological Processes: Innovative Hazardous Waste Technology Series,* Vol. 3, Freeman, H. M., and Sferra, P. R. (Eds.), Technomic Publication, Lancaster, Pa., pp. 143–162, 1991.

Villarreal, D. T., Turco, R. F., and Konopka, A., "Propachlor Degradation by a Soil Bacteria Community," *Appl. Environ. Microbiol.,* Vol. 57, pp. 2135–2140, 1991.

Vira, A., and Fogel, S., "Bioremediation: The Treatment for Tough Chlorinated Hydrocarbons," *Biotreatment News,* Vol. 1, p. 8, October 1991.

Vira, A., Fogel, S., Klink, L., Coho, J., and Campbell, M., "Bench-Scale Demonstration of a Methanotrophic Rotating Biological Contactor for Treatment of Groundwater Contaminated with Both Chlorinated and Non-Chlorinated Chemicals from Kelly Air Force Base in San Antonio, TX," *Water Environmental Federation 65th Annual Conference and Exposition,* New Orleans, La., September 1992.

Vogel, T. N., and Grbić-Galić, D., "Incorporation of Oxygen from Water into Toluene and Benzene during Anaerobic Fermentative Transformation," *Appl. Environ. Microbiol.,* Vol. 52, pp. 200–202, 1986.

Vogel, T. N., Criddle, G. S., and McCarty, P. L., "Transformation of Halogenated Aliphatic Compounds," *Environ. Sci. Technol.,* Vol. 21, No. 8, pp. 722–736, 1987.

Wackett, L. P., Brusseau, G. A., Householder, S. R., and Hansen, R. S., "Survey of Microbial Oxygenasses: Trichloroethylene Degradation by Propane-Oxidizing Bacteria," *Appl. Environ. Microbiol.,* Vol. 55, pp. 2960–2964, 1989.

Wickramanayake, G. B., Nack, H., and Allen, B. R., "Treatment of Trichloroethylene-Contaminated Groundwater Using Aerobic Bioreactors," *Biological Processes—Innovative Hazardous Waste Technology Series,* Vol. 3, Freeman, H. M., and Sferra, P. R. (Eds.), Technomic Publication, Lancaster, Pa., pp. 169–175, 1991.

Wilson, B. H., "Biotransformation of Selected Alkybenzenes and Halogenated Aliphatic Hydrocarbons in Methanogenic Aquifer Material," *Appl. Environ. Micro.,* Vol. 49, p. 242, 1985.

Wilson, S. B., "In-Situ Remediation of Groundwater and Soils: Seminar Outline," *Proceedings HMCRI's 10th National Conference,* Washington, D.C., Nov. 27–29, p. 227, 1989.

Yang, P. Y., and Li-Wang, M., "Immobilized Mixed Microbial Cell Process for Toxic-organic Water and Wastewater Treatment," presented at the *International Conference on Physicochemical and Biological Detoxification of Hazardous Wastes,* Wu, X. C. (Ed.), Technomic Publication, Lancaster, Pa., Vol. II, pp. 531–551, 1988.

Yare, B. S., "A Comparison of Soil-Phase and Slurry-Phase Bioremediation of PNA-Containing Soils," *On-Site Bioremediation: Processes for Xenobiotic and Hydrocarbon Treatment,* Hinchee, R. E., and Offenbuttel, R. E. (Eds.), pp. 173–187, Butterworth-Heinemann, Boston, Mass., 1991.

Young, J. C., and Yang, B. S., "Design Considerations for Full-Scale Anaerobic Filters," *Res. J. Water Pollut. Control Fed.,* Vol. 61, No. 9, pp. 1576–1587, 1989.

Young, L. Y., "Anaerobic Degradation of Aromatic Compounds," in Gibson, D. T. (Ed.), *Microbial Degradation of Organic Compounds,* Marcel Dekker, New York, pp. 487–523, 1984.

Zboinska, E., Lejczak, B., and Katarski, P., "Organophosphonate Utilization by the Wild Type Strain of *Pseudomonas fluorescens,*" *Appl. Environ. Microbiol.,* Vol. 58, pp. 2993–2999, 1992.

Zhang, X., and Wiegel, J., "Sequential Anaerobic Degradation of 2-4 Dichlorophenol in Freshwater Sediments," *Appl. Environ. Microbiol.,* Vol. 56, pp. 1119–1127, 1990.

Zheng, Z., Liu, S. Y, Freyer, A. J., and Bollag, J. M., "Transformation of Metalaxyl by the Fungus *Syncephalastrum racemosum,*" *Appl. Environ. Microbiol.,* Vol. 55, pp. 66–71, 1989.

Index

Index note: The *f.* after a page number refers to a figure, and the *t.* refers to a table.

ABOUT THE AUTHOR

John T. Cookson, Jr., has more than 30 years of experience in civil, sanitary, and environmental engineering. He holds a Ph.D. in environmental health engineering from the California Institute of Technology. He has served as chief engineer, vice president, president, and CEO of several environmental firms and currently is president of International Network for Environmental Training, Inc. He brings practical experience coupled with 10 years of teaching experience as a professor of civil engineering at the University of Maryland to the development of this book.